THE SOIL-ROOT INTERFACE

Proceedings of an international symposium
held in Oxford, England,
March 28 to 31, 1978

THE SOIL-ROOT INTERFACE

Edited by

J. L. HARLEY
Department of Forestry
Oxford

R. SCOTT RUSSELL
The Grange
East Hanney
Oxfordshire

1979

PUBLISHED
UNDER THE AEGIS OF THE

NEW PHYTOLOGIST

BY

ACADEMIC PRESS

London · New York · San Francisco
A Subsidiary of Harcourt Brace Jovanovich, Publishers

ACADEMIC PRESS INC. (LONDON) LTD.
24/28 Oval Road
London NW1

United States Edition published by
ACADEMIC PRESS INC.
111 Fifth Avenue
New York, New York 10003

Library of Congress Catalog Card Number: 78-73883
ISBN: 0-12-325550-3

Printed in Great Britain by
Galliard (Printers) Ltd, Great Yarmouth

PARTICIPANTS

Atkinson, D., East Malling Research Station, Maidstone, Kent
Bache, B.W., Macaulay Institute for Soil Research, Craigiebuckler, Aberdeen
Barber, S.A., Department of Agronomy, Purdue University, West Lafayette, IN 47907, USA
Barraclough, P.B., Rothamsted Experimental Station, Harpenden, Herts.
Bar-Yosef, B., Institute of Soils and Water, Agricultural Research Organisation, POB 6, Bet Dagan, Israel
Beusichem, M.L. van, Department of Soils and Fertilizers, Agricultural University, Wageningen, The Netherlands
Black, C.R., Physiology and Environmental Studies Department, Nottingham University, Sutton Bonington, Loughborough, Leics.
Brouwer, R., Botanical Laboratory, Lange Nieuwstraat 106, Utrecht 2501, The Netherlands
Brown, D.A., Agronomy Department, University of Arkansas, Fayetteville, ARK 72701, USA
Brown, M.E., Rothamsted Experimental Station, Harpenden, Herts.
Cabala-Rosand, P., Department of Soil Science, University of Reading, London Road, Reading, Berks.
Cabibel, M., INRA Centre de Recherches de Toulouse, Station d'Agronomie, B.P.12, Castanet, France
Campbell, G.S., University of Nottingham School of Agriculture, Sutton Bonington, Loughborough, Leics.
Clarkson, D.T., ARC Letcombe Laboratory, Wantage, Oxon
Coleman, D.C., Natural Resource Ecology Laboratory, Colorado State University, Fort Collins, Colorado 80523, USA
Cooke, G.W., Agricultural Research Council, 160 Great Portland Street, London W1N 6DT
Corbett, D.C.M., Agricultural Research Council, 160 Great Portland Street, London W1N 6DT
Crooke, W.M., Macaulay Institute for Soil Research, Craigiebuckler, Aberdeen
Cruz-Romero, G., ETS Ingenieros Agronomos, Universidad Politecnica, Valencia 10, Spain.
Darbyshire, J.F., Macaulay Institute for Soil Research, Craigiebuckler, Aberdeen
DeKock, P.C., Macaulay Institute for Soil Research, Craigiebuckler, Aberdeen
Dirksen, C., US Salinity Laboratory, Riverside, California 92501, USA
Dorp, F. van, Institute for Atomic Sciences in Agriculture, Association EURATOM-ITAL, Postbus 48, Wageningen, The Netherlands

PARTICIPANTS

Drew, M.C., ARC Letcombe Laboratory, Wantage, Oxon
Drews, M., Institut für Pflanzenernährung der Justus Liebig Universität, D-63 Giessen, Federal Republic of Germany
Egeraat, A.W.S.M. van, Laboratory Microbiology, Wageningen, The Netherlands
Elliott, L.F., Land Management and Water Conservation Research Unit, Washington State University, Pullman, Washington 99164, USA
Faizy, S., E-D. A., PO Box 62, Zamalek, Cairo, Egypt
Fitter, A.H., Department of Biology, University of York
Fleming, A.L., Plant Stress Laboratory, Beltsville Agricultural Research Centre, Beltsville, MD 20705, USA
Freckman, D.W., Department of Nematology, University of California, Riverside, CA 92521, USA
Gams, W., Centraalbureau voor Schimmelcultures, P.O.B. 273, Baarn, The Netherlands
Garrett, S.D., 179 Hills Road, Cambridge
Gaskins, M.H., Department of Agronomy, University of Florida, Gainesville, FL 32611, USA
Gasser, J.K.R., Agricultural Research Council, 160 Great Portland Street, London W1N 6DT
Gatehouse, D.M., ICI Plant Protection Division, Jealott's Hill, Bracknell, Berks.
Gilmour, C.M., Department of Bacteriology and Biochemistry, University of Idaho, Moscow, Idaho 83843, USA
Goodman, B.A., Macaulay Institute for Soil Research, Craigiebuckler, Aberdeen
Goodman, P.J., Welsh Plant Breeding Station, Plas Gogerddan, Nr. Aberystwyth
Goss, M.J., ARC Letcombe Laboratory, Wantage, Oxon
Greaves, M.P., ARC Weed Research Organisation, Begbroke Hill, Yarnton, Oxford
Greenland, D.J., Department of Soil Science, University of Reading, London Road, Reading, Berks.
Greenwood, D.J., National Vegetable Research Station, Wellesbourne, Warwick
Grimme, H., Landwirtschaftliche Forschungsanstalt Büntehof, Hannover, Federal Republic of Germany
Gundy, S.D. van, Department of Nematology, University of California, Riverside, CA 92521, USA
Harley, J.L., Department of Forestry, University of Oxford, Oxford
Hay, R.K.M., Department of Environmental Sciences, University of Lancaster, Bailrigg, Lancaster
Heap, A.J., Botany Department, University of Bristol, Bristol
Hendriks, L., Institut für Pflanzenernährung der Technischen Universität Hannover, Hannover, Federal Republic of Germany
Hopper, M.J., Grassland Research Institute, Hurley, Maidenhead, Berks.
Huck, M.G., Department of Agronomy and Soils, Auburn University, Auburn, Ala 36830, USA
Jackson, J.E., East Malling Research Station, Maidstone, Kent
Jager, A. de, Botanisch Laboratorium van de Ryksuniversiteit te Utrecht, Utrecht, The Netherlands
Jenkins, D.A., Department of Biochemistry and Soil Science, University College of North Wales, Bangor, Gwynedd

PARTICIPANTS

Johnen, B.G., ICI Plant Protection Ltd., Fernhurst, Haslemere, Surrey
Jones, L.H.P., Grassland Research Institute, Hurley, Maidenhead, Berks.
Jungk, A., Institut für Pflanzenernährung der Technischen Universität Hannover, Hannover, Federal Republic of Germany
Kafkafi, M.B., Agricultural Research Organisation, The Volcani Centre, P.O.B. 6, Bet Dagan, Israel
Keltjens, W., Department of Soils and Fertilizers, Agricultural University, Wageningen, The Netherlands
Kirkby, E.A., Department of Plant Sciences, The University, Leeds
Kirkham, M.B., Department of Agronomy, Oklahoma State University, Stillwater, OK 74074, USA
Lake, J.V., Agricultural Research Council, 160 Great Portland Street, London W1N 6DT
Lambert, J.R., Department of Agricultural Engineering, College of Agricultural Sciences, Clemson University, S.C. 29631, USA
Lawley, R.A., Department of Botany, Bristol University, Bristol
Leggett, J.E., Agronomy Department, University of Kentucky, Lexington, KY 40506, USA
Linehan, D.J., Macaulay Institute for Soil Research, Craigiebuckler, Aberdeen
Loneragan, J.F., School of Environmental and Life Sciences, Murdoch University, Perth, Western Australia 6153
Lynch, J.M., ARC Letcombe Laboratory, Wantage, Oxon
Maertens, C., INRA Centre de Recherches de Toulouse, Station d' Agronomie, B.P. 12, Castanet, France
Marschner, H., Institute of Plant Nutrition, Universität Hohenheim, P.O.B. 106, Stuttgart, Federal Republic of Germany
Milthorpe, F.L., School of Biological Sciences, Macquarie University, N.S.W. 2113, Australia
Moawad, M., Institut für Tropischen und Subtropischen Pflanzenbau, Göttingen, Federal Republic of Germany
Moorby, J., Glasshouse Crops Research Institute, Littlehampton, Sussex
Newman, E.I., Botany Department, The University, Bristol
Nicolson, T.H., Department of Biological Sciences, The University, Dundee
Nielsen, N.E., The Royal Veterinary and Agricultural University, Department of Soil Fertility and Plant Nutrition, Copenhagen, Denmark
Noordwijk, M. van, Institute for Soil Fertility, Haren, The Netherlands
Nye, P.H., Soil Science Laboratory, Department of Agricultural Science, University of Oxford, Oxford
Oertli, J.J., Institute of Botany, University of Basel, Basel, Switzerland
Olszta, W., AL PKWN 29, Lublin, Poland
Penning de Vries, F.W.T., Department Theoretical Production Ecology, Agricultural University, Wageningen, The Netherlands
Portas, C., Department of Crop Science, Instituto Universitario Evora, Portugal
Posner, A.M., Department of Soil Science and Plant Nutrition, University of Western Australia, Nedlands, Western Australia
Powell, D.B.B., Long Ashton Research Station, Long Ashton, Bristol

PARTICIPANTS

Prenzel, J., Institut für Bodenkunde und Waldernährung, Göttingen, Federal Republic of Germany
Quirk, J.P., FAA, Waite Agricultural Research Institute, University of Adelaide, Glen Osmond, South Australia
Raats, P.A.C., Institute for Soil Fertility, Haren, The Netherlands
Reid, C.P.P., Department of Forest and Wood Sciences, Colorado State University, Fort Collins, Colorado 80523, USA
Reid, J.B., Letcombe Laboratory, Wantage, Oxon OX12 9JT
Ritchie, J.T., Blackland Research Centre, PO Box 748, Temple, Texas, USA
Römheld, V., Institute of Plant Nutrition, University of Hohenheim, P.O.B. 106, Stuttgart, Federal Republic of Germany
Rowse, H.R., National Vegetable Research Station, Wellesbourne, Warwick
Rovira, A.D., CSIRO Division of Soils, Glen Osmond, South Australia
Sanders, F.E., Department of Plant Sciences, The University, Leeds
Sauerbeck, D., Agrikulturchemisches Institut der Universität Bonn, Bonn, Federal Republic of Germany
Schank, S.C., Department of Agronomy, University of Florida, Gainesville, Florida 32611, USA
Schippers, B., Phytopathology Laboratory 'W.C.S.', Baarn, The Netherlands
Scott Russell, R., ARC Letcombe Laboratory, Wantage, Oxon
Sewell, G.W.F., East Malling Research Station, Maidstone, Kent
Shipton, P.J., Plant Pathology Section, Agricultural Botany Division, School of Agriculture, Aberdeen
Smiley, R.W., Department of Plant Pathology, Cornell University, Ithaca, New York 14853, USA
Smucker, A.J.M., Department of Crop and Soil Sciences, Michigan State University, East Lansing, Michigan 48824, USA
So, Hwat Bing, Department of Agronomy and Soil Science, University of New England, Armidale, N.S.W. 2350, Australia
Sparling, G.P., Macaulay Institute for Soil Research, Craigiebuckler, Aberdeen
Stribley, D.P., Rothamsted Experimental Station, Harpenden, Herts.
Taylor, H.M., Department of Agronomy, Iowa State University, Ames, Iowa, USA
Thien, S.J., Department of Agronomy, Kansas State University, Manhattan, Kansas, USA
Tinker, P.B., Rothamsted Experimental Station, Harpenden, Herts.
Trolldenier, G., Landwirtschaftliche Forschungsanstalt Büntehof, Hannover, Federal Republic of Germany
Troughton, A., Welsh Plant Breeding Station, Plas Gogerddan, Aberystwyth
Vaughan, D., Macaulay Institute for Soil Research, Craigiebuckler, Aberdeen
Warembourg, F.R., CEPE, CNRS, BP 5051, Montpellier, France
Weatherley, P.E., Department of Botany, The University, Aberdeen
Welbank, P.J., Rothamsted Experimental Station, Harpenden, Herts.
Werff, M.M. van der, Biological Laboratory, Department of Ecology, Free University, Amsterdam, The Netherlands
White, R., Soil Science Laboratory, Department of Agricultural Science, University of Oxford, Oxford

Wieneke, D., Arbeitsgruppe Radioagronomie der Kernforschungsanlage Jülich, Jülich, Federal Republic of Germany
Wild, A., Department of Soil Science, The University, Reading, Berks.
Wilson, S.A., East Malling Research Station, Maidstone, Kent
Witty, J.F., Rothamsted Experimental Station, Harpenden, Herts.
Woodhead, T., Rothamsted Experimental Station, Harpenden, Herts.
Zentmyer, G.A., Department of Plant Pathology, University of California, Riverside, California 92521, USA

Symposium Planning Committee

Chairman: J.L. HARLEY

Joint Secretaries: P.B. Tinker and D.C. Clarkson

Members: R. Brouwer, D.J. Greenland, H. Marschner, A.D. Rovira, R. Scott Russell, H.M. Taylor, P.E. Weatherley.

FOREWORD

During the past decade interactions between the root systems of plants and the soil have received increasing attention from plant physiologists, microbiologists and soil scientists. Informal discussions and correspondence between a number of people engaged in these interdisciplinary studies in several countries led to the suggestion that a small symposium might be timely. A planning committee sent out a preliminary circular to find out who was interested in attending and contributing. The circular described the purpose of the meeting as follows:

> 'There is a growing awareness of the importance of processes at the soil-root interface which influence the water supply, mineral nutrition and pathology of plants. The study of this contact zone, in which root and soil interact directly, is technically difficult and necessarily involves a multi-disciplinary approach.
>
> The aim of the meeting is to bring together active research workers to discuss recent progress in research on the physical, chemical, biological and microbiological characteristics of this zone and to promote an exchange of information on techniques for its further exploration.'

The response was so great that plans had to be made for a much larger group of participants than had been expected. Even so, only a fraction of the contributions which were offered could be accommodated in the available sessions and poster displays.

Although the magnitude of the response emphasized the

importance and timeliness of the meeting, it involved the committee in the unhappy task of refusing many papers regardless of scientific merit. In addition it necessitated a change of plan about publication. The committee was originally not intending to publish a special symposium volume but had arranged with the Trustees and Editors of the NEW PHYTOLOGIST for them to accept symposium papers and publish them all together or in groups in the normal sequence of that journal. With the enlarged meeting a separate Symposium volume was more suitable and the committee were helped by the Trustees and Editors of the NEW PHYTOLOGIST in arranging for it to be published. The first five chapters contain the majority of papers presented at the meeting, as well as descriptions of posters and films which were displayed. Abstracts of other contributions offered by participants but not read at the meeting form the final chapter.

Some aspects of the interfacial zone have received attention for a century or more - for example, ectotropic mycorrhizal fungi, *Rhizobia* and other nitrogen fixing procaryotes, and pathogens. Others have until lately been relatively little considered, for example the extent of the loss or so-called 'exudation' which can occur from living roots, and the effects of the free-living rhizosphere organisms and vesicular arbuscular mycorrhizas on nutrient uptake. On the more physical side there have been important advances in understanding movement of nutrients through the soil towards roots and it has also been found that roots may not in all circumstances make uniform and continuous contact with the bulk soil through which they pass.

In large measure the new developments reflect the availability of new experimental techniques, but none the less it became clear during the meeting that improved methodology is a prime requirement for establishing quantitative relationships between many physical, chemical and biological parameters of the interfacial region; indeed some participants suggested that a further meeting devoted solely to this subject would soon be needed.

During formal and informal discussions it was realized that some of the terms used, especially by biologists, needed more careful definition. Dr A D Rovira was prevailed upon to write a short note on one set of terms, namely those used to describe the organic matter in the rhizosphere. The result of his thoughts, refined by discussion with his colleagues at CSIRO, Glen Osmond, S. Aust. are printed as the first paper in this book. The contributed papers have not of course been altered in the light of Dr Rovira's suggestions and the reader is left to decide how well these suggestions would fit any particular case.

It may be thought that the greatest value of this, the first attempt at a full-scale discussion meeting on the soil-root interface, lay in the clear demonstration it provided of current ignorance and inadequate comprehension of some aspects of the subject while others have been more fully explored. The meeting, and this volume, will have fulfilled their purpose if they encourage soil biologists, plant physiologists, soil chemists and soil physicists to become more fully acquainted with each other's work and to recognize the essential contribution which each discipline must make if the complexities of the interfacial zone are to be adequately understood.

The experimental work described and the opinions expressed are those of the individual authors.

J.L. Harley
R. Scott Russell

ACKNOWLEDGEMENTS

The participants in the Symposium formally acknowledged their indebtedness to all those who had brought them together or assisted their deliberations, especially the Royal Society, the Agricultural Research Council, Wadham College, the Forestry and Physiology Departments of Oxford University and the joint symposium secretaries, Dr. P.B. Tinker and Dr. D.C. Clarkson. The editors of this volume gratefully acknowledge the assistance of those who refereed and helped to edit the many contributions.

CONTENTS

CONTENTS

VI. **Abstracts of Papers not read in full**

In alphabetical order of senior authors.

CONTENTS

CONTENTS

NOTE ON TERMINOLOGY:
ORIGIN, NATURE AND NOMENCLATURE OF THE ORGANIC MATERIALS IN THE RHIZOSPHERE

A.D. ROVIRA, R.C. FOSTER and J.K. MARTIN

C.S.I.R.O., Division of Soils, Glen Osmond, South Africa

During this Conference on "The Soil-Root Interface" it became apparent that there is considerable confusion on the origin, nature and nomenclature of the organic materials found around plant roots. The organizers of the Conference requested that a short statement be prepared to help clarify the problem. A difficulty with such an assignment with limited space is the complexity of the topic and the fact that many of the materials form a continuum with intermediate forms rather than discrete, well-defined compounds of a single origin. Further details are given in papers listed at the end of this paper.

Our conclusions on the origin, nature and terminology of these compounds is illustrated in the diagram and summarized as follows:

1. *Exudates* Compounds of low molecular weight which leak from all cells into either the intercellular spaces and then to the soil via the cell junctions or directly through the epidermal cell walls into the soil. The release of these compounds is not metabolically mediated.
2. *Secretions* Compounds of low molecular weight as well as high molecular weight mucilages which are released as a result of metabolic processes.
3. *Plant Mucilages* Four sources of plant mucilages which

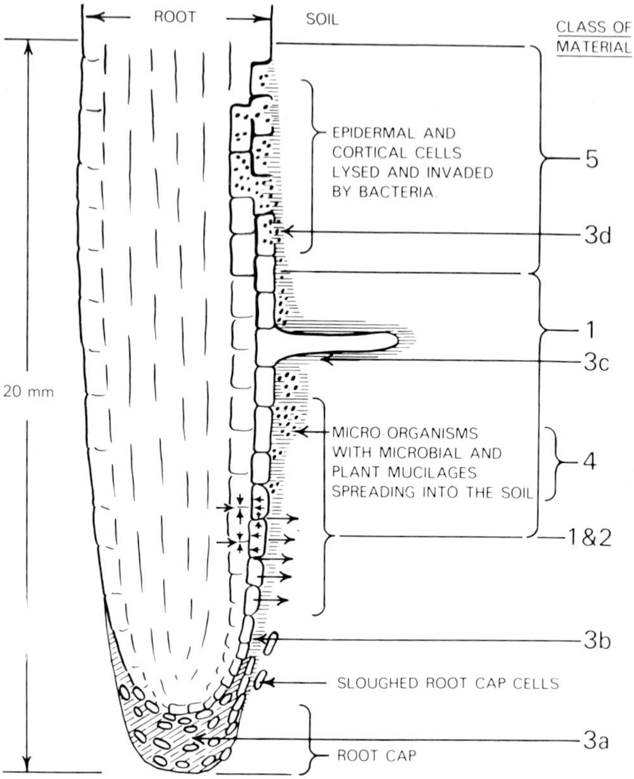

Fig. 1. Diagram of a root showing the origins of organic materials in the rhizosphere

contribute to the organic materials in the rhizosphere are:

(a) Mucilage originating in the root cap and secreted by Golgi.

(b) Hydrolysates of the polysaccharide of the primary cell wall between the epidermal cells and sloughed root cap cells.

(c) Mucilage secreted by the epidermal cells which still only have primary walls. This includes mucilages secreted by root hairs.
The development of secondary walls in these cells further from the tip means that no further secretions or mucilages are released, only exudates.

At this part of the root the primary plant mucilage is confined to the intercellular grooves where bacteria are unable to enter.

(d) Mucilage produced by bacterial degradation of the outer multilamellate primary cell walls of old, dead epidermal cells.

4. *Mucigel* Following Jenny & Grossenbacher (1963) we confine this term to mean the gelatinous material at the surface of roots grown in normal non-sterile soils. It includes natural and modified plant mucilages, bacterial cells and their metabolic products (such as capsules, glyocalyx materials and slimes) as well as colloidal mineral and organic matter from the soil. Whereas root mucilages entirely of plant origin and microbially produced mucilages can only be studied and characterized in axenic cultures, mucigel is a product of the entire root-soil-microbial complex with its own distinctive morphological and biochemical properties. The mucigel is important in maintaining contact between the root and the soil as the root shrinks during daytime water stress so permitting continuous uptake of nutrients and water.

5. *Lysates* These are the compounds released from autolysis of older epidermal cells when the plasmalemma fails. With further time the walls of these epidermal cells are digested by micro-organisms and the cells become heavily colonized releasing the products of microbial activity into the rhizosphere.

References

Thick Epidermal Cell Walls (Primary wall-root epidermis)

Foster, R.C. (1962). "Cell Wall Structure and Growth". Ph.D. Thesis, Leeds, 1962.

Leech, J.H., Mollenhauer, H.H. and Whaley, W.G. (1963). Ultrastructural changes in the root apex. *In* "Cell Differentiation" (Ed. G.E. Fogg), pp.74-84. Symp. Soc. Expl. Biol. XVII.

Preston, R.D. (1974). The Physical Biology of Plant Cell Walls. Chapman & Hall (London), p.396.

Root Cap Mucilage Secretions

Mollenhauer, H.H. and Morré, D.J. (1966). Golgi apparatus and plant secretions. *Ann. Rev. Pl. Physiol.*, 17, 27.

Mollenhauer, H.H., Whaley, W.G. and Leech, J.H. (1961). A function of the golgi apparatus in the outer root cap cells. *J. Ultra. Res.*, 5, 193.

Northcote, D.H. and Pickett-Heaps, J.D. (1966). A function of the golgi apparatus in polysaccharide synthesis and transport in root cap cells of wheat. *Biochem. J.*, 98, 159.

Lysis of Intercellular Substance at Root Cap

Juniper, B.E. and Roberts, R.M. (1966). Polysaccharide synthesis and the fine structure of root cap cells. *J. Roy. Micro. Soc.*, 85, 63.

Secretion of Mucilage by Epidermal Cells and Root Hairs

Sievers, A. (1963). Beteiligung des Golgi Apparates bei der Bildung der Zellwand von Wurzelhaaren. *Protoplasma*, 86, 188.

Lysates

Martin, J.K. (1977). Factors influencing the loss of organic carbon from wheat roots. *Soil Biol. Biochem.*, 9, 1.

Exudates

Rovira, A.D. (1969). Plant root exudates. *Bot. Rev.*, 35, 35.

Rovira, A.D. and McDougall, Barbara M. (1967). Microbiological and biochemical aspects of the rhizosphere. *In* "Soil Biochemistry" (Eds. A.D. McLaren and G.H. Peterson), p.417. Marcel Dekker, New York.

Mucigel

Dart, P.J. and Mercer, F.V. (1964). The legume rhizosphere. *Arch. Mikrobiol.*, 47, 344.

Bowen, G.D. and Foster, R.C. (1978). Dynamics of microbial colonization of plant roots. *In* "Symposium on Soil Microbiology and Plant Nutrition" (Eds. W.J. Broughton and C.K. Johns). Publ. Univ. Malaysia (in press).

Foster, R.C. (1979). Fine structure of mucigel. *New Phytol.* (in preparation).

Foster, R.C. and Rovira, A.D. (1976). Ultra-structure of wheat rhizosphere. *New Phytol.*, 76, 343.

Foster, R.C. and Rovira, A.D. (1978). The ultrastructure of the rhizosphere of *Trifolium subterraneum* L. *In* "Proc. Int. Symp. on Microbial Ecology" (Ed. M.J. Loutit), Springer Verlag, Berlin (in press).

Greaves, M.P. and Darbyshire, J.F. (1972). The ultrastructure of the mucilaginous layer on plant roots. *Soil Biol. Biochem.*, 4, 443.

Jenny, H. and Grossenbacher, K. (1963). Root-soil boundary zones as seen in the electron microscope. *Soil Sci. Soc. Amer. Proc.*, 27, 273.

Oades, J.M. (1978). Mucilages at the root surface. *J. Soil Sci.*, 29, 1.

GROWTH REQUIREMENTS FOR NUTRIENTS IN RELATION TO DEMAND AT THE ROOT SURFACE[1]

STANLEY A. BARBER

Department of Agronomy, Purdue University, Indiana, USA

Summary

Nutrient influx across the soil-root interface (rate of uptake per unit of root surface area) is influenced by the demand of the plant for the nutrients required for additional growth. Size of influx will depend on the relation between area of active root surface and the demand of the shoot. Nutrient demand per unit of root surface varies with many factors including plant species, cultivars within species, nutrient, plant age, root age, nutrient status of the plant, and light and temperature conditions. Interactions between these factors as well as changes within the plant to compensate for a low rate of supply of a particular nutrient make specific characterization of nutrient influx difficult. Average nutrient influx is usually greatest during the first two weeks of growth and then decreases as the plant matures. While effect of root age on influx differs with nutrient and plant species, influx of some nutrients continues even after root suberization. Split-root experiments indicate that increasing requirement of phosphate and potassium by the plant per unit of root does not increase influx until there is a reduction in the

---

[1] Journal Paper No. 7116. Purdue Univ. Agric. Expt. Stn., Lafayette, IN, 47907, U.S.A.

concentration of these nutrients in the shoot. Control of influx within the plant appears to reside in the shoot rather than the root. Limited data indicate that 60 per cent or more of the roots need to be absorbing at their maximum rate in order that influx of phosphate or potassium does not limit plant growth.

Plants have a capacity to compensate for deficiencies and when nutrients become limiting they may increase root size relative to shoot demand, increase root hair density and length.

Introduction

Plant demand for nutrients equals the increase in nutrient content of the plant with growth when nutrient supply to the root system is not growth limiting. Since mean nutrient concentration in the plant may decrease with plant age (Mengel and Barber, 1974) new growth may receive nutrients from older parts of the plant as well as those absorbed through the roots. The minimum demand for nutrients where growth rate is maintained is calculated using the lowest plant nutrient concentration that does not limit growth rate. This concentration has been called the "critical concentration" (Ulrich and Hills, 1973) by those using plant nutrient concentration as a measure of the nutritional status. Plant demand has frequently been expressed per plant per day or per unit of soil surface per day. A measurement which tests the ability of the soil to supply nutrients is the nutrient demand per unit of actively absorbing root per day. The product of the mean flux across the root surface times the total surface area of absorbing root determines the supply rate per plant. There is a relation between flux across the root surface and plant growth. An example of this is shown in Figure 1 where millet (*Panicum milaceum*) yield was related to mean P influx into the roots during growth (Adepetu and Barber, 1978). The paper presented here discusses plant factors

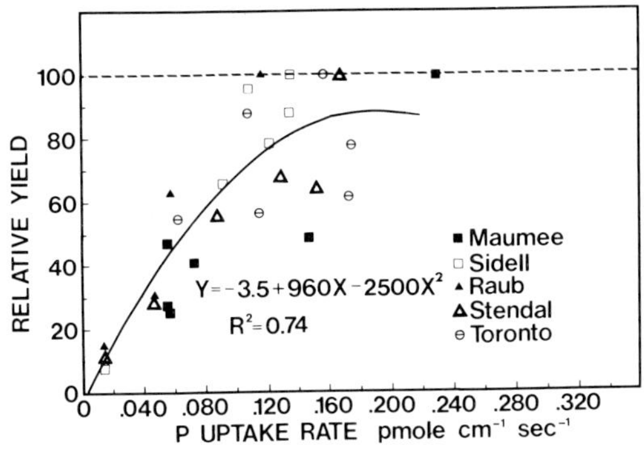

Fig. 1. Relation of mean phosphate uptake rate to relative yield of millet grown on five soils at five levels of P.

that affect plant demand for nutrients at the soil-root interface.

Characterizing Plant Demand

Nutrient absorption by plant roots requires respiration energy supplied by photosynthate transported to the root from the shoot. The rate of absorption is influenced by photosynthate supply plus a favourable temperature, aeration, and concentration of nutrients in the solution around the roots. The site of absorption is usually considered to be the plasmalemma of the root cortical cells. The relation between ion concentration in solution and ion influx into the root has been studied extensively using excised roots and to a lesser degree using intact plants. With intact plants both absorption and translocation occur and results may differ from those obtained with excised roots, hence in this discussion results with intact plants will be emphasized.

The isotherm of influx versus concentration of the ion in solution has been described by various modifications of the Michaelis-Menten equation, equation [1]

$$I = \frac{Imax\ C}{Km + C} \qquad [1]$$

where I is influx, Imax (also termed Vmax) is maximal influx, Km is the Michaelis-Menten constant and is the concentration where I = 1/2 Imax, and C is the concentration of ion in stirred solution at the root surface. This equation only fits the data over a limited concentration range and when the range is exceeded, different values of Km and Imax are needed for it to fit the experimental data (Nissen, 1977). Concentrations found for many nutrients in soil solution are low and usually in the range of the first phase. Hence the Michaelis-Menten equation is useful for describing the absorption isotherm. In addition, nutrient flux into intact plants usually changes less with increased nutrient concentration (Pitman, 1976) than nutrient flux into excised roots so that for intact plants one set of Michaelis-Menten parameters may often be sufficient for the range of nutrient concentrations usually found in soil.

Absorption isotherms that describe the demand for a nutrient as related to the concentration in solution have been obtained by measuring the rate of depletion from the solution supplying the plant (Claassen and Barber, 1974). When plants were allowed to deplete the nutrients from solution until net influx, In, reached zero, a minimum concentration, Cmin, was present. Because Cmin was not zero when In = 0, equation [1] had to be modified as shown in equation [2] in order that the equation would describe the absorption isotherm obtained.

$$In = \frac{Imax\ (C - Cmin)}{Km + (C - Cmin)} \qquad [2]$$

The values of Imax, Km, and Cmin describe the absorption characteristics of the root for each nutrient. At a particular level of C, In will be greater, the larger Imax, the smaller Km and the smaller Cmin.

Factors influencing Nutrient Demand

Nutrient

Demand by the plant for the various mineral nutrients varies. Requirement for NO_3^-, $H_2PO_4^-$ and K^+ by 18-day-old corn (*Zea mays* L.) plants are shown in Table 1 where data are the mean of several experiments conducted in a plant growth chamber at 28°C. The data for NO_3^- are from Edwards and Barber (1976); those for $H_2PO_4^-$ are from Jungk and Barber (1975) and those for K^+ from Claassen and Barber (1975).

TABLE 1

The influx characteristics of intact 18-day-old maize roots growing in solution culture at 28°C

Nutrient	Imax pmol cm^{-1} sec^{-1}	Km µM	Cmin µM
NO_3^-	1	12	4
$H_2PO_4^-$	0.4	3	0.2
K^+	2	17	2

The lower values for $H_2PO_4^-$ reflect both the lower requirement by the plant than for NO_3^- and K^+ and also the lower concentration of $H_2PO_4^-$ usually found in the soil solution which would require a higher efficiency of absorption in order to supply the demand of the plant root. Imax is described according to root length in this table, it can also be described in terms of root surface area and root fresh weight (root volume).

Root age

Average influx for the root system is frequently calculated to describe influx and it averages variation that may occur with root age. Where variation does occur the actual influx of actively absorbing roots may be several times larger than the mean value. Investigations of influx along plant roots have given results that vary with plant species and ion investigated. Ferguson and Clarkson (1975) measured $H_2PO_4^-$ and Ca^{++} absorption and translocation during 24 h along the roots of intact 10-day-old corn seedlings. Uptake of $H_2PO_4^-$ was relatively uniform in all parts of the seminal root axis even though suberization and thickening of the endodermis occurred. However Ca^{++} uptake and translocation reached a maximum 12 cm from the root tip and then decreased to a very low rate near the base of the root. Hence, the effect of root age on nutrient influx varied with nutrient.

Plant age

The mean influx per unit of root varies with plant age. Mengel and Barber (1974) measured mean influx per meter of root of corn growing in the field and obtained the relation between nutrient influx and plant age shown by the data in Table 2. Influx decreased markedly as plant age increased. Greenhouse experiments where corn was grown in solution culture showed a similar relationship of nutrient influx decrease with increasing plant age (Warncke and Barber,

TABLE 2

The effect of plant age on the mean net influx of nutrients into corn roots growing in a fertile Chalmers silt loam

Plant age	Mean influx of				
	N	P	K	Ca	Mg
days	μmol/m per day				
20	226.9	11.30	52.90	14.40	13.80
30	32.4	0.90	12.40	5.20	1.61
40	18.5	0.86	8.00	0.56	0.90
50	11.2	0.66	4.75	0.37	0.78
60	5.7	0.37	1.63	0.20	0.56
70	1.2	0.17	0.15	0.05	0.28
80	0.5	0.08	0.06	0.06	0.19
90	2.0	0.10	0.37	0.06	0.17
100	4.2	0.23	0.16	0.08	0.29

1974). The reduction in influx with plant age was greater than could be attributed to changes in mean root age. It is apparent that root growth rate was more rapid than the increase in nutrient demand per plant so that as the corn plant reached the reproductive stage average influx of only one to two per cent of that which occurred at 20 days was sufficient to meet plant demand for nutrients.

Since the foregoing experiments only measured mean influx, Jungk and Barber (1975) determined the relation between influx and concentration for corn plants of varying ages. They found, Figure 2, that Imax decreased with age just as the mean influx had in the experiments of Mengel and Barber (1974) and Warncke and Barber (1976). They found Km did not vary with plant age but Cmin decreased with plant age. How does plant demand determine influx?

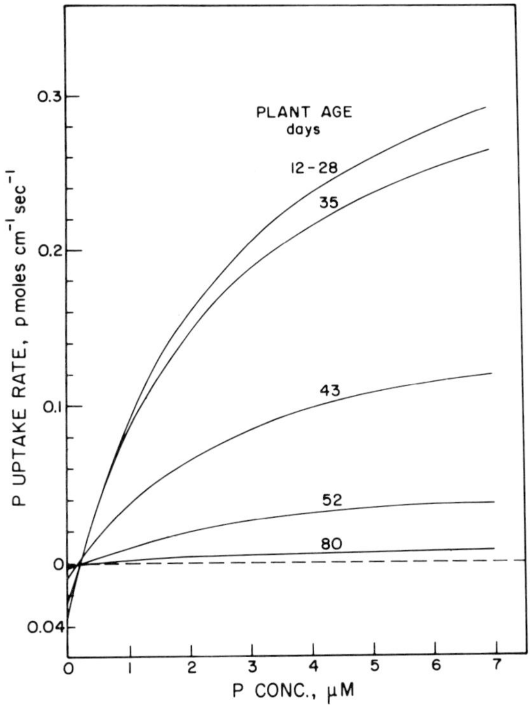

Fig. 2. Effect of plant age on phosphate uptake rate as influenced by P concentration in solution.

In order to alter plant demand per unit of root, Jungk and Barber (1974) used split root systems to supply phosphate to part of the root systems. They also removed part of the root system by trimming. When the treatments were imposed 24 h before influx measurements were made, they found influx was not influenced by plant demand per unit of root. Figure 3 shows phosphate uptake was proportional to root length per plant where plant size was constant.

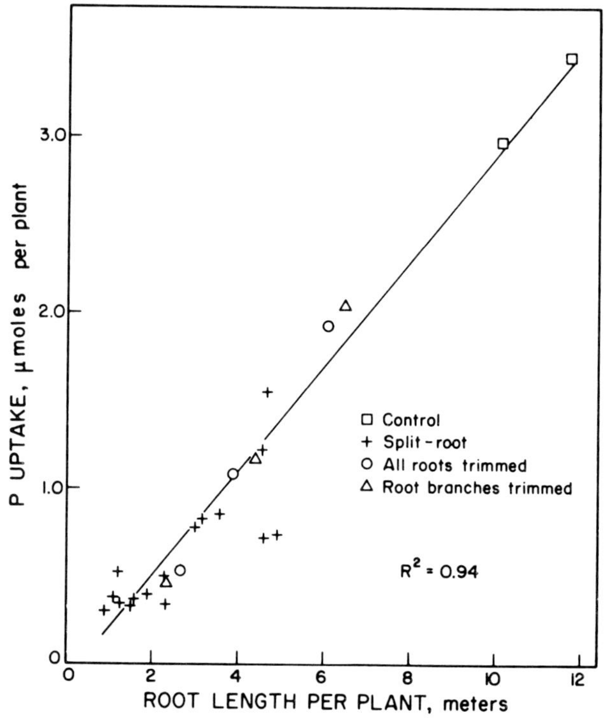

Fig. 3. Phosphate uptake in 4 h by 13-day-old corn plants as affected by root supplied with phosphate.

Plant nutrient status

In the trimmed-root experiments of Jungk and Barber (1974) the trimming was done only 24 h before influx measurement so the decrease in influx presumably was insufficient to decrease the nutrient status of the plant appreciably. Claassen and Barber (1977) studied the effect of potassium status of the shoot and of the root on the potassium influx characteristics of corn roots. Seventeen-day-old corn plants were produced which had a range in per cent K in the shoot dry matter of 2.25 to 9.66% by growing them in solutions differing in K concentration, by growing some in split-root systems where only part of the roots were supplied with K, and by growing them in nutrient solution without K during parts of the growth period. Corn plants that were lower in per cent K had higher Imax values for K influx by the roots. The data from six experiments shown in Figure 4 indicates a relation between K status of the plant and Imax. There was a 2.5 fold range in Imax values.

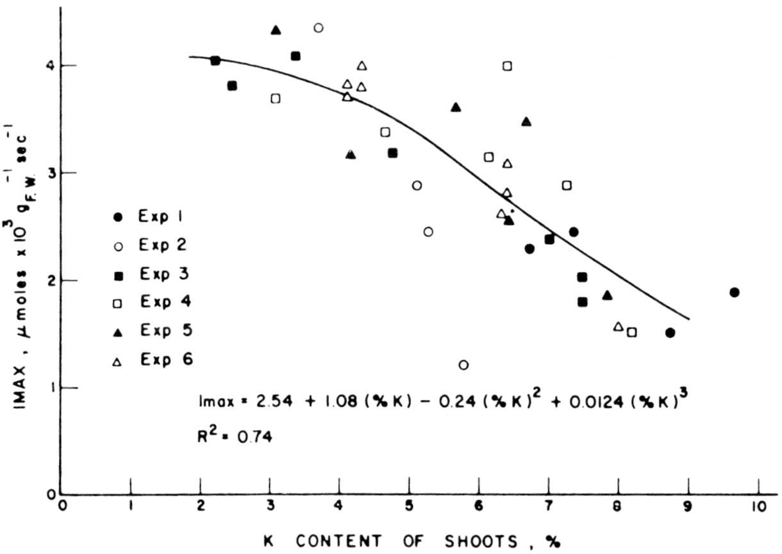

Fig. 4. Relation between K percentage in the shoot of 17-day-old corn plants and Imax for K influx into the roots using data from six separate experiments.

In corn plants grown in split-root containers it was possible to investigate the Imax for K uptake by roots which varied in K content and were attached to the same shoot since part of the roots were grown in K-free nutrient solution and part were in K-containing nutrient solution. The data shown in Table 3 show that Imax was about 10 per

TABLE 3

The effect of previous K treatment in a split-root solution culture experiment on corn shoot and root K concentration and Imax for subsequent K influx

Per cent of root exposed to K	K supplied	K in shoots %	K in roots %	Imax pmol cm^{-1} sec^{-1}
100	+	8.00	5.85	15.8
50	+	6.45	5.55	28.0
	-		1.88	30.8
25	+	4.35	4.99	33.8
	-		1.24	39.8
15	+	4.13	5.51	36.8
	-		1.20	38.0

cent higher for roots that had been grown in K-free nutrient solution than for those grown in K-containing nutrient solution. However the K influx for plants with only part of their roots supplied with K and hence a lower shoot K content had an influx 133 per cent more than plants that had been grown with all their roots in K-containing nutrient solution. These data indicate that there may be a feed-back mechanism between the shoot and root where the nutrient level in the shoot determines influx.

Species differences

Influx characteristics may vary with plant species. Loneragen and Asher (1967) have used flowing cultures to investigate the effect of phosphate concentration in solution on the relative growth rate of a number of plant species and have shown differences between species. Mean influx per gram fresh weight of root varied with species.

In my laboratory we have investigated the influx characteristics of NO_3^-, NH_4^+, $H_2PO_4^-$, and K^+ by maize and of $H_2PO_4^-$ and K^+ by soybeans (*Glycine max* L.). Values for NO_3^-, $H_2PO_4^-$ and K^+ uptake by corn roots were given in Table 1. Values for $H_2PO_4^-$ absorption by 18-day-old intact soybean roots were: Imax, 0.1 pmol $cm^{-1}sec^{-1}$; Km,

2 μM; and Cmin, 0.05 μM (Edwards and Barber, 1976). Values for K^+ absorption by 18-day-old intact soybean roots were: Imax, 0.4 pmol cm^{-1} sec^{-1}; Km, 16 μM; Cmin, 0.5 μM (Youngdahl and Barber, unpublished data). The difference between these two species was mainly in the lower Imax and Cmin values of the soybeans. Also there could be differences between varieties within each species as large as the differences between species shown here.

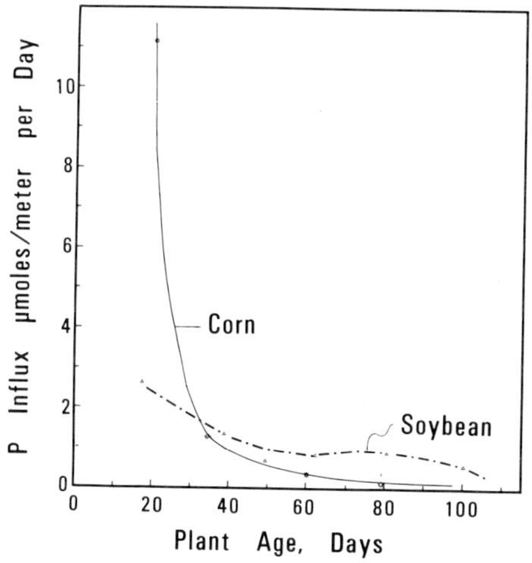

Fig. 5. Comparison of mean P influx by corn and soybean roots during their growth cycles.

There are differences between corn and soybeans in the change in nutrient influx with plant age. When mean P influx of plants grown in field experiments are compared, Figure 5 (Barber, 1978), we see that corn has a much greater influx at 20 days but at 80 days P influx into soybean roots is several times that found with corn. Part of the reason for this difference can be found in differences between corn and soybeans in the P requirement of the plant. The P concentration of corn plants continues to decrease throughout the life of the plant. The P concentration of soybean plants reaches a minimum at the time seed formation begins and increases after this time so that the plant requires a greater influx at the root surface to supply the plant requirement during this latter period of growth. The data shown in Figure 5 illustrate the wide differences that can occur between species in changes in nutrient demand at the root surface with plant age.

Variety differences

Influx varies with varieties within a plant species. Fortunately, this may allow plant breeders to develop varieties that are more efficient in nutrient absorption. Neilsen and Barber (1978) investigated the P influx characteristics of several inbreds and single-crosses of hybrid corn. The data for P influx by 5 inbreds and 4 single cross hybrids shown in Table 4 indicate a wide range in Imax, Km, and Cmin for P absorption by the root system of these plants measured when they were 21 days old. The data for the various single crosses indicate inheritance of P absorption properties.

Temperature

Changes in temperature of the rooting media of a plant changes the root growth rate relative to the shoot growth rate and also the rate of nutrient influx. Neilsen (1974) in a review on the effects of temperature on root growth found that for many species, increasing temperature from suboptimum levels reduced root-shoot ratio and hence changed the size of the absorption surface per unit of nutrient demand by the plant. Walker (1969) found that a temperature of 28°C was optimum for maximizing corn seedling growth. In the temperature range 10°C to 25°C increasing temperature usually increases influx with a Q_{10} of at least 2. Ching (unpublished data) working in our laboratory found Imax for K^+ uptake by 19-day-old corn seedlings was 2.41 pmol cm^{-1} sec^{-1} at 27°C and only 0.34 pmol cm^{-1} sec^{-1} at 13°C which gives a Q_{10} of 5. Increasing the temperature to 34°C caused a small decrease in K^+ influx over that measured at 27°C and influx at 22°C was

TABLE 4

Plant dry weight, root length, and P uptake characteristics of five inbreds and five single crosses of corn measured in solution culture at 20 to 22 days

Corn pedigree	Plant dry weight	Root length per g of		P uptake		
		root	plant	I_{max}*	Km	C_{min}
	g/plant	m/g			µM	
C103	2.04	173	26	0.26	2.3	0.12
H60	1.37	130	20	0.42	2.9	0.12
H84	3.10	118	22	0.25	3.1	0.09
H99	1.45	161	25	0.32	4.2	0.09
W64A	2.27	218	39	0.16	2.6	0.07
$lsd_{.05}$	0.67	54	4.6	0.065	1.3	----
n**	4	4	4	2	2	1
C103 x W64A	3.07	174	26	0.13	0.6	0.06
H60 x W64A	2.86	172	28	0.15	1.3	0.06
H60 x C103	2.82	150	22	0.23	1.5	0.04
H84 x H99	3.29	184	29	0.34	2.4	0.04
$lsd_{.05}$	0.40	38	4.0	0.13	0.44	0.07
n**	4	6	6	2	2	2

* pmol cm^{-1} sec^{-1}

** Number of experiments (replicates)

only slightly less than that at 27°C.

Light

When light intensity is reduced root growth decreases more than shoot growth. This would produce plants that would have a higher nutrient demand per unit of root surface.

Root morphology

The length and radius of plant roots varies according to the medium in which it grows. If the medium is difficult for the roots to penetrate then the roots will have a larger radius and may be shorter. In our laboratory W.R. Peterson (unpublished data) grew soybeans in solution culture and in sand through which a steady flow of nutrient solution was maintained. He found that the morphology of the roots was changed and some results are given in Table 5. Root radius was increased. The K^+ absorption characteristics of these two types of roots were examined. The sand grown roots were removed from the sand by washing immediately before measuring their influx characteristics.

TABLE 5

Influence of root morphology on potassium influx by roots of 18-day-old soybean plants

Rooting media	Root radius mm	Imax pmol cm^{-1} sec^{-1}	Root length per gram dry wt. meters
Solution	0.20	0.41	164
Sand	0.25	0.91	119

The data in Table 5 show that the coarser roots obtained in the sand had a higher Imax when measured per unit length of root. However when they were measured per unit of root fresh weight the influx values were not as different. Cross-section slices of the roots were examined microscopically and the radius of the stele was the same for both roots but the cortex while having the same number of cells was much thicker on the sand grown roots. Possibly the reason for the greater influx per unit of root was because of the greater surface area of plasmalemma per cm of root length that was available for nutrient absorption.

Nutrient demand as a proportion of Imax

Influx varies with concentration of the nutrient in the growth media. The influx as a fraction of Imax required to avoid growth limitation gives an indication of the level needed in solution. Claassen and Barber (1977), using split-root containers, found that approximately 50 per cent of the roots had to be supplied with K to avoid growth reduction. This would indicate that influx needs to be at 1/2 Imax when all plant roots are supplied to avoid growth limitation. Hence Km would measure the minimum concentration that would need to be maintained at the root surface so that growth would not be limited by lack of the nutrient measured.

Compensation in Plant Growth

When plants grow under conditions where a nutrient or an environmental condition is less than most favourable the plant responds by altering its root growth and demand for nutrients at the root in such a way as to provide more nutrients for the plant. In field experiments in Indiana, where corn was grown with various rates of application of N, P, and K we measured root length at the tasselling stage of growth (unpublished data). In experiments with high rates of N root density was 66 cm/cm^2 and with low N it was 96 cm/cm^2. In a separate experiment where all nutrients were at high levels except P or K, the root density with high levels of both P and K was 50 cm/cm^2, with low P and high K it was 72 cm/cm^2 and with low K and high P it was 53 cm/cm^2. Our observations are that when the corn plant had a deficiency of N or P, it produced more roots. A deficiency of K had little effect on root length and this agreed with observations of Claassen and Barber (1977).

References

Adepetu, J.A. and Barber, S.A. (1978). Soil phosphorus availability to millet roots. *Comm. in Soil Sci. and Plant Anal.*, 9 (in press).

Barber, S.A. (1978). Growth and nutrient uptake of soybean roots under field conditions. *Agron. J.*, 70 (in press).

Claassen, N. and Barber, S.A. (1974). A method for evaluating the influence of concentration on the uptake rate of nutrients. *Plant Phys.*, 54, 564.

Claassen, N. and Barber, S.A. (1977). Potassium influx characteristics of corn roots and interaction with N, P, Ca, and Mg influx. *Agron. J.*, 69, 860.

Edwards, J.H. and Barber, S.A. (1976). Nitrogen uptake characteristics of corn roots at low N concentrations as influenced by plant age. *Agron. J.*, 68, 17.

Ferguson, I.B. and Clarkson, D.T. (1975). Ion transport and endodermal suberization in the roots of *Zea mays*. *New Phytol.*, 75, 69.

Jungk, A. and Barber, S.A. (1974). Phosphate uptake rate of corn roots as related to the proportion of the roots exposed to phosphate. *Agron. J.*, 66, 554.

Jungk, A. and Barber, S.A. (1975). Plant age and the phosphorus uptake characteristics of trimmed and untrimmed corn root systems. *Plant and Soil*, 42, 227.

Loneragen, J.F. and Asher, C.J. (1967). Response of plants to phosphate concentration in solution culture: II. Rate of phosphate absorption and its relation to growth. *Soil Sci.* 103, 311.

Mengel, D.B. and Barber, S.A. (1974). Nutrient uptake rates per unit of root of corn grown under field conditions. *Agron. J.*, 66, 399.

Neilsen, K.F. (1974). Roots and root temperature. *In* "The Plant Root and its Environment" (Ed. E.W. Carson), p.293. The Univ. Press of Virginia, Charlottesville, Virginia.

Neilsen, N.E. and Barber, S.A. (1978). Differences between genotypes of corn in the kinetics of phosphorus uptake. *Agron.J.*, 70 (in press).

Nissen, P. (1977). Ion uptake in higher plants and KCl stimulation of plasmalemma adenosine triphosphatase: Comparison of models. *Physiol. Plant.*, 40, 205.

Pitman, M.G. (1976). Nutrient uptake by roots and transport to xylem: Uptake processes. *In* "Transport and Transfer Processes in Plants" (Ed. I.F. Wardlaw and J.B. Passioura), p.85. Academic Press, New York.

Ulrich, A. and Hills, F.J. (1973). Plant analysis as an aid in fertilizing sugar crops. I. Sugar beets. *In* "Soil Testing and Plant Analysis" (Ed. L.M. Walsh and J.D. Beaton), p.271. Soil Sci. Soc. Amer., Wisconsin.

Warncke, D.D. and Barber, S.A. (1974). Root development and nutrient uptake by corn grown in solution culture. *Agron. J.*, 66, 514.

Walker, J.M. (1969). One-degree increments in soil temperature affect maize seedling behaviors. *Soil Sci. Soc. Amer. Proc.*, 33, 729.

PROPERTIES OF ROOTS WHICH INFLUENCE RATES OF ABSORPTION

M.C. DREW

Agricultural Research Council Letcombe Laboratory, Wantage, England

Summary

A review is given of some of the structural and physiological factors which control the radial flow of nutrient ions and water across the root to the xylem. Consideration is given to the pathways of ion and water movement and the resistance provided by the development of the endodermis, the hypodermis and, in woody species, the periderm. Attention is drawn to the manner in which soil environmental factors - temperature, water potential, aeration, mechanical impedance and nutrient supply - can modify root structure and absorptive function, and the possible consequence in estimating the flow of materials across the soil-root interface is discussed.

Introduction

Absorption and transport of nutrient ions and water by the roots of intact plants are complex processes which depend not only on the properties of cell membranes, but also on the structural organization of the root as it ages, the requirements of the plant, and the numerous soil factors which continuously influence the growth and function of the root system. The aim of the present paper is to discuss some aspects of plant behaviour which can affect the ability of different parts of the root system to extract ions and water from the soil across the soil-root interface. To assess the soil properties which may limit the rates of uptake of nutrients and water by crops, it is necessary to know the average flux across root surfaces in soil, based on the length, weight or volume of roots and to assume that all parts of the root system function equally (Gardner, 1960; Newman, 1969; Brewster and Tinker, 1972). However, the anatomy and morphology in various parts of the root system varies greatly according to the age of the tissue and the soil environment; this has important consequences for root function.

Structural Changes in Roots which modify Rates of Absorption of Ions and Water

There are broadly two pathways available for the passage of ions and water across the cortex of the root to the xylem. The apoplast, or extracellular pathway, is comprised of the continuity of water filled micro-capillaries in the cell walls. Its properties are reviewed by Läuchli (1976a). The symplasm, or intracellular pathway, is from protoplast to protoplast, via plasmodesmata. Its general characteristics are reviewed by Spanswick (1976), and with particular reference to roots by Läuchli (1972, 1976b). There is now convincing evidence that the rapid movement of ions and of water across the root occurs mainly along low resistance pathways, probably in part both apoplasmic and symplasmic.

In cereals the cortical cells often comprise 80 per cent of the volume of the root and their walls are readily accessible to the flow of water and the diffusion of ions from the outer solution. A few millimeters behind the root tip the Casparian band is laid down in the radial and transverse walls of the endodermis, and the suberin and lignin incrustations of the primary wall at this point effectively block the passage of ions, and probably water, in the apoplast towards the stele (Clarkson, 1974;

Clarkson and Robards, 1975). The transport of all materials to, and from, the stele is therefore forced to occur through the protoplasts of the endodermal cells.

The presence of the Casparian band fulfils at least three functions. It acts as a barrier to the free diffusion out of the root *via* the apoplast, of ions previously transported to the xylem vessels; it allows the development of root pressure in the xylem vessels; it ensures selectivity in ion transport towards the xylem, as all ions must pass at least one set of cell membranes, either at the endodermis, while in its primary state, or in entering the symplasm. During further development of the root, a succession of potential barriers to the movement of ions and water are constituted by the deposition of suberin lamellae in the endodermis, hypodermis, and in woody roots in the cork cells (phellem) of the periderm. The suberin is laid down as continuous hydrophobic layers over the inner surfaces of the entire primary cell walls. Such water repellent barriers may confer some resistance to desiccation (Drew, 1978) and retard the invasion of soil microorganisms (Van Fleet, 1961; clarkson *et al.*, 1975), so allowing the conducting tissues within the stele to continue to function, but their influence on the passage of materials across the root to the xylem is appreciable.

In the apical zone of the root, cells are thin-walled, rich in cytoplasm and have a high rate of respiration. Absorption of ions and water is not restricted, however, to this zone. Relatively old parts of the seminal axes of barley, maize and marrow (*Cucurbita pepo*), located as much as 50 cm from the root tip, can be as effective as younger tissues in the absorption and transport of potassium and phosphate, despite the presence of a thick walled suberized endodermis with few passage cells (for reviews see Clarkson, 1974; Russell and Clarkson, 1976). Examination of the thickened inner cell walls of the endodermis (Fig. 1) reveals the presence of pits, in which numerous plas-

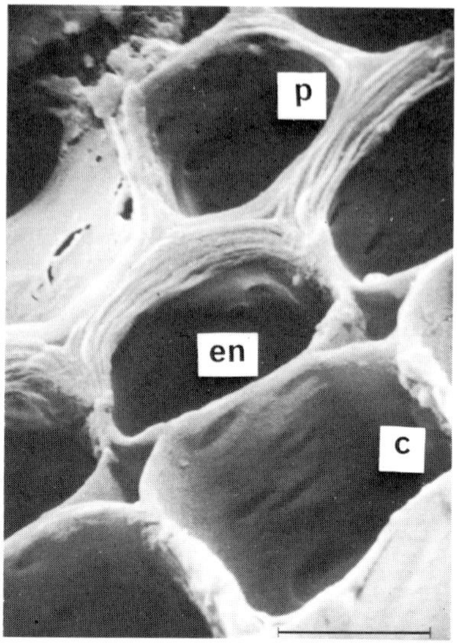

Fig. 1. Endodermis of maize root. The scanning electron micrograph is a transverse section of an adventitious root, about 20 cm from the tip. en, endodermis; c, cortical cell; p, pericycle. Note pits in walls between cortical cell, endodermis and pericycle and laminated thickenings of the wall of the endodermis (Drew, Chamel, Garrec and Fourcy, unpublished). Bar = 10 μm.

modesmata are located. These latter are thought to provide the symplasmic continuity between the cortex and stele for the transport of phosphate and potassium (Clarkson and Robards, 1975). The movement of calcium and magnesium into the xylem is, by contrast, greatly restricted by the development of suberized lamellae in the endodermis, suggesting that these ions are not readily transported in the symplasm, but normally cross the plasma membranes of the young endodermal cells (State 1) to reach the stele. The development of continuous lamellae of suberin (State 2) effectively cuts off the endodermis from access other than *via* the symplasm.

The region of maximum absorption of water by roots of monocots and dicots usually occurs from within a few mm of the tip, where xylem vessels are beginning to mature, to a zone about 10 cm behind the tip (Clarkson and Robards,

1975), that is, in tissues only 2-3 days old. In both barley and marrow the flux of water is progressively reduced to about 15 per cent of the maximum rate when suberin lamellae are laid down in the endodermis. With further increase in root age, or distance from the tip (up to 50 cm), the flux of water remains unchanged. The endodermis thus provides an important resistance to water flow across the root. However, under conditions of high transpiration the resistance to water flow appears to decrease, and flux may locally increase, the region of maximum water uptake becoming displaced towards the basal zones of the root where suberization of the endodermis is more advanced. The mechanisms regulating enhanced water flow in these basal zones are unknown (Weatherley, 1975).

In some species in which the root cortex persists, a hypodermis may develop in the cortical cells immediately beneath the epidermis. Suberin lamellae become deposited in the walls in much the same way as in the endodermis. In maize the hypodermis, unlike the endodermis, can act as a barrier to the uptake of phosphate (Ferguson and Clarkson, 1975, 1976). Micro-autoradiography revealed that ^{32}P-labelled phosphate failed to penetrate beyond the epidermis. Access of ions to the stele *via* the symplasm may be restricted by the paucity of plasmodesmata in the tangential walls linking epidermis and hypodermis, or perhaps by the blocking of pits which can occur in cork cells.

In the roots of woody angiosperms and gymnosperms, and in most herbaceous dicots, a periderm arises and the epidermal and cortical tissues exterior to it are eventually shed. The outer layer of the root is then constituted by the cork cells, which when mature are dead. Kramer and co-workers (Kramer and Bullock, 1966; Kramer, 1969; Chung and Kramer, 1975) have shown that such roots can display a highly variable but appreciable permeability to water in response to hydrostratic pressure gradients, and because they comprise most of the root system in mature plants the combined contribution to the water supply of the plant must often overwhelm that of the younger unsuberized tissues. The results of Chung and Kramer (1975) and references to the older literature cited by Kramer and Bullock (1966), indicate that despite the periderm, roots are able to conduct ions to the xylem. Water and solutes are believed to enter through lenticels and splits in the periderm, but the pathway towards the xylem in such roots has not been investigated by modern techniques. The extent to which the phellogen, the secondary phloem and the cambium resist water flow, or exert a selectivity over the entry of ions, is largely unknown.

In some cereal and grass species and in some dicots the cortex is not persistent and degenerates at an early stage in the development of the root leaving the endodermis or pericycle as the outermost layer of intact cells (Head, 1973; Waid, 1974). Although the root environment influences the degeneration of the cortex, the tendency appears to be constant within species, oats and barley for example having a more persistent cortex than wheat or rye (Troughton, 1957, 1962). Apart from any influence the breakdown of the cortex may have on the nature of the contact between the root and the soil, and on the activities of soil microflora and fauna, the absorptive properties of the root might be expected to be modified. However, experiments in which the cortex or part of it have been removed from the stele show that some degree of control of ion and water absorption is retained provided that the endodermis remains intact (Anderson and Reilly, 1968; Clarkson, 1974). Whether the above conclusion can be applied to soil grown roots is not clear, for the degradation of the cortex in older roots is accompanied by a considerable lignification of the endodermis, pericycle and stelar parenchyma, and no assumptions concerning the properties of such roots can be made at present.

Modifications to Root Structure and Absorptive Function in Response to Soil Conditions

The primary edaphic factors which influence the growth and function of roots in soil can be identified as temperature, soil water potential, oxygen concentration, mechanical resistance, and the concentrations of nutrient ions and other solutes. For each factor, the range favourable for root metabolism is narrow, so that roots are almost always subject to some degree of stress. In considering absorption by plants, whether in agricultural or natural conditions, the influence of edaphic factors on root growth is often of predominant importance, for these largely determine the extent to which the soil is exploited by the root system (see reviews by Shaw, 1952; Kramer, 1969; E.W. Russell, 1973; Drew and Goss, 1973; R.S. Russell, 1977). The manner in which these same soil factors can also influence the ability of roots to absorb and transport materials to the stele is examined in this section.

In attempting to quantify the degree of stress which affects any root function, it is important to distinguish between direct effects on the behaviour of the root, and effects determined mainly by the response of the shoot and its control of root activities. Transient changes can take place in the rate of transport of ions to the xylem in response to water stress in the shoot, without direct injury to roots (Pitman *et al.*, 1974). The temperature of the rooting medium may also influence absorption indirectly, through effects of root-produced hormones on the growth of the shoot and its capacity as a "sink" for ions. In maize, root temperatures below the optimum range for shoot growth cause a decrease in the amounts of cytokinins, and an increase in abscisic acid, in xylem exudates (Atkin *et al.*, 1973). In *Nicotiana rustica* and *Phaseolus vulgaris*, heat treatments (46-47°C) to roots of only 2 min duration led to similar changes in the pattern of hormones in the xylem sap to those described above, which were correlated with reductions in shoot growth (Itai *et al.*, 1973).

It is also worthwhile bearing in mind what are probably general responses to adverse soil conditions. It is a common observation that when the extension of roots is retarded, differentiation is slowed less than extension, so that cell maturation occurs much closer to the apex than in rapidly extending roots. The hypodermis and endodermis are thus found to develop closer to the apex when soil conditions are unfavourable (Van Fleet, 1961). A second general response concerns the behaviour of root systems growing in non-uniform media, in which growth is accelerated in parts of the root system receiving the most favourable conditions, often at the expense of the remainder (see R.S. Russell, 1977). This compensatory response may also extend to rates of ion uptake, per unit length, weight or volume of roots. The overall effect is thus for both root growth and absorption to become concentrated, transiently, in the more favourable zones of the soil.

In what follows we consider some of the mechanisms by which individual factors of the soil environment influence root function.

Temperature

Plant species originating from contrasting climates display wide differences in the optimum temperature for root function, and relatively small departures from the optimum temperature can depress rates of ion uptake and transport, and permeability to water, particularly in short-term experiments (Nielsen and Humphries, 1966; Kramer, 1969). In bean, shifts in temperature above or below the optimum range (25-30°C) caused the endodermis to become suberized closer to the root tip, with concomitant changes in the rate of xylem exudation, and development of symptoms of leaf water stress (Brouwer and Hoogland, 1964). When root temperatures are gradually reduced, allowing a period of adaptation, effects on permeability to water may become much less marked (Kramer, 1969). This type of adaptation has been studied in detail with barley and rye (Clarkson, 1976). Pretreatment of the root system at low temperature ($8^{o}C$) raised the subsequent permeability to water and stimulated transport of potassium, calcium and phosphate ions to the xylem when roots were returned to $20^{o}C$, compared with plants continuously maintained at $20^{o}C$. Pretreatment at $8^{o}C$ increased the unsaturation of the fatty acids associated with phospholipids in the roots, particularly in rye. These changes might influence transport processes across cell membranes by maintaining them in a fluid condition at the lower temperature (Clarkson and Hall, 1977). Such observations provide some insight into mechanisms acting at the cell membrane level by which roots may adapt to seasonal changes in soil temperature, and to temperature gradients within the soil profile.

Soil water potential

Even in temperate climates, plant roots are often subjected to prolonged periods of water stress, followed by re-wetting, particularly in the upper layers of the soil. Mechanisms by which roots combine a degree of resistance to desiccation with continued ability to absorb ions and water probably owe much to the role of suberin lamellae in the endodermis, hypodermis and phellem. When water stress is allowed to develop gradually in the soil, roots show a decreased permeability to water for some days after re-wetting (Kramer, 1950; Hansen, 1974), an effect attributed to retardation of root extension and suberization closer to the root apex. The suberization of dormant roots (metacutinisation) is a particularly striking example. In roots of *Pinus halepensis* and *P. resinosa* in response to dry soil conditions, root extension ceases and cells just beneath the root cap become suberized forming a continuous layer with the endodermis (Leshem, 1965, 1970; Wilcox,

1968). When the soil is re-wetted the root apices are able to rupture the suberized layer and extension resumes within only a few days.

When seminal root axes of barley, which do not form a hypodermis, were subjected to prolonged desiccation, leading to collapse of the cortex, the rate of transport of ^{32}P-labelled phosphate to the xylem on subsequent re-wetting was reduced to about 5 per cent of that in un-stressed roots (Clarkson *et al.*, 1968). Autoradiography showed that phosphate did not penetrate as far as the stele, suggesting that damage to the symplasm prevented circumvention of the suberized endodermis.

Mucigel, produced largely by the root cap, invests the outer surface of the root tip and the epidermis for about 1 cm from the root tip (Oades, 1978). It is possible that mucigel retards desiccation of the apical meristem and young tissues before suberized lamellae develop in the endodermis or hypodermis, but there is no direct evidence on this. In dry soil there may be increased production of mucigel by the roots of crop plants and native species, resulting in an annulus of soil tightly adhering to the root (see Drew, 1978). Because of its pectic constituents, mucigel readily binds cations and may therefore serve a protective function by retarding the penetration of heavy metal cations towards the apical meristem prior to development of the Casparian band (Barlow, 1975). The mucigel may also help maintain hydraulic conductivity between root and soil.

When young root tissues are rapidly exposed to water stress, changes in the ability of the cells to accumulate ions can be greatly reduced by mechanisms unrelated to suberization. Cole and Alston (1974) exposed root tips of wheat to dry air, causing tissue water potentials to fall as low as -24 bar. The ability to accumulate ^{36}Cl immediately upon re-wetting was greatly impaired, but after 18 h the rate of accumulation was restored to the same level as in unstressed roots. No observations were reported in the above study of the effect of water stress on cell fine-structure, but Nir *et al.* (1969) found that root tips of maize were able to tolerate loss of up to 70 per cent of their water content, causing disintegration of cell membranes and sub-cellular organelles (mitochondria, plastids, nuclei), without irreversible damage. On rehydration, normal cell structures were reassembled. The above results suggest considerable capacity for recovery of function, at least in relatively young tissues.

Soil aeration

In roots not especially adapted to oxygen stress, ion absorption and water permeability depend on adequate supplies of oxygen in the rooting medium (Grable, 1966; Kramer, 1969). In well-stirred solution cultures, dissolved oxygen concentrations in equilibrium with oxygen partial pressures of 0.01 to 0.02 bar are usually adequate for normal growth and function, the affinity of terminal oxidases for oxygen being high (Greenwood, 1969; Greenwood and Goodman, 1971). When the oxygen concentration falls beneath the above threshold, the structure or morphology of the root system is profoundly altered. The original roots cease growth and may degenerate, while newly initiated adventitious roots emerge from the base of the stem with considerable modification to their structure (Fig. 2). Lacunae develop in the cortex by the enlargement of the inter-cellular spaces, followed by degeneration of cell walls (McPherson, 1939), leaving only occasional bands of intact cells between cortex and stele. Despite these changes, the passage of ions across the roots is little affected (Drew, Chamel, Garrec and Fourcy, unpublished). The predominant pathway of ions and water (apoplast or symplast) across the cortex of such roots remains on open question, but if transport is predominantly symplasmic, then the flux through the bands of intact cells must be

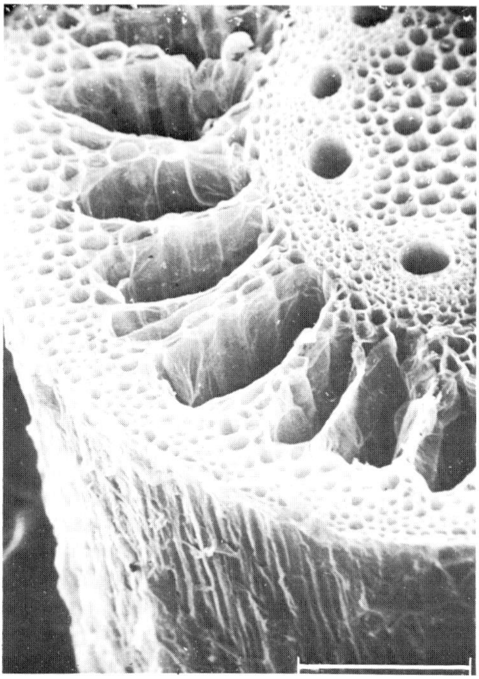

Fig. 2. Air-spaces in maize root. The scanning electron micrograph is a transverse section of an adventitious root developed under anoxic conditions. The distance from the root tip was about 10 cm. Note strands of collapsed and of intact cells between outer cortex and stele (Drew, Chamel, Garrec and Fourcy, unpublished). Bar = 200 μm.

far higher than in the normal cortex of aerated roots.

Mechanical impedance

Apart from those roots which extend in fissures and pores larger than their own diameter, the passage of the root tip through the soil requires the expansion of pores initially smaller than itself. Soil particles are radially displaced by the tip and the cells of the cap and meristematic region receive a mechanical stress, the magnitude of which depends on the compressibility of the soil. Even in soils of only moderate strength, the rate of root extension is often slowed by mechanical resistance (Barley *et al.*, 1965). Studies under well controlled conditions in artificial media show that relatively small stresses, about 0.2 bar, can halve the rate of root extension in cereals (Drew and Goss, 1973; R.S. Russell, 1977) with accompanying changes in patterns of cell division and extension. Impeded roots are found to develop an unusual type of biseriate endodermis (Wilson *et al.*, 1978) in which suberized lamellae are formed much closer to the root tip; at 15 mm from the tip the proportion of endodermal cells with suberized walls (State 2) was more than 80 per cent. In unimpeded roots, about 10 per cent of the cells were in State 2 at 20 mm from the tip. Cells of the pericycle also matured much closer to the apex. Goss (1974) found that in barley seedlings in which the unbranched root axes were subjected to a stress of 0.27 bar, the uptake of labelled potassium in 24 h was not different from that in unimpeded roots. Early maturation of the endodermis would be expected, however, to interfere in the transfer of calcium and magnesium across the root.

Inorganic nutrients

In soils, the concentrations of inorganic nutrients can vary widely within the rooting zone (E.W. Russell, 1973). Studies in solution culture have shown that when the supply of either inorganic nitrogen or phosphate is non-uniform, there is a stimulation of the absorption and transport of ions in parts of the root system which come in contact with favourably high concentrations (Drew and Saker, 1975, 1978). This effect, together with a localized promotion of the growth of lateral roots, largely compensates for the deficient supply to the remainder of the root system, since the concentration of phosphate in the shoots and their relative growth rate were restored to levels similar to those in plants receiving a uniform supply of phosphate. The enhanced rate of ion absorption is reminiscent of that observed when plants starved of specific nutrients are re-introduced to a supply (Bowen, 1970), or in split-root experiments in which only part of the root system is deprived of phosphate (Jungk and Barber, 1974). The stimulation of phosphate uptake following phosphate-deprivation or root excision is not immediate but requires an induction period of several days, during which the concentration of phosphate in the shoot decreases appreciably (Jungk and Barber, 1974). This suggests that the phosphate status of the shoot in some way controls the rate of net absorption of phosphate and its translocation.

The rate of uptake in any one part of the root system thus depends on the previous nutritional history of the plant and on the rate of uptake by other parts of the root system. These interactions are a reflection of the co-ordination maintained between ion uptake and the growth rate of the plant (for further discussion see Nye *et al.*, 1975; Pitman, 1976; Drew and Saker, 1978).

Conclusions

The age of the root tissue, and its previous history, are seen to exert a profound influence on absorptive function. Modifications to the structure of roots, and their contact with the soil, are likely to influence the radial movement of ions and water across the cortex of the root towards the xylem. Degeneration of the root cortex, or early development of suberized lamellae, would be expected greatly to increase resistance to the radial flow of water, and delay the re-establishment of continuous films of water when roots are moistened after a period of desiccation. There is relatively little information, however, concerning the extent to which roots are able to withstand stresses from the soil environment without irreversible damage to their normal function.

The variable conditions within the soil, both with respect to space and time, present special problems, making it difficult to predict the rate at which absorptive function in any root segment will decline with age. The prediction of root growth is posed a similar problem (Drew and Goss, 1973). A further complication arises from the fact that changes in absorptive ability occur at various levels of organization. Suberization and cortical degeneration are readily detectable even in soil grown roots, but this will not be true of damage to cell membranes and to sub-cellular structures essential to symplasmic transport.

The foregoing must lead to the conclusion that at any one time, only a fraction of the total root system in soil-grown plants, with the exception of seedlings, has the potential to absorb water and some nutrient ions at an appreciable rate. The flux of materials across the soil/root interface therefore may locally differ greatly from average values calculated by assuming that the entire root system is equally functional. The value of mechanistic models of the transfer of materials from soil to plants, and their use for predictive purposes, may depend upon recognition of this limitation.

References

Anderson, W.D. and Reilly, E.J. (1968). A study of the exudation of excised maize roots after removal of the epidermis and outer cortex. *J. exp. Bot.*, 19, 19.

Atkin, R.K., Barton, G.E. and Robinson, D.K. (1973). Effect of root-growing temperature on growth substances in xylem exudates of *Zea mays*. *J. exp. Bot.*, 24, 475.

Barley, K.P., Farrell, D.A. and Greacen, E.L. (1965). The influence of soil strength on the penetration of a loam by plant roots. *Austr. J. Soil Res.*, 3, 69.

Barlow, P.W. (1975). The root cap. *In* "The Development and Function of Roots" (Ed. by J.G. Torrey and D.T. Clarkson). p.21. Academic Press, London.

Bowen, G.D. (1970). Early detection of phosphate deficiency in plants. *Soil Sci. Pl. Anal.*, 1, 293.

Brewster, J.L. and Tinker, P.B.H. (1972). Nutrient flow rates into roots. *Soils Fertilizers*, 35, 355.

Brouwer, R. and Hoogland, A. (1964). Responses of bean plants to root temperatures. 2. Anatomical aspects. *Jaarb. Inst. biol. scheik. Onderz. LandbGewass. 1964*, 23.

Chung, H.H. and Kramer, P.J. (1975). Absorption of water and ^{32}P through suberized and unsuberized roots of Loblolly pine. *Can. J. For. Res.*, 5, 229.

Clarkson, D.T. (1974). "Ion Transport and Cell Structure in Plants".

McGraw-Hill, London.
Clarkson, D.T. (1976). The influence of temperature on the exudation of the xylem sap from detached root systems of rye (*Secale cereale*) and barley (*Hordeum vulgare*). *Planta*, 132, 297.
Clarkson, D.T., Drew, M.C., Ferguson, I.B. and Sanderson, J. (1975). The effect of the "Take-all" fungus, *Gaeumannomyces graminis*, on the transport of ions by wheat plants. *Physiol. Pl. Path.*, 6, 75.
Clarkson, D.T. and Hall, K.C. (1977). Membrane lipids of roots of rye and barley in relation to cooling. *Agric. Res. Coun. Letcombe Lab. Ann. Rep. 1976*, 18.
Clarkson, D.T. and Robards, A.W. (1975). The endodermis, its structural development and physiological role. *In* "The Development and Function of Roots" (Ed. by J.G. Torrey and D.T. Clarkson). p.415. Academic Press, London.
Clarkson, D.T., Sanderson, J. and Russell, R.S. (1968). Ion uptake and root age. *Nature*, 220, 805.
Cole, P.J. and Alston, A.M. (1974). Effect of transient dehydration on absorption of chloride by wheat roots. *Pl. Soil*, 40, 243.
Drew, M.C. (1978). Root development and activities. *In* "Arid-land Ecosystems" (Ed. by R.A. Perry and D.W. Goodall), Vol. 1. International Biological Programme, Cambridge University Press.
Drew, M.C. and Goss, M.J. (1973). Effect of soil physical factors on root growth. *Chemistry and Industry* No. 14, 679.
Drew, M.C. and Saker, L.R. (1975). Nutrient supply and the growth of the seminal root system in barley, 2. Localized, compensatory increases in lateral root growth and rates of nitrate uptake when nitrate supply is restricted to only part of the root system. *J. exp. Bot.*, 26, 79.
Drew, M.C. and Saker, L.R. (1978). Nutrient supply and the growth of the seminal root system in barley. 3. Compensatory increases in growth of lateral roots, and in rates of phosphate uptake, in response to a localized supply of phosphate. *J. exp. Bot.* In press.
Ferguson, I.B. and Clarkson, D.T. (1975). Ion transport and endodermal suberization in the roots of *Zea mays*. *New Phytol.*, 75, 69.
Ferguson, I.B. and Clarkson, D.T. (1976). Ion uptake in relation to the development of a root hypodermis. *New Phytol.*, 77, 11.
Gardner, W.R. (1960). Dynamic aspects of water availability to plants. *Soil Sci.*, 89, 63.
Goss, M.J. (1974). Effects of mechanical impedance on root growth. Ph.D. Thesis, University of Reading.
Grable, A.R. (1966). Soil aeration and plant growth. *Adv. Agron.*, 18, 57.
Greenwood, D.J. (1969). Effect of oxygen distribution in the soil on plant growth. *In* "Root Growth" (Ed. by W.J. Whittington), pp.202-223. Butterworth, London.
Greenwood, D.J. and Goodman, D. (1971). Studies on the supply of oxygen to the roots of mustard seedlings (*Sinapis alba*). *New Phytol.*, 70, 85.
Hansen, G.K. (1974). Resistance to water transport in soil and young wheat plants. *Acta Agric. Scand.*, 24, 37.
Head, G.C. (1973). Shedding of roots. *In* "Shedding of Plant Parts" (Ed. by T.T. Kozlowski), pp.237-293. Academic Press, New York.

Itai, C., Ben-Zioni, A. and Ordin, L. (1973). Correlative changes in endogenous hormone levels and shoot growth induced by short heat treatments to the root. *Physiol. Plant*, 29, 355.

Jungk, A. and Barber, S.A. (1974). Phosphate uptake rate of corn roots as related to the proportion of the roots exposed to phosphate. *Agron. J.*, 66, 554.

Kramer, P.J. (1950). Effects of wilting on the subsequent intake of water by plants. *Amer. J. Bot.*, 37, 280.

Kramer, P.J. (1969). "Plant and Soil Water Relationships". McGraw-Hill, New York.

Kramer, P.J. and Bullock, H.C. (1966). Seasonal variations in the proportions of suberized and unsuberized roots of trees in relation to the absorption of water. *Amer. J. Bot.*, 53, 200.

Läuchli, A. (1972). Translocation of inorganic solutes. *Ann. Rev. Pl. Physiol.*, 23, 197.

Läuchli, A. (1976a). Apoplasmic transport in tissues. *In* "Encycl. Pl. Physiol.: Transport in Plants, 2B" (Ed. by U. Lüttge and M.G. Pitman), pp.3-34. Springer Verlag, Berlin.

Läuchli, A. (1976b). Symplasmic transport and ion release to the xylem. *In* "Transport and Transfer Processes in Plants" (Ed. by I.F. Wardlaw and J.B. Passioura), pp.101-112. Academic Press, New York.

Leshem, B. (1965). The annual activity of intermediary roots of the Aleppo pine. *Forest. Sci.*, 11, 291.

Leshem, B. (1970). Resting roots of *Pinus halepensis*: structure, function and reaction to water stress. *Bot. Gaz.*, 131, 99.

McPherson, D.C. (1939). Cortical air spaces in the roots of *Zea mays*. *New Phytol.*, 38, 190.

Newman, E.I. (1969). Resistance to water flow in soil and plant. 1. Soil resistance in relation to amounts of root: theoretical estimates. *J. appl. Ecol.*, 6, 1.

Nielsen, K.F. and Humphries, E.C. (1966). Effects of root temperature on growth. *Soils and Fertilizers*, 29, 1.

Nir, I., Klein, S. and Poljakoff-Mayber, A. (1969). Effect of moisture stress on sub-microscopic structure of maize roots. *Aust. J. biol. Sci.*, 22, 17.

Nye, P.H., Brewster, J.L. and Bhat, K.K.S. (1975). The possibility of predicting solute uptake and plant growth response from independently measured soil and plant characteristics. 1. The theoretical basis of the experiments. *Pl. Soil.*, 42, 161.

Oades, J.M. (1978). Mucilages at the root surface. *J. Soil Sci.*, 29, 1.

Pitman, M.G. (1976). Ion uptake by plant roots. *In* "Encycl. Pl. Physiol.: Transport in Plants, 2B" (Ed. by U. Lüttge and M.G. Pitman), pp.95-128. Springer Verlag, Berlin.

Pitman, M.G., Lüttge, U., Läuchli, A. and Ball, E. (1974). Effect of previous water stress on ion uptake and transport in barley seedlings. *Aust. J. Pl. Physiol.*, 1, 377.

Russell, E.W. (1973). "Soil Conditions and Plant Growth" 10th ed. Longman, London.

Russell, R.S. (1977). "Plant Root Systems: Their Function and Interaction with the Soil". McGraw-Hill, London.

Russell, R.S. and Clarkson, D.T. (1976). Ion transport in root systems. *In* "Perspectives in Experimental Biology" (Ed. by N. Sunderland), Vol. 2, 401-11. Pergamon, Oxford.

Shaw, B.T. (1952). Ed. "Soil Physical Conditions and Plant Growth". Amer. Soc. Agron. Monograph No. 2, Academic Press, New York.

Spanswick, R.M. (1976). Symplasmic transport in plants. *In* "Encycl. Pl. Physiol.: Transport in Plants, 2B" (Ed. by U. Lüttge and M.G. Pitman), pp.35-53. Springer Verlag, Berlin.

Troughton, A. (1957). "The Underground Organs of Herbage Grasses". Bull No. 44, Commonwealth Agric. Bureaux, Farnham Royal, Bucks.

Troughton, A. (1962). "The Roots of Temperate Cereals (wheat, barley, oats and rye)". Mimeographed publication No. 2. Commonwealth Agric. Bureaux, Farnham, Bucks.

Van Fleet, D.S. (1961). Histochemistry and function of the endodermis. *Bot. Rev.*, 27, 165.

Waid, J.S. (1974). Decomposition of roots. *In* "Biology of Plant Litter Decomposition" (Ed. by C.H. Dickinson and G.J.F. Pugh), Vol. 1, pp.175-211. Academic Press, London.

Weatherley, P.E. (1975). Water relations of the root system. *In* "The Development and Function of Roots" (Ed. by J.G. Torrey and D.T. Clarkson), pp.397-413. Academic Press, London.

Wilcox, H.E. (1968). Morphological studies of the root of red pine, *Pinus resinosa* 1. Growth characteristics and patterns of branching. *Amer. J. Bot.*, 55, 247.

Wilson, A.J., Robards, A.W. and Goss, M.J. (1978). Effects of mechanical impedance on barley roots - changes in ultrastructure. *Agric. Res. Coun. Letcombe Lab. Report 1977*, 25.

SOIL PROPERTIES CONTROLLING THE SUPPLY OF NUTRIENTS TO THE ROOT SURFACE

P.H. NYE

Department of Agricultural Science, University of Oxford, England

Summary

Assuming average values for absorption and steady state conditions for uptake a mathematical model is derived for the uptake of nutrients by root systems growing in uniformly mixed soil. Laboratory experiments with young plants grown under these conditions show that the model leads to reasonable predictions of uptake of nitrate and potassium, but not of insoluble nutrients such as phosphate.

Introduction

Inorganic plant nutrients move to the surfaces of absorbing roots by diffusion and mass flow. The rate of nutrient uptake by a plant depends upon the demand imposed by its roots on the soil at the root surface, the size of the root system, and the mobility of the nutrient in the soil. In the simple instances of unabsorbed anions and exchangeable cations it may be sufficient to treat the root merely as an absorbing sink. For relatively insoluble nutrients the active part played by the root in solubilization has also to be taken into account.

Model of Uptake by a developing Root System

The rate of uptake from unit volume of soil will be influenced by the distribution of roots within it (Fig. 1), and the time each root has been exploiting it. Since it is

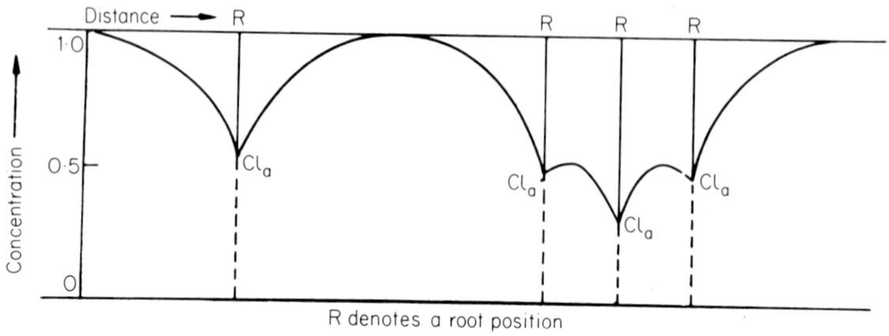

Fig. 1. Concentration-distance profile to show effect of overlap of diffusion zones on concentration at the root surface.

difficult to take into account the variation in uptake performance along and between individual roots, or the different degrees of competition between them, it is necessary to set up a model in which the root properties and the mean inter-root distance have average values. Likewise it is necessary to average over the soil heterogeneity that exists within the unit soil volume. The consequences of such averaging can be roughly assessed, and it is only dangerous if it is forgotten that it has been done. The model makes two main assumptions. The first is that each root, of radius a, may be assigned a concentric zone of exploitation of radius x. In fact the zone exploited by a root will be of irregular shape, but may be approximated by an equivalent cylindrical volume. If the rooting density (length per unit volume) is L_v, then $x = 1/\sqrt{\pi L_v}$. The second assumption concerns the concentration profile around each root. Because uptake of nutrients, relative to their concentrations in the soil solution, is usually greater than uptake of water, the concentration at the root surface C_{1a} will be lower than the average soil solution concentration $\bar{C}_1$. The ratio $C_{1a}/\bar{C}_1$ is a measure of the reduction in uptake imposed by the need for nutrient to move through the soil. It is assumed that the form of the concentration profile around each root (as in Fig. 1) corresponds to a steady state in which nutrient and water are fed into the zone of exploitation at x as fast as they are removed by the root at a. It may then be shown (Baldwin *et al.*, 1973) that the relation between C_{1a} and $\bar{C}_1$ is given by

$$\frac{C_{1a}}{\bar{C}_1} = \frac{1}{1 + \frac{a\alpha - av}{D_1^* \theta f} \ln \frac{x}{1.65a}} \tag{1}$$

Here α = root absorbing power (cm s^{-1}), defined by flux = $\alpha\, C_{1a}$

v = flux of water across root surface ($cm^3\ cm^{-2}\ s^{-1}$)

D_1^* = dispersion coefficient of the solute in the soil solution ($cm^2\ s^{-1}$)

θ = volumetric soil moisture content

f = soil impedance factor

α is related to the root inflow (uptake rate per unit root length), I, used by Barber (this volume, p. 5), by $I = 2\pi\ a\alpha\ C_{1a}$. D_1^* differs from the molecular diffusion coefficient in allowing for dispersion of solute caused by non-uniform flow of water through the soil pores. At values of v commonly induced by transpiration D_1^* is unlikely to exceed the molecular diffusion coefficient by more than a factor of two. The impedance factor, f, allows for the tortuous pathway followed by solutes through the soil, and for the increased viscosity of water within a few molecular layers of soil solid surfaces.

When the nutrient concentration in the soil solution is high compared with the plant's needs, α may be less than v, and the concentration at the root surface will exceed the average concentration. Equation (1) remains a reasonable approximation provided $av \ll D_1^* \theta f$, which is usually the case except possibly in dry soil.

The uptake rate per unit volume of soil, U_v, is $2\pi\ a\alpha\ C_{1a}\ L_v$. Substituting for C_{1a} from equation (1)

$$U_v = \frac{2\pi\ a\alpha\ L_v\ \bar{C}_1}{1 + \frac{a\alpha - av}{D_1^* \theta f} \ln \frac{x}{1.65a}} \tag{2}$$

In this expression the concentration at the root surface C_{1a}, which cannot be measured directly, has been eliminated, and the uptake rate expressed entirely in terms of quantities that are measurable in principle. $\bar{C}_1$ is the concentration that would be measured in a solution expressed from the soil pores.

Equation (2) shows the relative importance of the plant terms α, v, L_v, a and x; and the soil terms D_1^*, θ and f, in regulating the uptake rate. If the plant demand αa is large and av is small compared with soil transmission $D_1^* \theta f$

$$U_v \simeq \frac{2\pi\ D_1^* \theta f\ L_v\ \bar{C}_1}{\ln \dfrac{x}{1.65a}} \qquad (3)$$

so that the plant does not increase its uptake rate by increasing its demand per unit root length, since the rate is limited by transmission through the soil. The uptake rate can however be increased by increase in root density. These conditions prevail when growth of young plants is restricted by low concentration of nitrate, phosphate or potassium in the soil solution (Nye, 1977).

If both $a\alpha$ and av are very small compared with $D_1^* \theta f$, the denominator in equation (2) approaches unity and

$$U_v \simeq 2\pi a\alpha\ L_v\ \bar{C}_1 \qquad (4)$$

the value it would attain in a stirred solution of the same concentration as the soil solution. These conditions prevail at high nutrient levels in moist soil.

Transmission through the soil is very sensitive to the moisture level, θ, since, as shown in Figure 2, the impedance factor decreases steeply and approaches zero as the moisture level decreases to the wilting point. When transmission is rate limiting the accuracy of predictions of plant uptake from soil is restricted by the difficulty of knowing θ over the whole root system and the consequent uncertainty of the value of the product θf.

So far I have considered only the rate of uptake from a unit volume of soil at a particular time. As uptake pro-

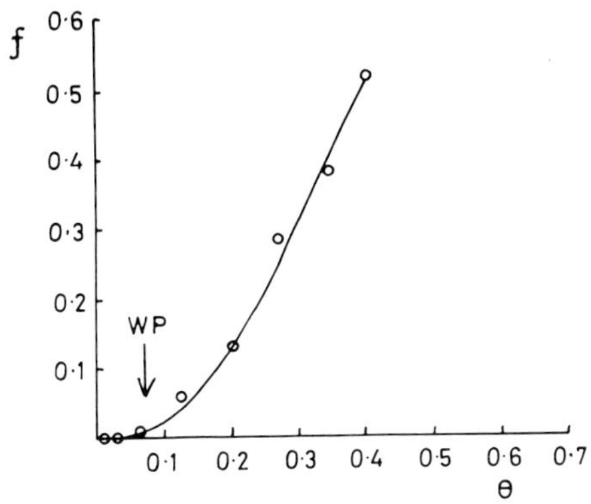

Fig. 2. The relation between the diffusion ion impedance factor f and volumetric moisture fraction θ (after Rowell *et al.*, 1967).

ceeds, the soil's ability to maintain the solution concentration depends on its buffer power, b, defined as the gradient dC/dC_1, where C is the total concentration of diffusible solute in the soil (solid + solution). Since the rate of decrease in total solute per unit volume of soil, $- dC/dt$, must equal the rate of uptake by the plant from unit volume, U_v, we have

$$- \frac{dC}{dt} = - \frac{b d\bar{C}_1}{dt} = U_v = \beta \bar{C}_1 \quad (5)$$

where $\beta \bar{C}_1$ is the right hand side of equation (2). Thus

$$\frac{d\bar{C}_1}{\bar{C}_1} = \frac{\beta}{b} dt \quad (6)$$

If β and b are constant, or if average values for them are used, equation (6) may be integrated to give

$$\frac{C_{1i} - C_1}{C_{1i}} = \frac{C_i - C}{C_i} = \frac{\text{Uptake from soil}}{\text{Initial content of soil}} = 1 - e^{-\beta t/b} \quad (7)$$

where subscript i represents the initial value at t = o. Thus, the time taken to exhaust a given fraction of solute in the soil is proportional to the buffer power. It is more accurate to solve equation (6) numerically by a time step procedure, using values of β and b appropriate to each time step, than to use an analytical solution and average values of β and b.

The details of the development of Equations (2) and (6) are given by Baldwin *et al.* (1973). Their predictions have been tested against the output of an electrical analogue under conditions in which β increases rapidly with time and b increases rapidly with decreasing concentration (Baldwin and Nye, 1974). The two methods agreed very well.

Experimental Validation of the Model

Transport equations have been extensively used to predict solute uptake by single roots from soil (Nye and Tinker, 1977, Chapter 6). There are few experiments in which the uptake of solutes by whole root systems have been predicted from independently measured plant and soil parameters, and compared with experimental results. Bhat *et al.* (unpublished) have determined the uptake of nitrogen by rape growing in a constant environment cabinet from a soil at two initial levels of nitrate. In the same environment the uptake of nitrate from stirred nutrient culture solution at a range of nitrate levels was also determined. The predicted uptakes from the soil are compared with the observed uptakes in Figure 3. In these experiments root length increases exponentially with time. Nitrate was also taken up exponentially from the soil until nearly exhausted, when further uptake was related to the slow rate of mineralisation.

Claassen and Barber (1976) have made a similar and more extensive comparison for potassium uptake by maize. They studied four soils, one of which started at 5 levels of K and the remainder at a single level. The soils covered a wide range of buffer power, and the initial concentrations of K in the soil solution ranged from 0.1×10^{-3} to 5.3×10^{-3} M. Their results are summarized in Figure 4. Considering the difficulty of measuring the soil and plant parameters accurately, the predicted and observed uptakes agree remarkably well over this very wide range of condi-

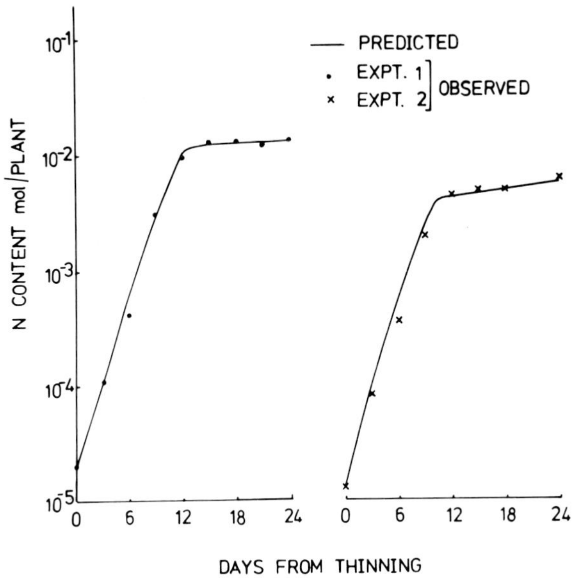

Fig. 3. Comparison of experimental and predicted uptake of nitrogen by rape from soil in a constant environment (Bhat, Nye and Brereton, unpublished data).

tions. To predict uptake they used the theory for diffusion of solute to a root in an infinite volume of soil. This assumes there will be no competition between roots. In fact there was some overlap of depletion zones, and they suggest competition as a reason why the observed uptakes were somewhat less than those they predicted.

These two sets of experiments, taken in conjunction with the results of single root experiments, show that it is possible to predict uptake of solutes by uniformly growing plants in a well defined environment from homogeneous soils provided they have well defined buffer powers. However, in the development of equation (2) it has been assumed that the root is a simple cylindrical sink. In practice, its effective geometry may be altered by root hairs, and by mycorrhizal hyphae; and the effective buffer power of the soil may be altered by changes in pH and redox potential in the rhizosphere, and excretion of solubilizing chelating ions. These effects make it difficult to predict the uptake of ions maintained at a low level in the soil solution by sparingly soluble complexes.

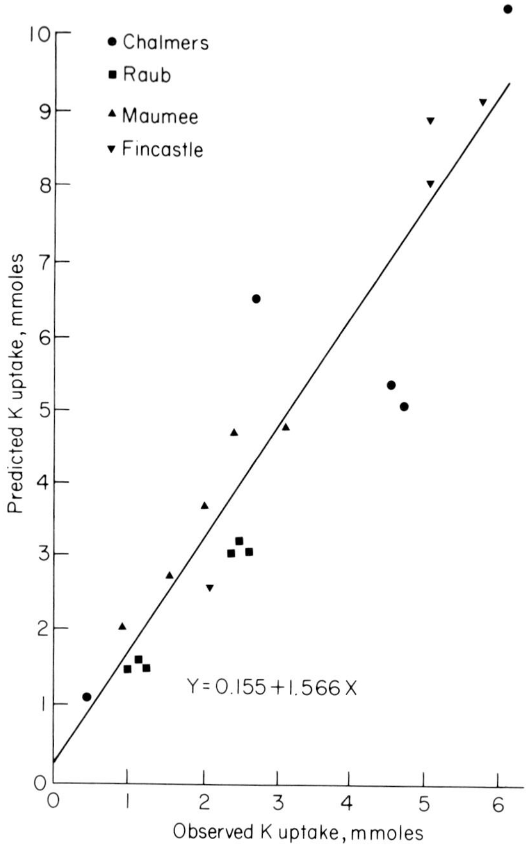

Fig. 4. Comparison of experimental and predicted uptake of potassium by maize from four soils. For the Chalmers soil there were 5 levels of potassium. For the Raab and Maumes soils there were two uptake periods (after Claassen and Barber, 1976).

Some of these difficulties have been revealed by intensive studies of phosphorus uptake. Brewster *et al.* (1975, 1976) compared the phosphate uptake by onion and rape from a soil at 5 levels of phosphate, with the uptake predicted from independent measurements on the plants and the soils. For onion plants, in the soil at the lowest phosphate level the simulated uptake of phosphorus was less than the experimental; with increasing additions of phosphate the simulated uptake became steadily larger than the experimental, and at the higher level the predicted uptake was greater than the experimental. At intermediate levels agreement was good. The discrepancy at the highest level was explained by indications that the effective buffer power in the mechanically undisturbed aggregated soil around the roots was lower than in samples shaken with solution - the method used for the independent soil measurement. The discrepancy at the lowest level would have been explained by solubilization of phosphate in the rhizosphere. Much stronger evidence for solubilization was obtained from the experiments with rape. This plant absorbed far more phosphorus than predicted from soils with a low level of phosphate. The concentration profiles of exchangeable phosphorus near onion and rape roots, revealed by densitometer

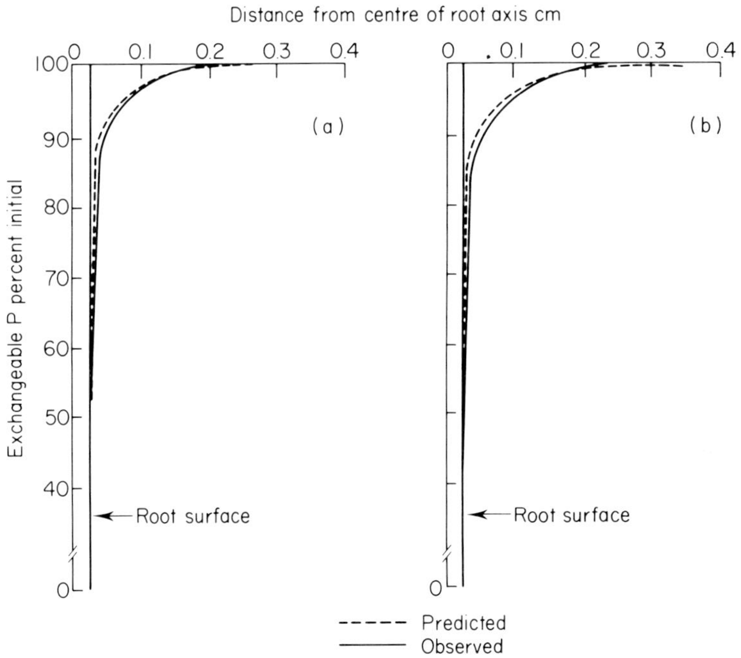

Fig. 5. Concentration profiles of exchangeable P near an onion seedling root in a low-P soil labelled with ^{33}P (a) 5 days, (b) 10 days old (after Bhat and Nye, 1974).

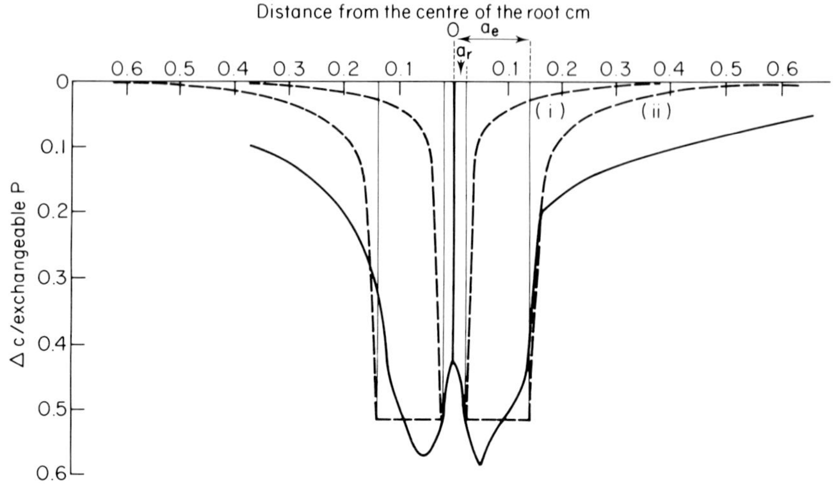

Fig. 6. Concentration profile of exchangeable P near a rape seedling root in a low-P soil labelled with ^{33}P. Broken line (i): calculated assuming the root hairs as inactive. Broken line (ii): calculated assuming intense root hair activity and uniform depletion from within the root hair cylinder. Solid line: observed in the experiments (after Bhat and Nye, 1973).

scans across autoradiographs of ^{33}P labelled soil, are shown in Figures 5 and 6 (Bhat and Nye, 1973, 1974). The depletion zone around the rape roots is much deeper and wider than predicted. The very much greater absorption by rape roots than onion roots in these experiments is shown in Table 1. The most likely explanation for these effects is that rape roots release organic chelating anions which exchange with surfaces bonding phosphate. Moghimi, Tate

TABLE 1

P uptake by seedlings from soil

	Onion		Rape	
	P_0	P_1	P_0	P_1
Soil solution P (x 10^{-6}M)	2.6	26	2.6	26
Day	12	11	5	7
Root length (cm)	14	18	26	30
P uptake (mols x 10^7)	0.25	1.4	10	31

and Oades (1978) have isolated 2-ketogluconate from the rhizosphere of wheat roots in amounts that could solubilize significant amounts of phosphate from hydroxyapatite. If roots can release chelates from the tips of their root hairs they have a particularly efficient mechanism for solubilizing phosphorus.

References

Baldwin, J.P. and Nye, P.H. (1974). A model to calculate the uptake by a developing root system or root hair system of solutes with concentration variable diffusion coefficients. *Pl. Soil*, 40, 703.

Baldwin, J.P., Nye, P.H. and Tinker, P.B. (1973). Uptake of solutes by multiple root systems from soil. III. A model for calculating the solute uptake by a randomly dispersed root system developing in a finite volume of soil. *Pl. Soil*, 38, 621.

Bhat, K.K.S. and Nye, P.H. (1973). Diffusion of phosphate to plant roots in soil. I. Quantitative autoradiography of the depletion zone. *Pl. Soil*, 38, 161.

Bhat, K.K.S. and Nye, P.H. (1974). Diffusion of phosphate to plant roots in soil. III. Depletion around onion roots without root hairs. *Pl. Soil*, 41, 383.

Brewster, J.L., Bhat, K.K.S. and Nye, P.H. (1975). The possibility of predicting solute uptake and plant growth response from independently measured soil and plant characteristics. III. The growth and uptake of onions in a soil fertilized to different initial levels of phosphate and a comparison of the results with model predictions. *Pl. Soil*, 42, 197.

Brewster, J.L., Bhat, K.K.S. and Nye, P.H. (1976). The possibility of predicting solute uptake and plant growth response from independently measured soil and plant characteristics. V. The growth and phosphorus uptake of rape in soil at a range of phosphorus concentrations and a comparison of results with the predictions of a simulation model. *Pl. Soil*, 44, 295.

Claassen, N. and Barber, S.A. (1976). Simulation model for nutrient uptake from soil by a growing plant root system. *Agronomy J.*, 68, 961.

Moghimi, A., Tate, M.E. and Oades, J.M. (1978). Phosphate dissolution by rhizosphere products. II. Characterization of rhizosphere products especially 2-ketogluconic acid. *Soil Biol. Biochem.* (in press).

Nye, P.H. (1977). The rate limiting step in plant nutrient absorption from soil. *Soil Sci.*, 123, 292.

Nye, P.H. and Tinker, P.B. (1977). "Solute Movement in the Soil-Root System". Blackwell's Scientific Pub., Oxford.

Rowell, D.L., Martin, M.W. and Nye, P.H. (1967). The measurement and mechanism of ion diffusion in soils. III. The effect of moisture content and soil solution concentration on the self-diffusion of ions in soils. *J. Soil Sci.*, 18, 204.

LIMITATIONS TO CURRENT MODELS DESCRIBING WATER UPTAKE BY PLANT ROOT SYSTEMS

BETTY KLEPPER and H.M. TAYLOR

Federal Research, Science and Education Administration, United States Department of Agriculture, Pendleton, Oregon and Ames, Iowa, USA

Summary

Most models which describe water movement from soil into the plant xylem at "crown" level are at present limited in their usefulness by their lack of field validation for a variety of crop species over a range of soil conditions. Major limitations include uncertainties about extrapolation of laboratory data on soil physical properties to field situations; lack of understanding of conditions at the root-soil interface; inability to incorporate known effects of root age, time of day, temperature, aeration, and other factors that can affect tissue permeability into quantitative equations; also, inadequate descriptions of root morphology (especially branching patterns) and root anatomy (especially vessel properties) under field conditions. Regardless of these limitations, progress is being made in the elucidation of many of these factors.

Introduction

Soil water flow into roots has been described by a number of models which either are concerned with microscale processes moving water into single roots or view the function of the root sink as a whole with little specification of processes (macroscale models). Most microscale models (*e.g.*, Philip, 1957; Gardner, 1960; Cowan, 1965) do not lend themselves to interpretation of field data gathered under heterogeneous conditions of soil texture, chemistry, and moisture, and over long periods of shoot and root development. Furthermore, different parts of this root-soil system interact with one another through xylem potential gradients. Earlier macroscale models (Rose and Stern, 1967; Whisler, Klute and Millington, 1968; van Bavel, Stirk and Brust, 1968; Molz and Remson, 1971; Feddes and Rijtema, 1972) were very general and physiological parameters were so poorly specified that their use in interpreting field data or in predicting crop water use was limited by insufficient specificity or versatility.

Recently, several researchers (Nimah and Hanks, 1973; Hansen, 1975; Rose, Byrne and Hansen, 1976; van Bavel and Ahmed, 1976; Hillel, Talpaz and van Keulen, 1976; Taylor and Klepper, 1978) have modelled soil water extraction by root systems using a somewhat "hybrid" approach, *i.e.*, by incorporating several time-specific plant and soil parameters into a calculation system so that various parts and coefficients interact at any one time and change with the passage of time. These systems predict the time-course of transpiration given certain initial conditions for a particular crop and the time course of meteorological or plant conditions.

Most of these models are internally consistent and will compute a value for water uptake commensurate with measured transpiration. Some have been partially verified so that temporal and spatial arrangements of computed uptake rates approximate measured rates for the data used to generate model coefficients. However, there is as yet no model validated for data sets generated under different conditions at different locations using different crops; most are crop-specific, location-specific, and even season-specific.

This paper describes three models of water extraction

and delivery to the crown by root systems and assesses obstacles to such simulations. The models by Hillel *et al.* (1976), Taylor and Klepper (1978), and Busscher and Fritton (1978) conveniently differ in emphasis. Their selection is no indication of our opinion of validity, applicability, value or convenience. We naturally emphasize the model we devised because it is recent, and close to our present interests.

Three Models compared

Most comprehensive is the soil-plant-air model of Hillel *et al.* (1976). Inputs are soil hydraulic resistances and potentials as functions of time and depth, hydraulic resistance of the root system, including a "depth factor", initial soil water and solute concentrations, relative distribution of active roots within the profile and daily evapotranspiration (including diurnal dynamics). From this information, the model calculates the xylem water potential at ground level ("crown") needed to supply daily transpiration, predicted soil water at each depth each day, and exchange of water and salts across the lower boundary of the root zone. Uptake from all horizontal layers is equated to transpiration.

Taylor and Klepper (1978) also sum uptake from horizontal layers, but their model is limited to water movement from bulk soil to, into, and up root xylem to the "crown". Their model varies from that of Hillel *et al.* (1976) in two significant ways. First, Taylor and Klepper (1978) use hourly values of "crown" water potential as an input variable because present information is insufficient to predict plant water potential from soil and meteorological factors, especially when partial canopies (Ritchie, 1972) or substantial plant water stresses (Ritchie, Burnett and Henderson, 1972) are involved. Second, Taylor and Klepper (1978) consider that axial resistance to conduction of water is a function of both depth and flow rate while Hillel *et al.*

(1976) specify that axial resistance is a function only of depth.

The model by Taylor and Klepper can be stated

$$U_i = (V_i)(L_{vi})(q_i)(\psi_{si} - \psi_p + \psi_{zi} + \Sigma\psi_{fi})$$

where

U_i = water uptake rate from soil volume V_i (m^3/min)

V_i = the rooted volume of soil in layer i (m^3)

L_{vi} = root length density in V_i (m root/m^3 soil)

q_i = average root water uptake in V_i (m^3 H_2O/m root/MPa/min)

ψ_{si} = total soil water potential in V_i (MPa)

ψ_p = total water potential in plant xylem at the ground surface (MPa)

ψ_{zi} = water potential loss due to elevation differences between the ground surface and the midpoint of V_i (MPa)

$\Sigma\psi_{fi}$ = frictional energy loss from water flow through xylem vessels between the midpoint of V_i and the ground surface (MPa).

Input parameters are root length density, "root-soil-system" permeability, water potential of bulk soil and of plant crown, including correcting terms for elevation and for potential losses in narrow xylem vessels due to friction during water flow. The equation is simply an Ohm's Law statement that equates water uptake in each layer to the product of root length ($V_i L_{vi}$), root-soil system permeability (q_i), and the water potential gradient from bulk soil to root xylem in the midpoint of the layer (ψ_{si} - ψ_{ri}), where $\psi_{ri} = \psi_p - \psi_{zi} - \Sigma\psi_{fi}$. Thus, xylem water flow in lateral roots is included in this permeability term. The frictional term is used only for main roots where water accumulated from laterals and from roots in deep layers flows to the crown. In this model, layers interact through the $\Delta\psi_{fi}$ term, which is itself flow-dependent. Upper taproots carry water accumulated by distal roots as well as water taken up in the immediate vicinity.

The least comprehensive model (Busscher and Fritton, 1978) calculates water flow in laterals and up the main root of a cereal (oats) using the Poiseuille equation. Input parameters are number and radii of xylem vessels in main and lateral roots, segment lengths and branching patterns of the root system, and pressure drops from xylem in root rips to plant crowns. The model calculates water delivered to the "crown". It does not include resistance to water flow into xylem at root tips.

None of these models accounts for plant water "capacitance". Taylor and Klepper (1978) claim that the error is not important for good-sized plants under crop conditions. Thus, the xylem is an "inelastic pipe" system capable of undergoing reductions in pressure to cause water potential gradients which move water.

The first two models either incorporate or assume that layer to layer soil water flow can be accounted for by present theoretical equations. In fact, large errors can result in estimates of soil water flow during a time interval long enough for most field experiments to be conducted (Flühler, Ardakani and Stolzy, 1976). Spatial variability of field soils and plant canopies can also introduce substantial errors (Nielsen, Biggar and Erh, 1973). These limitations, although real and significant, will not be discussed in detail because they are outside the subject matter of this conference. Instead, we will discuss how these models deal with processes involved in water movement to, into, and up the root xylem.

The Soil-Root Interface

The surface area of root-soil contact seems to be an important input parameter in any physically rigorous model because most conductivities are stated on an areal basis. Note, however, that both Hillel *et al.* (1976) and Taylor and Klepper (1978) ignore root surface area and use, instead, root length density, *i.e.*, root length per unit soil volume. This parameter is presently preferred because validated techniques are available for assessing it (Newman, 1966; Rowse and Phillips, 1974). Furthermore, root length is preferred because location of controlling shell layer(s) in the soil-to-root xylem pathway is still uncertain. Thus, we do not know the correct diameter to use in measuring surface area. For example, if the endodermis were the *only* important resistance in the radial path, then it would be important to know the total endodermal surface area. At any rate, one can calculate relevant surface area from root length density if average surface area of the controlling tissue layer can be related to root length. Thus root length density and root surface area models are readily inter-convertible and it is presently more convenient to use root length density.

Experimentally, root length measurements are difficult and time-consuming (Böhm, Maduakor and Taylor, 1977). Thus, it is tempting to assume that root length density varies with depth in some mathematically convenient way. For example, root length might decrease linearly (van Bavel and Ahmed, 1976) or exponentially (Feddes and Rijtema, 1972) from the land surface to its maximum depth. Then root densities at all depths could be predicted from values near the soil surface and depth of root penetration.

In fact, rooting patterns do at times follow such a relationship in field soils. Gerwitz and Page (1974) found that root dry weight declined exponentially with depth in the majority of cases. Since dry weight measurements are greatly increased by larger, older, woodier and presumably less absorptive roots commonly found near the surface, it is probably not valid to assume that root length density or root surface area will also show an exponential decline with depth.

For 57-day-old soybean (*Glycine max* L. (Merr.)) plants, Sivakumar, Taylor and Shaw (1977) found that root length density was about 110 m/m^3 in the surface 0.075 m and decreased exponentially with depth to zero at 1.05 m. However, after a 17-day period of moderate drought, these root systems changed dramatically. Almost 80% of roots in surface layers disappeared and root length density decreased at all depths above 0.6 m so that a linear relationship of root length density with depth more appropriately described the distribution.

Proportions of "new" and "old" roots presumably changed substantially during the 17-day drought discussed above. It is commonly assumed that root permeability (and therefore our q_i term) decreases as roots age. We can incorporate these permeability changes in our model by assigning different q_i values among various fractions of root length density (L_v). Methods are needed both to discriminate among various root ages and to measure average water uptake rates of the various age classes.

Substantial changes also can occur over short time intervals in the total volume of soil explored by roots. For example, over the 17-day growth period discussed above, soybean rooting depth increased from 1.05 m to 1.80 m, an average elongation rate of 0.04 m/day (Sivakumar *et al.*, 1977). In a horizontally layered model, the soil volume available to the root system would have increased by more than 70% in 17 days. This root system explored new soil and exploited "new" water as it grew. A model initially based on root densities at the beginning of this period would have been woefully wrong 17 days later.

Of course, one can model root length densities from root initiation and extension rates (Hackett and Rose, 1972; Lungley, 1973) and death rates (Hillel and Talpaz, 1976). Such models are not yet field-tested. It might be more appropriate to develop models to predict root growth rates as functions of measurable or estimable soil properties, using relationships among selected soil factors and root elongation rates. Then a model relating the two data sets could be used to estimate changes in root length density from soil parameters. Unfortunately, root elongation rate is a complex variable affected by many factors, all of which are likely to vary with time in field situations. In a laboratory experiment, Taylor, Huck and Klepper (1972) found that 100 hours after germination, cotton taproot length could be predicted from temperature, aluminium ion activity and soil strength with eight coefficients which accounted for only 68% of the variance. Except in very simple cases where one soil factor overrides all others, root extension estimates based on physical and chemical soil factors will contain intolerable errors. In view of this limitation, we believe that water extraction models for field crops must be validated with measured root length densities as an input parameter, generally on a bi-weekly to monthly basis.

Sample-to-sample variability at any one time and depth is sizable for root-length densities (Böhm, 1977). To these sampling variabilities, we must add real seasonal changes and real discontinuities in soil properties (worm holes, old root channels, hard pans, rocks, and minor chance variations caused by peculiarities of geological history). Samples must be taken not only at rather frequent intervals, but also with sufficient sample size and replication to overcome high sample variance.

Our lack of knowledge about the root-soil contact areas are well-described (Tinker, 1976). Diurnal root-soil gaps have been experimentally demonstrated (Huck, Klepper and Taylor, 1970; Faiz, 1973). Mucilage layers and hairs may mitigate effects of this gap, but the fact remains that roots shrink about 50% as they dry from -2 to -10 bars (Cole and Alston, 1974) and soils probably do not shrink in perfect tandem with roots. If vapour gaps were to occur during a three- to five-hour period in midday, effects on plant water potential might be imperceptible if midday stomatal closure stopped water loss soon enough to compensate for the sudden loss of water supply.

Even if root length densities were measured, water flow patterns in the rhizosphere are not well investigated. There are no field measurements of soil water potential gradients near root surfaces of transpiring plants, but laboratory measurements show that water content gradients do occur near areas of concentrated roots in dry soil (Dunham and Nye, 1973). Techniques are available for measuring water potentials of root surfaces (Fiscus, 1972; So, Aylmore and Quirk, 1976; So, this volume), but we are not certain that root surface water potential accurately tracks diurnal changes in root xylem potential.

Several models, including the one by Hillel *et al.* (1976) describe flow of water to root surfaces using Gardner's (1964) equation relating soil resistance to the inverse of BKL, where B is an empirical root-length activity constant, K is soil hydraulic conductivity, and L is the root length present. This approach assumes that accurate water content-water potential and water content-hydraulic conductivity relationships are available for the soil. Present techniques for measuring soil water potential with thermocouple psychrometry have a precision of about 0.1 MPa and diurnal temperature fluctuations and point-to-point variances in field soils further reduce the accuracy. Often, tensiometers or pressure membrane equipment is used; these measure matric potential but osmotic potential in some soils can be substantial. The hysteresis between wetting and drying portions of these curves is not taken into account. Presumably soil near the root will dry during the day and rewet at night, but along different pathways because of hysteresis effects. Finally, laboratory measurements are sometimes made on disturbed soil samples from the ploughed layer and frequently are not done separately for undisturbed samples from separate layers of the profile.

Taylor and Klepper (1978) avoided some of these problems by resorting to an empirical "engineer's" approach. Because of uncertainties about path length, soil water potential gradients near roots, root-soil contact efficiency, and radial root resistance to water movement, they used a "system permeability", q_i, which is physically not very rigorous. This term, q_i, specifies the flow rate from bulk soil to root xylem at the mid-point of a layer given the water potential difference between bulk soil and root xylem and the root length available in the soil volume. Data for cotton roots in a loamy fine sand show that this term decreases, sometimes linearly, by about an order of magnitude as the soil dries from near field capacity to near the wilting point (Taylor and Klepper, 1975). Herkelrath, Miller and Gardner (1977) have proposed that this decrease of water uptake rate can be modelled by assuming that q_i is proportional to relative saturation of the soil layer. Our 1978 model provides for altering q_i as soil water content changes but effects of soil salinity, unfavourable temperature and deficient aeration on permeability of the soil-root system are not included. It should be possible to account for endogenous diurnal rhythms (Parsons and Kramer, 1974) and effects of flow rate on permeability when consistent quantitative data are available. At present, root resistance seems to remain constant with changing flow rates in some experiments (Hailey, Hiler, Jordan and van Bavel, 1973) and to decrease with increasing flow rates in other experiments (Stoker and Weatherley, 1971). Root-shoot ratios possibly will need to be specified because the q_i term at a given transpiration rate seems to increase as the root-shoot ratio decreases in soybeans (unpublished data, B.W. Eavis, Ministry of Overseas Development, Surbiton, England).

Root Tissue-Root Xylem Relationships

Both Taylor and Klepper (1978) and Busscher and Fritton (1976) assume that the Poiseuille equation (Greacen, Ponsana and Barley, 1976) is appropriate for calculating axial resistances in xylem vessels. Because xylem pressure potential measurements are not available, this assumption is valid as any other. Note that this equation predicts hydraulic potential (not total potential) differences that are required to move a volume of fluid along a small-diameter, impervious, rigid tube as a function of tube length, radius (4th power) and fluid viscosity. Low concentrations of osmotica in xylem vessels probably have no effect on viscosity. Both of these models, however, neglect any effects of exchange of salts between the xylem sap and surrounding tissues (Klepper, 1967; Klepper and Kaufmann, 1966). Both models assume that xylem sap moves unaltered from root tip to crown. Although changes in osmotic potential of xylem sap make no difference in use of the Poiseuille equation, such changes could lead to unexpected water exchange between the xylem sap and surrounding tissues. These exchanges would modify flow velocity along the length of the vessel, which is far from impervious.

The main limitations in the use of the Poiseuille equation to calculate axial root resistance are inability to measure xylem vessel radii sufficiently accurately, lack of knowledge about effects of growth conditions on longitudinal variations in vessel radii, lack of knowledge about numbers of vessels that carry water from one layer to another in field crops, and inability to account for water exchange along the vessel length. The most serious of these limitations are those associated with vessel radii, especially in view of the fact that vessel radii enter the equation as r^4. Most investigators measure vessel diameter at one depth within the profile, but Meyer (1976) found that radii of metaxylem vessels of wheat increased from less than 30 μm near the soil surface to about 45 μm at 1 m depth. If all factors other than the radius remained constant, a metaxylem vessel of that wheat plant could carry five times the water at 1 m depth as at the surface. We cannot be sure that xylem vessel radii always are altered in such a predictable manner.

Busscher (1976) has carefully defined a root system, including the segment length and branching patterns so that quantitative information is available on the numbers of vessels for moving water from one soil volume to another. Calculations by Busscher and Fritton (1978) show that a pressure differential of only a few millibars between root tip xylem and crown xylem of oats will provide sufficient water to replace transpired water. In contrast, Hellkvist, Richards and Jarvis (1974) and H.B. So (this symposium) have shown that axial resistances can be substantial in other species.

The Future

Tremendous progress has occurred in describing water uptake by root systems in the 21 years since the model of Philip (1957) was published. Many groups now are working to create more realistic (and hopefully simpler) models. Many other groups are conducting experiments to help resolve the uncertainties that presently exist, especially at or near the root-soil interface. Conferences such as this one should help clarify our concepts about the relevant parameters to be included in future models. The day surely will come when, with appropriate identification of soil, climate and plant characteristics, we can predict the time courses for transpiration and for water content changes within the profile.

References

Böhm, W. (1977). Development of soybean root systems as affected by plant spacing. "Z. Acker- und Pflanzenbau". *J. Agron. and Crop Sci.*, 144, 103.

Böhm, W., Maduakor, H. and Taylor, H.M. (1977). Comparison of five methods for characterizing soybean rooting density and development. *Agron. J.*, 69, 415.

Busscher, W.J. (1976). A simulation of soil water flow to and through a growing root system. Ph.D. Thesis, Pennsylvania State University.

Busscher, W.J. and Fritton, D.D. (1978). Simulated flow through the

root xylem. *Soil Sci.*, 125, 1.
Cole, P.J. and Alston, A.M. (1974). Effect of transient dehydration on absorption of chloride by wheat roots. *Plant and Soil*, 40, 243.
Cowan, I.R. (1965). Transport of water in the soil-plant-atmosphere system. *J. Appl. Ecol.*, 2, 221.
Dunham, R.J. and Nye, P.H. (1973). The influence of water content on the uptake of ions by roots. I. Soil water content gradients near a plane of onion roots. *J. Appl. Ecol.*, 10, 585.
Faiz, S.M.A. (1973). Soil-root-water relations. Ph.D. Thesis, University of Aberdeen, Scotland, U.K. 176 pp.
Feddes, R.A. and Rijtema, P.E. (1972). Water withdrawal by plant roots. *J. of Hydrology*, 17, 33.
Fiscus, E.L. (1972). *In situ* measurement of root-water potential. *Pl. Physiol.*, 50, 191.
Flühler, H., Ardakani, M.S. and Stolzy, L.H. (1976). Error propagation in determining hydraulic conductivities from successive water content and pressure head profiles. *Soil Sci. Soc. Amer. J.*, 40, 830.
Gardner, W.R. (1960). Dynamic aspects of water availability to plants. *Soil Sci.*, 89, 63.
Gardner, W.R. (1964). Relation of root distribution to water uptake and availability. *Agron. J.*, 56, 41.
Gerwitz, A. and Page, E.R. (1974). An empirical mathematical model to describe plant root systems. *J. Appl. Ecol.*, 11, 773.
Greacen, E.L., Ponsana, P. and Barley, K.P. (1976). Resistance to water flow in the roots of cereals. *In* "Water and Plant Life" (Ed. by O.L. Lange *et al.*), p.86. Springer Verlag, Berlin.
Hackett, C. and Rose, D.A. (1972). A model of the extension and branching of a seminal root of barley, and its use in studying relations between root dimensions. *Aust. J. Biol. Sci.*, 25, 669.
Hailey, J.L., Hiler, E.A., Jordan, W.R. and van Bavel, C.H.M. (1973). Resistance to water flow in *Vigna sinensis* L. (Endl.) at high rates of transpiration. *Crop Sci.*, 13, 264.
Hansen, G.K. (1975). A dynamic continuous simulation model of water state and transportation in the soil-plant-atmosphere system. I. The model and its sensitivity. *Acta Agriculturae Scandinavica*, 25, 129.
Hellkvist, J., Richards, G.P. and Jarvis, P.G. (1974). Vertical gradients of water potential and tissue water relations in Sitka spruce trees measured with the pressure chamber. *J. Appl. Ecol.*, 11, 637.
Herkelrath, W.N., Miller, E.E. and Gardner, W.R. (1977). Water uptake by plants. II. The root contact model. *Soil Sci. Soc. J.*, 41, 1039.
Hillel, D. and Talpaz, H. (1976). Simulation of root-growth and its effect on the pattern of soil water uptake by a non-uniform root system. *Soil Sci.*, 121, 307.
Hillel, D., Talpaz, H. and van Keulen, H. (1976). A macroscopic-scale model of water uptake by a non-uniform root system and of water and salt movement in the soil profile. *Soil Sci.*, 121, 242.
Huck, M.G., Klepper, B. and Taylor, H.M. (1970). Diurnal variations in root diameter. *Pl. Physiol.*, 45, 529.
Klepper, Betty. (1967). Effects of osmotic pressure on exudation

from corn roots. *Aust. J. Biol. Sci.*, 20, 723.
Klepper, Betty and Kaufmann, M.R. (1966). Removal of salt from xylem sap by leaves and stems of guttating plants. *Pl. Physiol.*, 41, 1743.
Lungley, D.R. (1973). The growth of root systems. A numerical computer simulation. *Plant and Soil*, 38, 145.
Molz, F.J. and Remson, I. (1971). Application of an extraction-term model to the study of moisture flow to plant roots. *Agron. J.*, 63, 72.
Newman, E.I. (1966). A method of estimating the total length of root in a sample. *J. Appl. Ecol.*, 3, 139.
Meyer, W.S. (1976). Seminal roots of wheat: Manipulation of their geometry to increase the availability of soil water and to improve the efficiency of water use. Ph.D. Thesis, University of Adelaide, Australia.
Nielsen, D.R., Biggar, J.W. and Erh, K.T. (1973). Spatial variability of field-measured soil-water properties. *Hilgardia*, 42, 215.
Nimah, M.N. and Hanks, R.J. (1973). Model for estimating soil water, plant and atmospheric interrelations. I and II. *Soil Sci. Soc. Amer. Proc.*, 37, 522.
Parsons, L.R. and Kramer, P.J. (1974). Diurnal cycling in root resistance to water movement. *Physiol. Plant.*, 30, 19.
Philip, J.R. (1957). The physical principles of soil water movement during the irrigation cycle. 3rd Congr. Inter. Comm. on Irrigation and Drainage. Quest. 8, pp.8.125-8.154.
Ritchie, J.T. (1972). Model for predicting evaporation from a row crop with incomplete cover. *Water Resources Research*, 8, 1204.
Ritchie, J.T., Burnett, E. and Henderson, R.C. (1972). Dryland evaporative flux in a subhumid climate. III. Soil water influences. *Agron. J.*, 64, 168.
Rose, C.W., Byrne, G.F. and Hansen, G.K. (1976). Water transport from soil through plant to atmosphere: A lumped-parameter model. *Ag. Meteor.*, 16, 171.
Rose, C.W. and Stern, W.R. (1967). Determination of withdrawal of water from soil by crop roots as a function of depth and time. *Aust. J. Soil Res.*, 5, 11.
Rowse, H.R. and Phillips, D.A. (1974). An instrument for estimating the total length of root in a sample. *J. Appl. Ecol.*, 11, 309.
Sivakumar, M.V.K., Taylor, H.M. and Shaw, R.H. (1977). Top and root relations of field-grown soybeans. *Agron. J.*, 69, 470.
So, H.B. (1978). This volume, p.99.
So, H.B., Aylmore, L.A.G. and Quirk, J.P. (1976). The resistance of intact maize roots to water flow. *Soil Sci. Soc. Amer. J.*, 40, 222.
Stoker, R. and Weatherley, P.E. (1977). The influence of the root system on the relationship between the rate of transpiration and depression of leaf water potential. *New Phytol.*, 70, 547.
Taylor, H.M., Huck, M.G. and Klepper, B. (1972). Root development in relation to soil physical conditions. *In* "Optimizing the Soil Physical Environment Toward Greater Crop Yields" (Ed. by D.I. Hillel), pp.57-77. Academic Press, New York.
Taylor, H.M. and Klepper, Betty (1975). Water uptake by cotton root systems: An examination of assumptions in the single root model. *Soil Sci.*, 120, 57.

Taylor, H.M. and Klepper, Betty (1978). The role of rooting characteristics in the supply of water to plants. *Adv. Agron.*, 30 (in press). Academic Press, New York.
Tinker, P.B. (1976). Transport of water to plant roots in soil. *Phil. Trans. R. Soc. London B*, 273, 445.
van Bavel, C.H.M. and Ahmed, J. (1976). Dynamic simulation of water depletion in the root zone. *Ecol. Modelling*, 2, 189.
van Bavel, C.H.M., Stirk, G.B. and Brust, K.J. (1968). Hydraulic properties of a clay loam soil and the field measurement of water uptake by roots. I. Interpretation of water content and pressure profiles. *Soil Sci. Soc. Amer. Proc.*, 32, 310.
Whisler, F.D., Klute, A. and Millington, R.J. (1968). Analysis of steady-state evapotranspiration from a soil column. *Soil Sci. Soc. Amer. Proc.*, 32, 167.

AN ELECTRON PARAMAGNETIC RESONANCE STUDY OF THE UPTAKE OF Mn (II) AND Cu (II) BY WHEAT ROOTS

B.A. GOODMAN and D.J. LINEHAN

The Macaulay Institute for Soil Research, Craigiebuckler, Aberdeen, U.K.

Summary

Electron paramagnetic resonance (EPR) spectroscopy was used to study the uptake of manganese and copper by living excised wheat roots. Investigations of uptake were made from solutions in which the cations were present as free ions or as EDTA complexes. Little progress was made with manganese because the solutions of manganous complexes gave only the spectrum of free Mn^{2+} ions arising from dissociation of the complexes; the undissociated complexes giving no spectrum in solution at room temperature. With copper the uptake was shown to have little dependence on the form of the copper in the uptake solution and to involve an initial immobilization with the subsequent progressive appearance of a soluble complex having EPR parameters similar to those of some copper-amino acid complexes.

Introduction

There is little general agreement about the involvement of metabolism in the transport of micronutrient cations (Moore, 1972; Pitman, 1976). This results, at least in part, because the rather complex and poorly understood solution chemistry of many micronutrients, such as copper and manganese (Baes and Mesmer, 1976), makes the interpretation of physiological experiments somewhat uncertain.

In the study of some micronutrient reactions, EPR spectroscopy is a potentially useful technique because of its great sensitivity to paramagnetic species and its ability to provide information on their chemical and structural properties. EPR is a branch of microwave spectroscopy which is applied to the study of molecules or ions with unpaired electron spins. Under the influence of a magnetic field the energies of the two spin states of an unpaired electron are different, and resonance between these two energy states is produced by the introduction of microwave radiation of the appropriate energy. The resonance condition is $\Delta E = h\nu = g\beta H$, where ΔE is the energy separation between the two spin states, h is Planck's constant, υ is the microwave frequency, β is the Bohr magneton, H is the magnetic field strength and g is a constant. The value of g is equal to 2.0023 for a free electron and similar values are obtained for many organic and inorganic free radicals. With transition metal complexes, however, appreciable deviations from this "free spin" value are often observed.

An electron under no magnetic influence except the applied external field produces only a single peak EPR spectrum, but, if it is able to interact with a nucleus having non-zero spin, I, the peak is split into 2I + 1 components, known as the hyperfine structure, with their separation proportional to the unpaired electron density.

The magnitudes of the g-value and the hyperfine structure are dependent on the orientations of the principal axes of the paramagnetic species relative to the direction of the applied magnetic field, so that in a solid sample consisting of randomly-oriented molecules the spectrum consists of a summation of all possible molecular orientations. Figure 1a illustrates such an anisotropic spectrum where the nucleus has I = 3/2 and the molecule has axial symmetry. In fluid solution, if the paramagnetic molecules are able to tumble rapidly compared to the time scale of

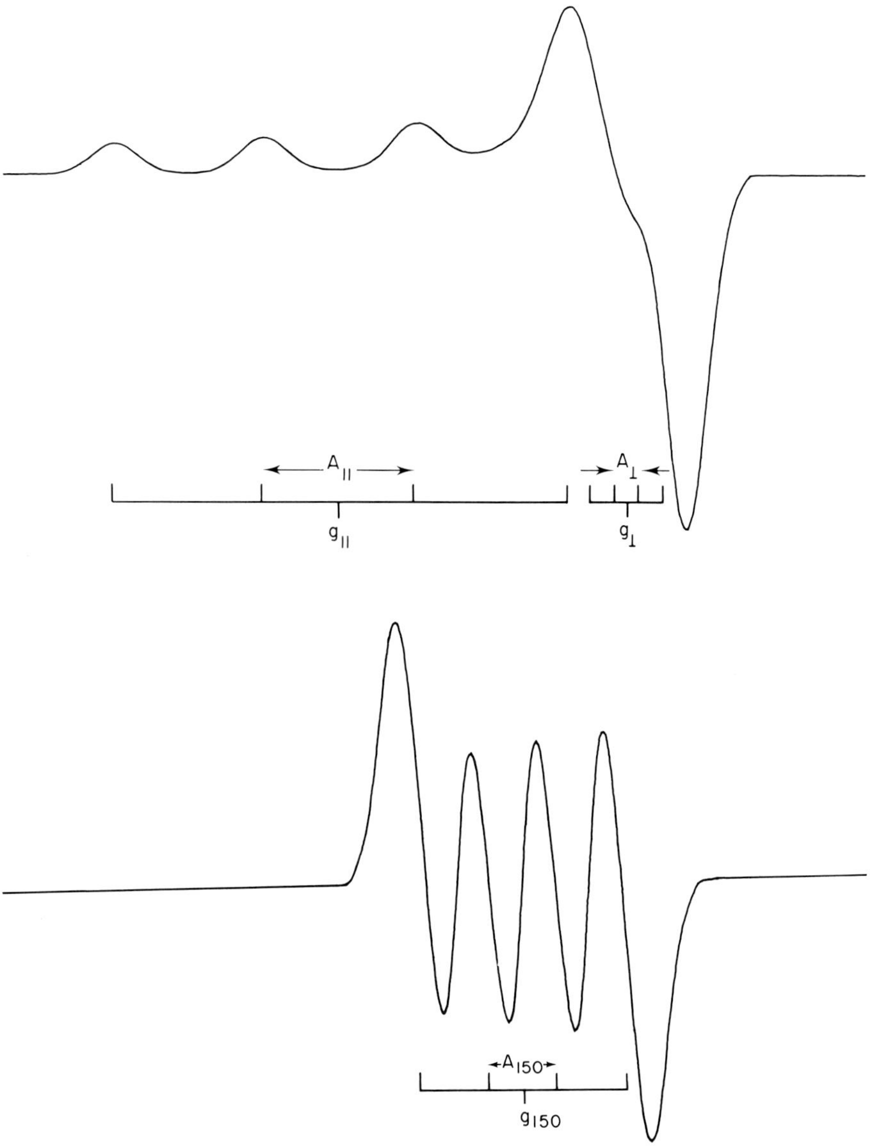

Fig. 1. Computer simulated EPR spectra of a species with axial symmetry and nuclear spin I = 3/2, (a) anisotropic spectrum with $g_{\perp}$ = 2.06, $g_{||}$ = 2.27, $A_{\perp}$ = 2.0 mT, $A_{||}$ = 17.0 mT, (b) isotropic spectrum of the same species showing the averaging of parameters when it undergoes rapid motion.

the EPR transition, an averaged isotropic spectrum is obtained (Fig. 1b). With large molecules or in a viscous medium, however, the motion of the paramagnetic molecules may be slower and an anisotropic spectrum may again be observed. An EPR spectrum can, therefore, provide information on the structure (e.g. degree of symmetry of the environment of the nucleus containing the unpaired electron), bonding (e.g. degree of covalency of molecular orbitals containing the unpaired electron(s)) and the size of any paramagnetic species occurring in solution.

In the present work experiments have been performed with the object of assessing the extent to which EPR spectroscopy can be used to provide information on the uptake of paramagnetic species by plant roots and any chemical changes that may subsequently take place.

Materials and Methods

Growth of wheat plants

Wheat (*Triticum aestivum* L c.v. Capelle) seeds were surface sterilized with 0.4% w/v peracetic acid (Laporte Industries Ltd., Luton, U.K.) and germinated on perforated polypropylene rafts in two litre tall-form lipless beakers containing 500 ml of autoclaved half-strength Hoagland's nutrient solution with Cu and Mn excluded. The tall-form beakers were covered by two litre squat-form beakers to maintain sterile conditions within the beaker. After 72 h in the dark at 25°, samples of nutrient solution were tested for sterility and non-sterile beakers rejected. After 4 days in light at 25°, uniform plants were selected for ion uptake as described below.

Ion uptake

The wheat seedlings were placed in crystallizing basins with the roots bathed for 30 min in 10^{-4} M solutions of manganese or copper as the sulphate or EDTA complex. After rinsing for 1 min in distilled water the roots were excised and threaded into silica EPR sample tubes, with the root tips immersed in water to prevent them from drying out during the experiment.

EPR spectra

The EPR spectra, obtained with a Varian E104 spectrometer operating at X-band frequencies, were displayed as the 1st derivative of the absorption. All spectra were recorded from regions of the roots at least 3 cm above the water surface.

Results and Discussion

The spectrum obtained from a root grown in Hoagland's nutrient solution with copper and manganese excluded is shown in Figure 2. This consists of a single narrow peak with a g-value close to 2, and must, therefore, be assigned to a free radical species. The absence of hyperfine struc-

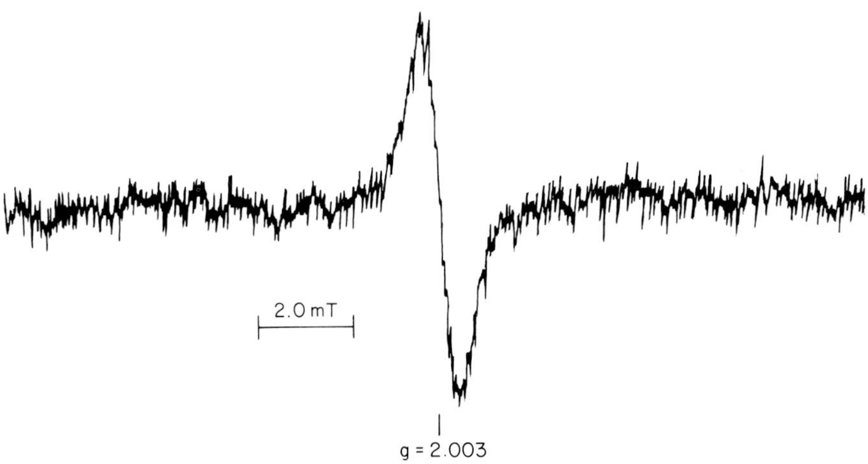

Fig. 2. EPR spectrum of an untreated wheat root. Microwave power = 20 mW, modulation frequency = 100 kHz, modulation amplitude = 0.5 mT and receiver gain = 8.0×10^4.

ture prevents identification of this species and consequently it will not be considered further in this paper, although, as subsequent results will show, it can undergo quite dramatic changes in intensity.

A typical spectrum obtained as a result of uptake from a manganese sulphate solution is shown in Figure 3. In addition to a weak central peak from the free radical species described above, this spectrum consists of 6 peaks which arise from coupling of the unpaired electrons with the ^{55}Mn nucleus (I = 5/2). The g- and hyperfine values from this species are the same as those obtained from the $Mn(H_2O)_6^{2+}$ ion. Identical spectra were obtained when the manganese was present in the uptake solution either as the free ion or as the EDTA complex. However, many manganese complexes show an EPR spectrum only from the dissociated component in solution at room temperature. The technique, therefore, is limited to measuring the amount of $Mn(H_2O)_6^{2+}$, i.e. free

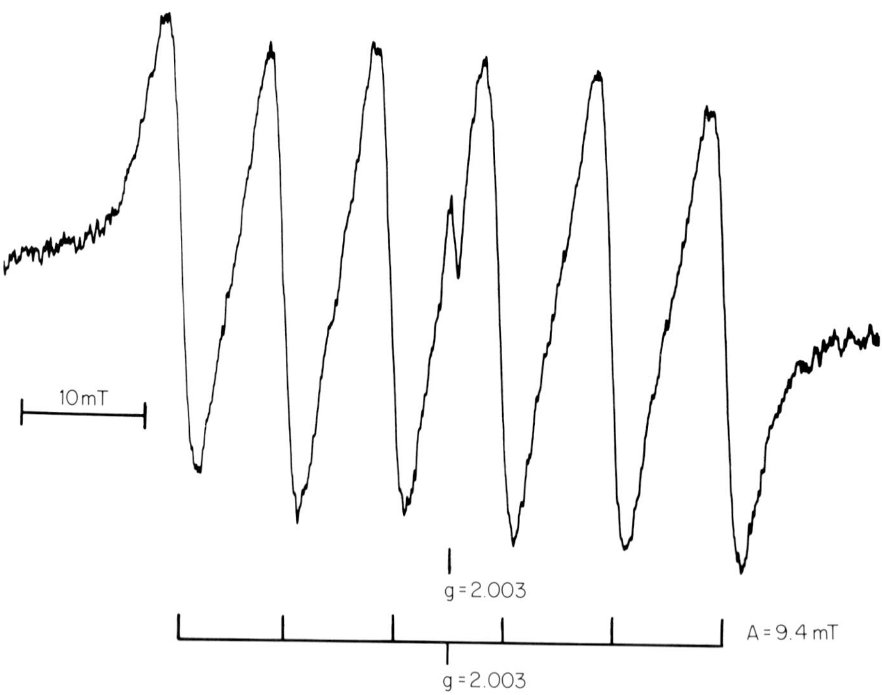

Fig. 3. EPR spectrum of a wheat root, treated with 10^{-4} M manganese EDTA for 0.5 h, 1.5 h after removal from solution. Microwave power = 20 mW, modulation frequency = 100 kHz, modulation amplitude = 0.5 mT and receiver gain = 4.0 x 10^4.

Mn^{2+}, present in the roots, and its variation with time. In fact it is much easier to measure relative than absolute quantities and, by studying the variation of peak height with time, it was shown that the concentration of $Mn(H_2O)_6^{2+}$ remained essentially constant over a period of several hours after removal of the roots from the Mn-uptake solution.

In contrast, complexed forms of copper usually give satisfactory EPR spectra and many results have been reported in the literature (e.g. Goodman and Raynor, 1970). Spectra obtained as a result of copper uptake by wheat roots from copper sulphate solution are shown in Figure 4. Shortly after removal from the copper-uptake solution it can be seen that the root produces a spectrum which contains an anisotropic component superimposed on a component similar to that obtained from the parent solution. There is also some evidence for the presence of an isotropic component. With time the free copper component decreased in intensity quite rapidly so that after 2.5 h (Fig. 4d) it had virtually disappeared. The intensity of the anisotropic component also decreased. The most notable phenomenon, however, is the intensity of the isotropic component, which though weak at first, increased rapidly and was still increasing after 23 h (Fig. 4h). The only other feature in these spectra is the sharp peak previously assigned to a free radical. A comparable experiment where the uptake of copper was from a copper-EDTA solution was also carried out (results not shown). Spectra were similar to these obtained from the $CuSO_4$ uptake with only minor differences. Not surprisingly the component assigned to free Cu^{2+} in the early spectra was appreciably weaker. Also, although the parameters of the final isotropic species were identical to those obtained as a result of uptake from the $CuSO_4$ solution, the increase in intensity with time was much more rapid being well developed only 3 h after uptake.

At this stage it is worthwhile to consider the possible

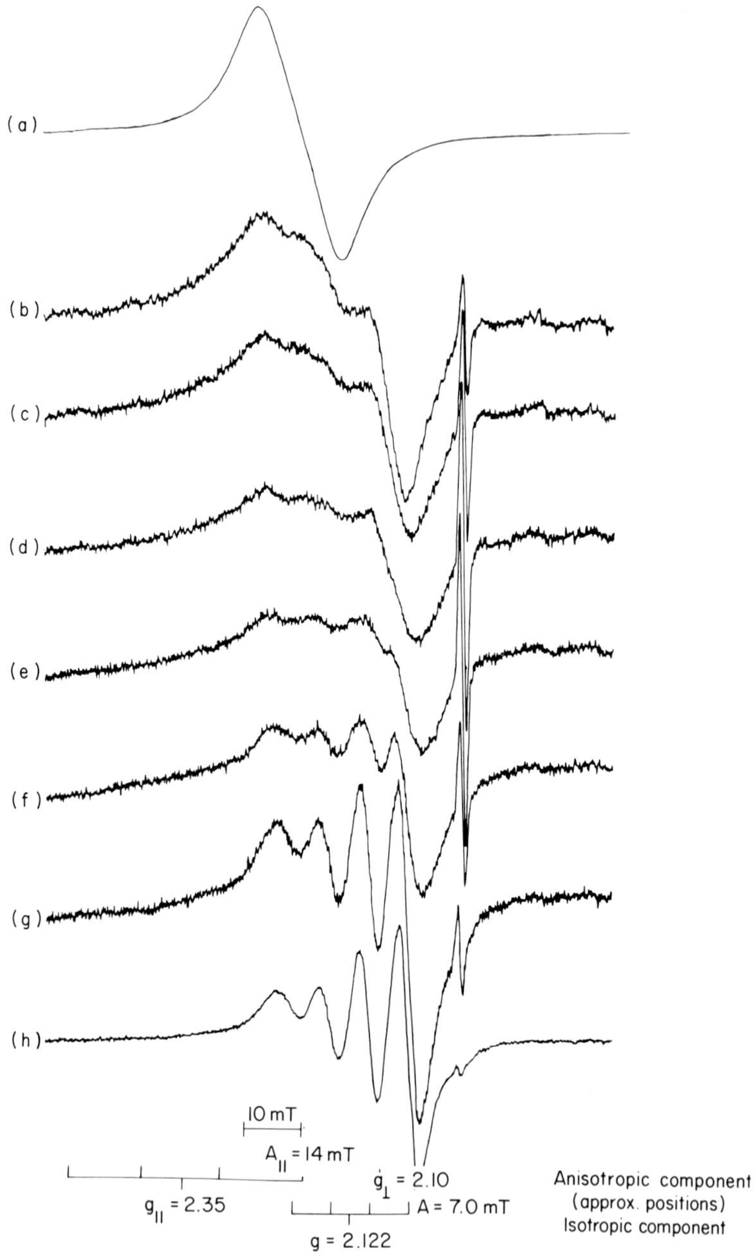

Fig. 4. EPR spectra of (a) 10^{-2} M copper sulphate and (b) - (h) wheat roots treated with 10^{-4} M copper sulphate for 0.5 h. After removal from solution spectra taken at (b) 0.5 h, (c) 1.5 h, (d) 2.5 h, (e)

nature of the various copper species which produce these spectra. Firstly, there is evidence for the presence of some copper as the free Cu^{2+} ion, $Cu(H_2O)_6^{2+}$, in the roots for a short time after removal from the uptake solution. This species is much more abundant when chelating agents are absent from the uptake solution, thus suggesting that these complexes limit the supply of the free ion to the root. The anisotropic component must arise from Cu^{2+} bound either in the solid state, possibly to cell wall material, or in high molecular weight complexes in solution, e.g. copper proteins. However, because of the decrease in intensity of this component with time, the former explanation is the more likely, since it would be surprising if copper were to be lost from high molecular weight complexes during the course of the experiment. The possibility that some of the copper is present as high molecular weight complexes in solution cannot, however, be excluded. The isotropic component can only be assigned to low molecular weight complexes in solution within the root, and, because of the delay before appreciable amounts of this complex were produced, the complexing agent is probably one that was produced by the root during the course of the experiment or which diffused slowly to the sites of copper uptake. Table 1 lists the parameters obtained from a number of copper complexes together with those of the species present in the root. It is apparent that one possible explanation for the complex in the root is that it arises from a copper-amino acid complex. Spectra were obtained from complexes of copper with several different amino acids for various concentration ratios and at pH 5 and 6. The results for the copper-glutamic acid system are illustrated in Figure 5.

Fig. 4. (opposite) (cont'd)
4.0 h, (f) 6.0 h, (g) 11.5 h, (h) 23.0 h. For all spectra microwave power = 20 mW, modulation frequency = 100 kHz, modulation amplitude = 1 mT. Receiver gain = 1.25×10^3 for (a), 3.2×10^4 for (b) - (g) and 10×10^3 for (h).

TABLE 1

EPR parameters for wheat roots and some copper complexes in aqueous solution

Complexing groups	g	A (mT)	Reference
Wheat root - isotropic component	2.124	7.0	Figs 4 & 5, this work
Amino acid			
Cu glycine	2.148	5.14	Dezor (1977)
Cu (glycine)$_2$	2.125	6.25	
Cu (glycine)$_3$	2.13	7.96	
Cu glutamic acid 1:1 complex	2.155	5.9	Fig 6, this work
Cu glutamic acid 1:2 complex	2.126	7.0	
Tetrapyrrole derivatives			
Cu etioporphyrin	2.0975	9.12	Roberts & Koski (1960)
Cu Na chlorophyllin	2.1077	9.8	Schoffa (1968)
Simple carboxylic acid			
Cu (acetate)$_2$	single broad peak		Unpublished results
Phenolic acid			
Cu (salicylate)$_2$	single broad peak		Unpublished results
Proteins	anisotropic spectra obtained because of slow tumbling $g_\perp$ = 2.05, $g_{\parallel}$ = 2.3		Ingram (1969)
Citric acid (pH 2-7)	single peak because of formation of dimeric species		Dunhill *et al.* (1966)

At least 3 different species are present with there being an increasing tendency to progress from free Cu^{2+} → 1:1 → 1:2 Cu:amino acid complex with increasing concentration of amino acid and increasing pH. Other amino acids exhibited a similar behaviour with some forming the 1:2 complex more readily (e.g. histidine) and others less readily (e.g. glycine) than glutamic acid. The isotropic spectra obtained from the copper-treated roots are similar to those from the 1:2 copper-amino acid complexes.

Confirmation of the nature of the copper complex formed

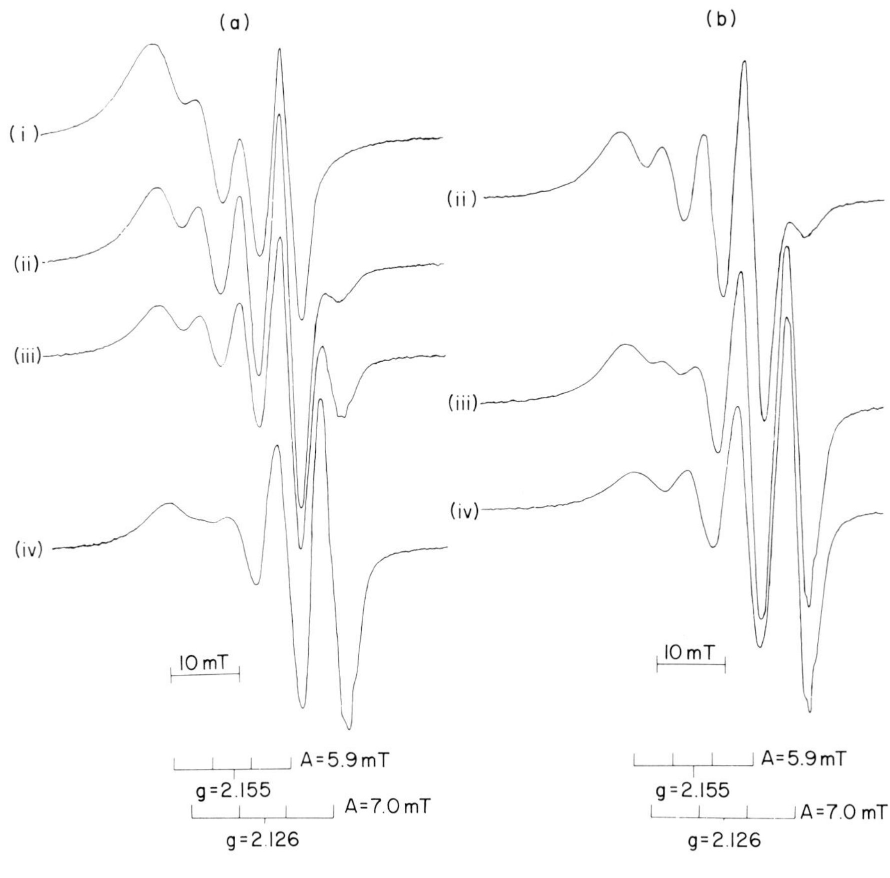

Fig. 5. EPR spectra of copper glutamic acid complexes in aqueous solution with (a) pH 5 and (b) pH 6, and with Cu^{2+}: glutamic acid ratios of (i) 2:1, (ii) 1:1, (iii) 1:2 and (iv) 1:5. For all spectra microwave power = 20 mW, modulation frequency = 100 kHz, modulation amplitude = 0.4 mT. Receiver gain = 2.0×10^3 for (a) (i) - (iii), (b) (ii), 1.6×10^3 for (a) (iv), (b) (iii), 1.25×10^3 for (b) (iv).

in wheat roots was sought by allowing uptake from a solution containing copper as an amino acid complex. However, the resulting series of spectra were very similar to those obtained after uptake from copper sulphate or copper-EDTA except that the isotropic component increased in intensity more rapidly. It thus appears that the copper and amino acid are taken up separately by the root and that the formation of the complex occurs in a compartment isolated from that in which initial immobilization of the copper occurs.

In order to check whether or not the changes observed in the EPR spectra after the uptake of copper were related to a physiological rather than a pathological process two experiments were performed on roots which were killed by immersion in liquid nitrogen for a few minutes. With a root killed before treatment with the copper-uptake solution, only a very weak spectrum was observed if the copper was present as the EDTA complex. If the copper was present as copper sulphate, a more intense, but largely anisotropic spectrum was observed (Fig. 6a). No significant changes in this spectrum were observed over an extended period. In the second experiment in which the root was killed immediately after removal from the copper-uptake solution the spectrum (Fig. 6b) has an intense isotropic component, similar to that observed with the live roots several hours after removal from the copper-uptake solution. However, the intensity did not increase further and remained essentially unchanged over a period of 4 days. The use of liquid nitrogen to kill the roots results in very substantial tissue damage so that metabolites are free to diffuse within the tissue or, where it is immersed in solution, away from the root. It is thus likely that in the case of the root killed prior to uptake, the free amino acids, present in the root, would have diffused into the uptake solution and thus not be available to form copper complexes within the tissue. In the case of the root killed after copper uptake the amino acids would be free to form copper complexes immediately the tissue thawed.

Very little has hitherto been clearly established about the forms in which trace elements are translocated in plants or even of the forms in which they occur in different parts of the plant. Manganese is thought to occur in

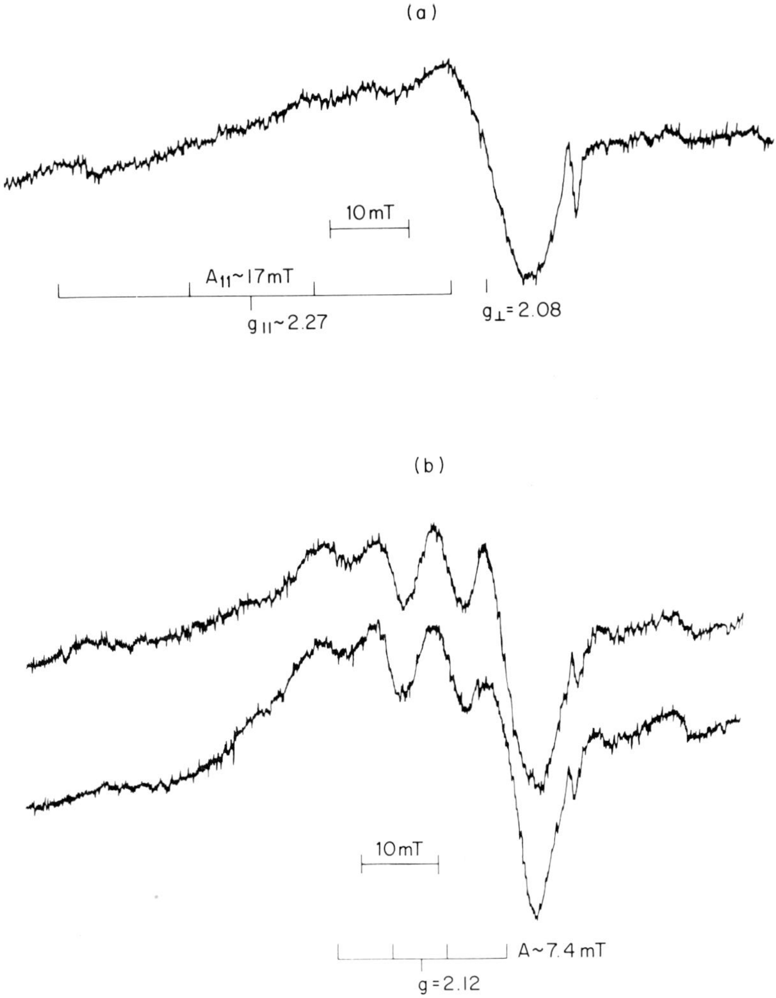

Fig. 6. EPR spectra from dead roots; (a) root killed before uptake of copper sulphate, (b) root killed immediately after uptake of copper sulphate (i) spectrum after 1.5 h, (ii) spectrum after 96 h. For all spectra microwave power = 20 mW, modulation frequency = 100 kHz, modulation amplitude = 1 mT. Receiver gain = 5.0×10^4 for (a) and 4.0×10^4 for (b).

plant tissues mainly as an uncomplexed cationic species (Bremner and Knight, 1970) and to be translocated in this form or complexed by citrate (Tiffin, 1972). Our finding of the $Mn(H_2O)_6^{2+}$ ion in all of our manganese treated wheat roots, whilst tending to lend some support to the former of these suggestions, is certainly not inconsistent with the latter, since it is likely that a complex formed between Mn^{2+} and citrate would be of the outer-sphere type and would, therefore, give the same spectrum as the uncomplexed solvated ion. Also, as pointed out earlier, many manganese complexes do not give EPR spectra so it is not possible to estimate what fraction of the manganese taken up by the root is occurring as $Mn(H_2O)_6^{2+}$.

Copper in plant tissues is known to occur as protein and amino acid complexes (Reilly, 1969, 1970). Tiffin (1972) showed that copper is translocated within the xylem of peanut plants mainly as anionic species and suggested that these were likely to be copper-amino acid complexes. Similarly, Hill (1973) speculated that copper is translocated through the plant complexed with low molecular weight carriers, probably amino acids, and that these complexes would be formed at the site of copper absorption in the root. Our finding in wheat roots of a soluble copper complex having EPR parameters consistent with it being a copper-amino acid complex thus supports the speculations of these earlier workers. Conclusive evidence regarding the nature of the copper species translocated within the plant should be available from an EPR study of xylem exudates. Detailed comparisons with the root spectra will then be possible. Successful application of EPR methods to other micronutrients will depend on the extent to which distinctive spectra are obtained from the various species produced within the root.

Conclusions

A number of conclusions regarding the uptake of copper, by roots, are possible even from these preliminary investigations. Complexed forms of copper, even highly stable chelates such as copper-EDTA, dissociate prior to uptake. The soluble copper complex, which appears in the root after uptake, is almost certainly an amino acid complex. The progressive increase in the concentration of this complex depends on the structural integrity of the root and is thus likely to be dependent on metabolism. The persistence of the complex over extended periods indicates that it is formed in a protected environment probably within the cytoplasm.

More extensive investigation of the uptake of copper by both excised and attached roots, under conditions likely to affect metabolism, are currently underway and should provide more conclusive evidence regarding the mechanisms involved in the uptake and transport of copper by plants.

References

Baes, C.F. and Mesmer, R.E. (1976). "The Hydrolysis of Cations". Wiley Interscience, New York.

Bremner, I. and Knight, A.H. (1970). The complexes of zinc, copper and manganese present in ryegrass. *Brit. J. Nutr.*, 24, 279.

Dezor, A.M. (1977). EPR study of glycine-copper complexes in aqueous solution. *Acta Phys. Pol.* A51, 417.

Dunhill, R.H., Pilbrow, J.R. and Smith, T.D. (1966). Electron spin resonance of copper (II) citrate chelates. *J. Chem. Phys.*, 45, 1474.

Goodman, B.A. and Raynor, J.P. (1970). Electron spin resonance of transition metal complexes. *Adv. Inorg. Chem. Radiochem.*, 13, 135.

Hill, J.M. (1973). The changes with age in the distribution of copper and some copper-containing oxidases in red clover. *J. exp. Bot.*, 24, 525.

Ingram, D.J.E. (1969). "Biological and Biochemical Applications of Electron Spin Resonance". Adam Hilger, London, England.

Moore, D.P. (1972). Mechanisms of micronutrient uptake by plants. *In* "Micronutrients in Agriculture" (Eds. J.J. Mortvedt, P.M. Giordano and W.L. Lindsay), p. 171. Soil Sci. Soc. Amer. Inc., Madison, U.S.A.

Pitman, M.G. (1976). Ion uptake by plant roots. *In* "Encyclopedia of Plant Physiology: Transport in Plants, IIB" (Eds. V. Lüttge and M.G. Pitman), New Series Vol. 2, p.95. Springer Verlag, Berlin.

Reilly, C. (1969). The uptake and accumulation of copper by *Becium homblei*. *New Phytol.*, 68, 1081.

Reilly, C. (1972). Amino acids and amino acid copper complexes in water soluble extracts of copper tolerant and non-tolerent *Becium homblei*. *Z. Pflanzenphysiol.*, 66, 294.

Roberts, E.M. and Koski, W.S. (1960). An electron spin resonance study of copper etioporphyrin II. *J. Amer. Chem. Soc.*, 82, 3006.

Schoffa, G. (1968). Hyperfine and superhyperfine structure of electron spin resonance and chemical bonding in copper (II) - sodium chlorophyllin. *Z. Naturforsch.*, 23a, 550.

Tiffin, L.O. (1972). Translocation of micronutrients in plants. *In* "Micronutrients in Agriculture" (Eds. J.J. Mortvedt, P.M. Giordano and W.L. Lindsay), p.199. Soil Sci. Soc. Amer. Inc., Madison, U.S.A.

THE PHYSICS AND CHEMISTRY OF THE SOIL-ROOT INTERFACE: SOME COMMENTS

D.J. GREENLAND

Department of Soil Science, University of Reading, U.K.

Summary

Important aspects of soil structure relevant to root growth and the contact between roots and the soil are reviewed and the inadequacy of present knowledge stressed.

Introduction

For many years soil and plant scientists have studied factors affecting the uptake of nutrient ions by plants, and there has recently been keen interest in soil factors controlling the movement of nutrient ions to plant roots. Work on both plant and soil aspects has been described in recent books (e.g. Lüttge and Pitman, 1976; Nye and Tinker, 1977; Russell, 1977). Although some success has been achieved in developing mathematical models to describe the movement and uptake of nutrient ions such as nitrate and potassium from soils by roots, these are successful only insofar as soil conditions conform to an "idealised structure", and roots behave in a consistent manner. Nye and Tinker (1977) conclude their book with the comment "The major factors impeding further progress are the inaccuracy of soil physical measurements; the natural heterogeneity of soils; and the difficulty of predicting root development and properties ...". Our ability to produce accurate and mathematically useful descriptions of soil structure is still very limited, although it is essential to be able to do so if we are to develop a full understanding of the relationship between soil properties and root growth, and particularly of ion transport to the root-soil interface. An appreciation of soil structure is also necessary if we are to define the root soil interface and its influence on plant growth.

This paper comments on the inadequacy of our knowledge of the physical condition of soils, and particularly of soil porosity as it relates to root development, and on the physics and chemistry of the soil surfaces at the interface. Our knowledge of the plant side of the root-soil interface is also inadequate (Oades, 1978) and some comments are made relating to the interaction of the root surface with soil particles.

Soil Porosity and Root Growth

Roots grow unrestrictedly only in pores of diameters greater than their own, and only enter those of at least similar size to the root tips and which are easily enlarged. No precise size can be ascribed to root tips, but they can certainly enter pores of the order of 10 μm diameter. Provided that these offer little mechanical resistance to enlargement, the roots will continue to grow in them (Russell and Goss, 1974). More commonly roots will grow into larger pores, and these will normally be air-filled. Root diameters of cereals for instance are of the order of 200 to 400 μm (Russell, 1977), and pores wider than about 50 μm are drained of water at field capacity. Total porosity of soils is normally greater than 40%, and frequently over 50%, but except in sandy soils the greater part of the pore space is contained in pores less than 1 μm (10^3 nm) equivalent diameter. In aggregates from loams or clays the proportion of pores or voids larger than 50 μm is frequently very small indeed (Fig. 1, which refers to aggregates of about 5 cm^3). In such soils root growth is largely re-

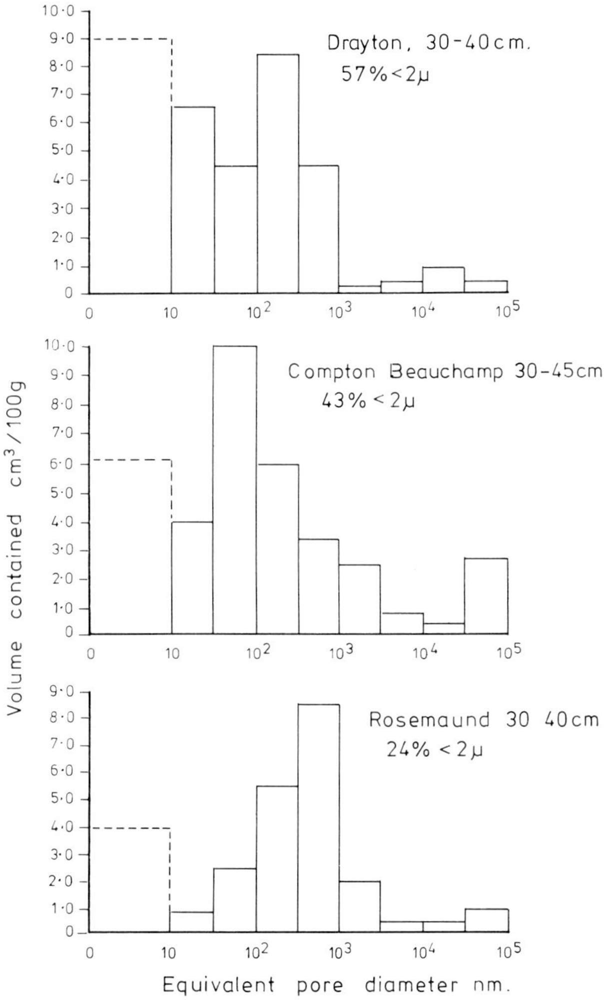

Fig. 1. Pore size distributions of 5 cm^3 aggregates from subsurface horizons of a clay, loamy clay and silty loam, determined by mercury injection porosimetry after critical point drying. (Pores <10 nm radius were not measured but assumed to account for the residual porosity to bring the total to that given by the initial aggregate densities, and so include porosity lost in critical point drying, which is predominantly due to interlamellar water within swelling clays).
(a) Denchworth series soil from Drayton E.H.F., U.K. 57% <2μ clay, 2.8% organic matter, aggregate density ρa = 1.19
(b) Denchworth series soil from Letcombe Laboratory site, Compton Beauchamp, U.K. 42% <2μ, 5.8% organic matter, ρa = 1.25
(c) Bromyard series soil from Rosemaund, E.H.F., U.K. 24% <2μ, 1.0% organic matter, ρa = 1.59. (Unpublished data of Lawrence, Payne and Greenland.)

stricted to the fissures between aggregates. Some root hair penetration of pores less than 50 μm diameter probably occurs. Russell and Goss (1974) and others have shown that a pressure as low as 100 mbars or 10^{-2} MPa is sufficient to reduce the rate of root elongation by 50%. The tension of water in pores of around 50 μm diameter is of corresponding magnitude. Thus while roots may enter some of the larger pores which occur within aggregates, they will have difficulty enlarging them to allow further growth unless the soil is close to field capacity, when the pressure the root must exert to enlarge the void in which it is growing will be less than 10^{-2} MPa.

Aggregate Structure and the Root-Soil Interface

Inter-ped pores, fissures or channels are easily lost by compression (Fig. 2). In many clay soils their formation is closely related to the shrinkage behaviour of the soil associated with wetting and drying. Aggregates from some soils such as the Denchworth series clay from Drayton follow a normal shrinkage curve, i.e. withdrawal of water during drying does not remove water from successively smaller pores, but merely causes the aggregate to shrink. The loss in volume causes the soil surface to be lowered, and cracks or fissures to appear within the soil (Fig. 3). The development of critical point drying methods allied to mercury injection porosimetry has enabled the pore size distribution of aggregates at different water contents to be determined with reasonable confidence that the information obtained relates to the actual conditions in the aggregate (Lawrence, Payne and Greenland, in press). Thus it is now known how the porosity of the soil aggregates changes as their water content changes. In the Drayton soil the few pores larger than 1 μm are quickly lost as the sample dries, so that little or no entry of air into the aggregates occurs during drying under field conditions.

The Drayton soil is an extreme example as it contains

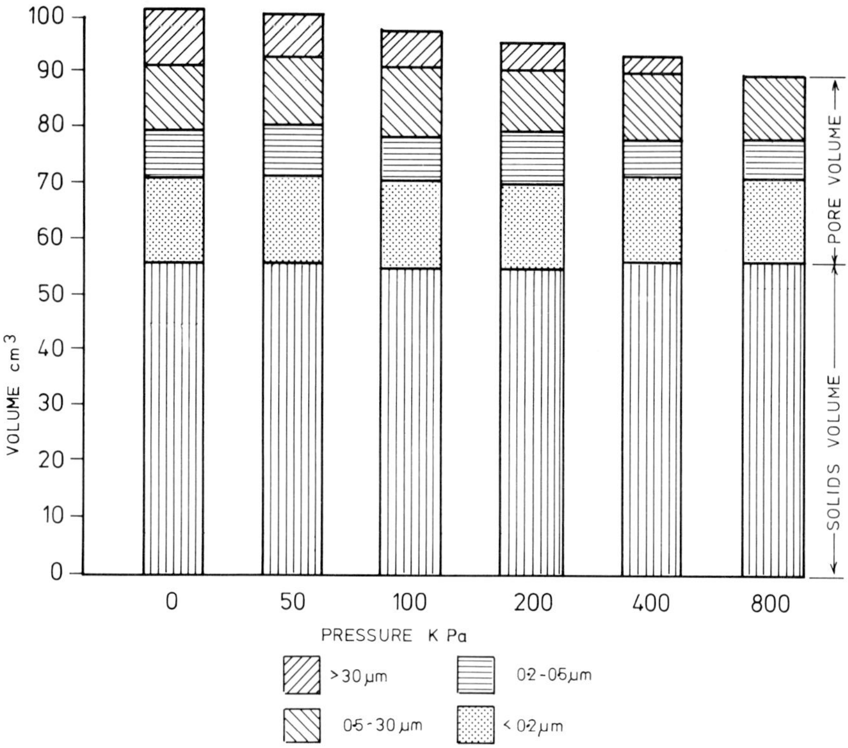

Fig. 2. Change in porosity of a clay loam from Sweden, due to compressive stress (reproduced with permission from Eriksson, Håkensson and Danfors, 1974).

almost 60 per cent of <2 μm clay. More typical examples of non-calcareous clays such as that from Compton Beauchamp (Fig. 1) have around 40 per cent clay, and often higher organic matter levels than the Drayton soil. The coarser texture does not appear to produce a general shift in the pore size distribution to larger pores, although the few extra pores in the Compton Beauchamp soil in the size range 30-100 μm (3×10^4 to 10^5 nm) may be particularly important. In the field root growth into aggregates is seen in this soil, but only very rarely in aggregates of the Drayton soil.

The data quoted refer to subsurface horizons. Aggregates from the surface horizon often contain more organic matter content, and may contain a greater proportion of transmission pores (those greater than 50 μm e.c.d.) although this depends on their agronomic history (Currie, 1965).

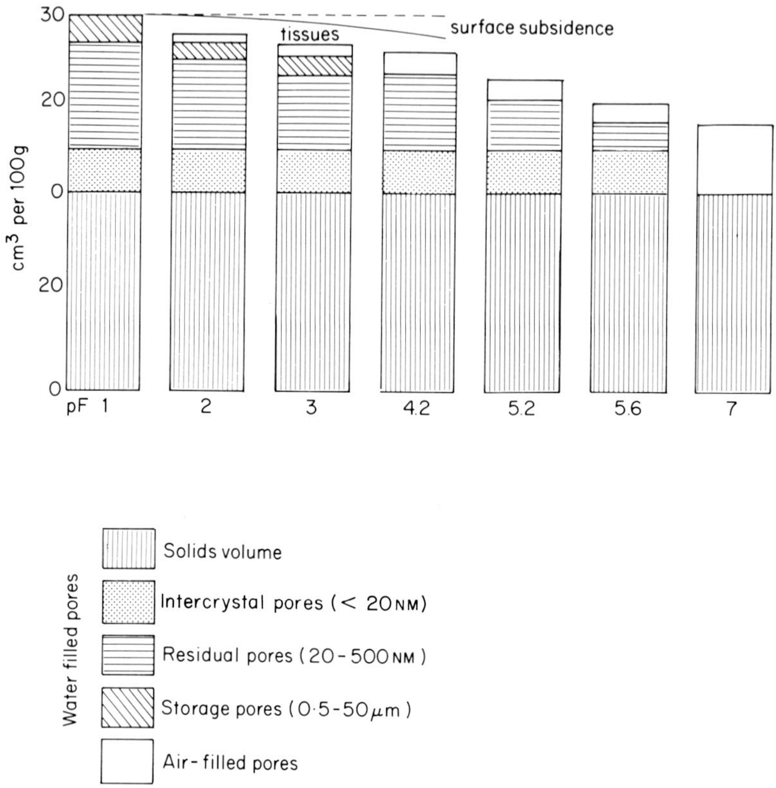

Fig. 3. Shrinkage and changes in pore size distribution of 6 cm^3 aggregates of the Denchworth series soil from Drayton, U.K. Water was removed from aggregates equilibrated at different water suctions by critical point drying, and the pore size distribution then determined by mercury injection porosimetry. (Unpublished data of Lawrence, Payne and Greenland.)

While the structure of aggregates is partly defined by the pore size distribution within them, their physical and chemical properties can also be influenced by the presence of clay coatings or cutans on their surfaces (Brewer, 1960). The importance of such coatings in relation to the movement of ions and water to plant roots has been a matter of some controversy. Brewer and others have suggested, without experimental support, that such coatings would in effect provide a barrier to ion and water transport in and out of the aggregate, whereas Nye and Tinker (1977) have discounted their significance. There is now no doubt of their widespread occurrence in subsurface horizons in a very wide range of environments. The term 'argillic horizon' is used in 'Soil Taxonomy' (USDA, 1975) as a differentiating property for several higher categories. Recently Kane and Mott (unpublished) have succeeded in isolating the cutanic material from B horizons of some British soils (Fig. 4). This material often has a root-like morphology,

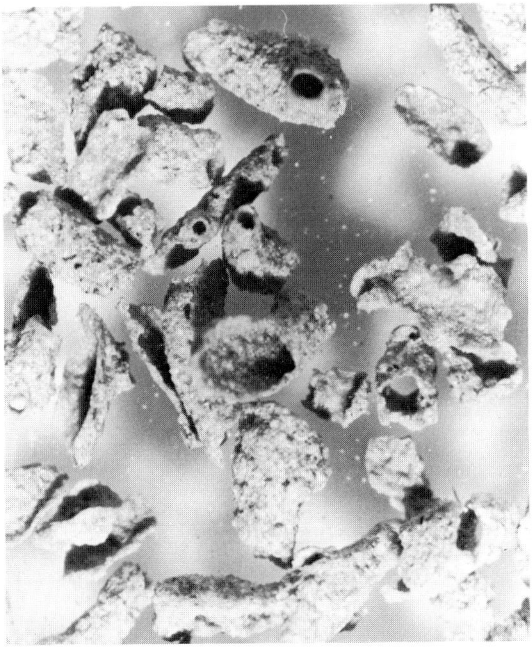

Fig. 4. Photomicrograph of clay skins isolated from the B horizon of a loessal soil from Burghfield Common, near Reading, U.K. The diameter of the clay tubes is approximately 250 μm. (Photo kindly supplied by Mr. P. Kane.)

which probably arises because the clay was initially deposited in an old root channel, but could also be produced by root pressure on cutans at the surfaces of soil aggregates, the effect being reinforced by subsequent root growth which follows the old channel. The importance of the work of Kane and Mott in isolating the cutans is that it becomes possible to study the physical chemistry of the soil side of the root-soil interface.

Cutans are less common in surface soils, because of the disturbing action of faunal and mechanical manipulation, and because most are formed by mobilisation and deposition of clay particles at a lower level in the soil. Nevertheless they may still occur, as coatings on peds or larger particles. Pressures exerted by the growing root can help to orient particles, and create a layer of higher bulk density at the surfaces of pores in which the root grows. Bradfield (1937) described the formation of well lined "non-capillary" pores in surface soils in very clear terms many years ago. If clay coatings are continuous, and the particles platy and well aligned, then the pores separating them are likely to be very small. Gas adsorption studies with soils and clays (Aylmore and Quirk, 1967; Mott, unpublished results) have indicated that mean clay particle separations are usually close to 3 nm. Mass flow in such pores will be negligible, and diffusion restricted (see below).

However cutans are seldom continuous. More commonly the soil side of the root-soil interface will consist of clay domains, or surfaces of soil aggregates (those > 1 mm diameter) or microaggregates (5-1,000 μm diameter). Between these structural units much larger pores occur. For example in the Denchworth soils at Drayton and Compton Beauchamp, when close to field capacity there is a major peak in the pore size distribution curve near 10^2 nm (Fig. 1). This probably represents the pores between domains of clay particles. Mass flow in such pores will be greater by a factor of $(10^2)^4=10^8$ than in the pores between clay particles with radii close to 1 nm, because flow rates in capillaries are proportional to the fourth power of the radius.

Clay skins can therefore be expected to impede mass flow very considerably. Their influence on diffusion is likely to be much less. Water within 1 nm of the surface of calcium montmorillonite has a viscosity three times greater than that of bulk water, and diffusion of electrolytes will be impeded to a rather greater extent (Kemper, Maasland and Porter, 1964). However Gerber, Wilding and Franklin (1974) who have measured diffusion across cutans report relatively small effects.

Surfaces of Mineral Particles and Root-Soil Contacts

Surfaces of mineral particles fall into two major categories, those of constant charge and those of constant potential. Those with constant charge include those of the micas, smectites and vermiculites and those with constant potential the oxides and hydrous oxides. At constant charge surfaces there is a high concentration of counterions. The charge density ranges from about 0.3 to 0.1 C m^{-2}. The counterions form a double layer at the surface. Calcium, magnesium and aluminium are usually dominant. They are hydrated, and most exist in a surface layer two or three water molecules thick. In this layer water has a high viscosity as already mentioned and other properties similar to those of water in a salt hydrate.

For hydrous oxide surfaces the negative charge density is dependent on pH and electrolyte concentration. In common soil conditions, pH 5 to 8 and approximately 0.01 M solution, it lies between 0.3 and 0.4 C m^{-2} for silica or quartz, but is much lower for the oxides of iron and aluminium. At these surfaces water molecules associate more strongly with the surface hydroxyls, and this association gives rise to a similar hydrophilicity as for constant charge surfaces, so that movement of water and electrolytes close to such surfaces is likely to be similarly restricted.

All mineral surfaces other than those which have been rigorously dehydrated by strong heating will tend to conform to one or other of these surface types. As they are hydrophilic water will condense on them according to its partial pressure in the atmosphere to which they are exposed. At around 20 per cent relative humidity (R.H.) a monolayer is formed, and at 98.6 per cent R.H. a multilayer many molecules thick has formed (Quirk, 1957). This R.H. corresponds to a soil water suction of 15 bars, pF 4.2 or 1.5 MPa, the so-called "wilting point". Thus in soils containing water available to plants there will always be a thick water film on particle surfaces, even if the pore is not water filled. Pores up to an equivalent diameter of 200 nm are filled at the "wilting point" and up to a diameter of about 5 x 10^4 nm at field capacity.

Although roots may normally grow into air-filled pores larger than 5 x 10^4 nm diameter, they will not contact exposed particle surfaces, but the soil solution. In soils of weak structure clay particles may be dispersed within the soil solution so that pore surfaces are ill defined. More generally the plant will produce root hairs that may penetrate aggregates adjacent to the root. The response of the root to soils of weak and stable structure is different. Figure 5 illustrates roots growing in a poorly structured soil, before and after stabilisation with a soil conditioner, and clearly showing the effect of structural condition on root surface characteristics. Soileau (p.c.) has obtained similar photographs for cotton plants in a soil treated or not treated with a chemically quite different conditioner, krilium, to provide coarsely aggregated and finely aggregated soil. He found root hair development to be prolific in coarsely aggregated (4.8 - 3.3 mm) soil, and much less in the <1.2 mm aggregate size fraction.

The Root Surface

The external surface of the root epidermis is probably

Fig. 5. Roots of tomato in the structurally unstable fine sandy loam of Rowland series soil from the University farm, Sonning, near Reading. (a) untreated soil (b) soil to which 0.12% by weight of polyvinyl acetate soil conditioner had been added to stabilize the structure (from Carr and Greenland, 1975).

not the immediate surface of contact between plant material and soil, because it appears that it is commonly covered by a layer of mucilage (Oades, 1978). This may not be true of older, and particularly of suberized, roots, but as such roots may be less active in absorption, so that soil contact is of less importance, they will be omitted from this discussion. Most root mucilage is of plant origin, but as it invariably becomes heavily populated by microorganisms it is normally contaminated with materials of microbial origin. Examination by staining methods suggests a predominantly pectin-like composition. More recent work has shown that while polysaccharides are a major constituent, lipids and peptides are also present. The possibility exists that a membrane surrounds the mucilagenous layer, at least close to the root tip. If such membranes do exist, they do not persist far from the root tip, as in this region electron micrographs have shown the mucilage to be rather freely penetrated by microorganisms and soil particles (Foster and Rovira, 1976).

Thus in the regions where most absorption of water and nutrients is occurring, the root side of the interface is commonly represented by the 'mucigel' layer, as Jenny and Grossenbacher (1963) termed the mucilage. It appears to be a true gel, allowing water and ions to diffuse through it in a manner not significantly different from that in free water. It would however, be expected to restrict mass flow. Again it may be speculated that in common with cutans on the soil side of the interface there may be channels through it.

Where the organic constituents of the mucigel layer meet the interface, they are liable as a result of their thermal motion to approach the surfaces of mineral particles. There have been no direct studies to establish the interaction which arises. However where the sugar and uronic acid composition of the root mucilages have been determined the results indicate that the pectin-like components are closely similar to the polysaccharides isolated from soils. These are readily adsorbed by clays (Fig. 6) and so it is to be expected that the mucigel materials will be adsorbed. Turchenek and Oades (1976) consider that the adsorption of fine clay particles to mucigel constituents is the first step in the formation of stable microaggregates around plant roots, a part of the process described much earlier by Bradfield (1937). However it is hazardous to predict the adsorption behaviour of polymeric materials, and direct studies of mucilage adsorption are needed.

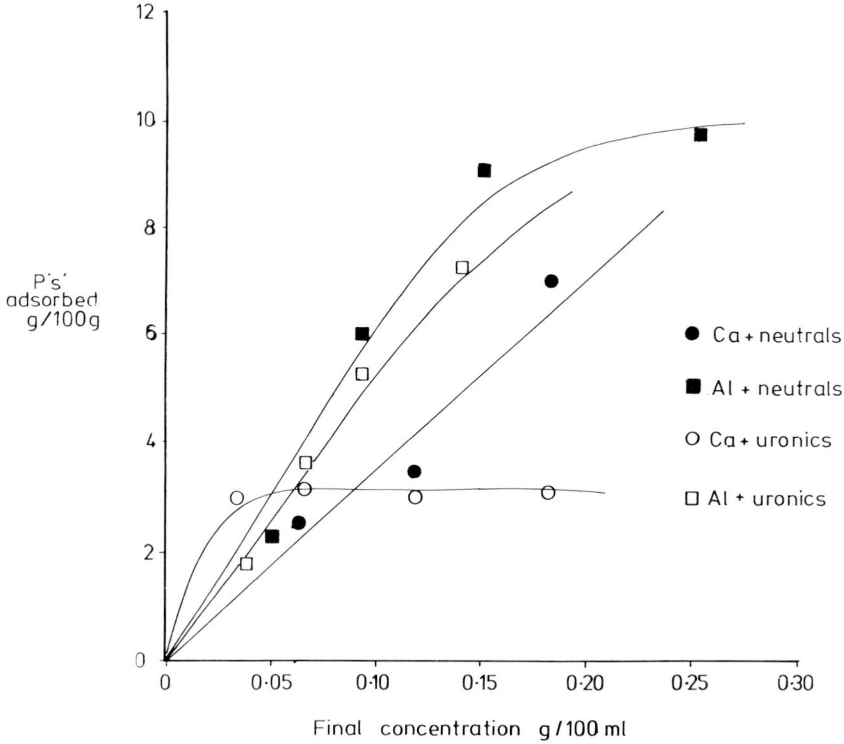

Fig. 6. Sorption by calcium and aluminium saturated montmorillonite of neutral and uronic components from a soil polysaccharide (from Parfitt and Greenland, 1970).

The continuum which may be formed by mucilage from soil particles to root may be of significance to plants as well as soils. This arises because the binding of inorganic particles to the root implies that as the soil dries, and the root shrinks, a soil layer is likely to be drawn close to the root (Fig. 7). This soil layer will normally contain many fine water-filled pores which form a protective buffer against further desiccation and help to preserve a moisture film around the root. Russell (1977) has suggested that the 'root-soil gap' may be significant in restricting the uptake of water and nutrients in drying soils. Tinker (1976) quotes others who have calculated that a 'gap' is likely to develop, and suggested that the root will probably stick to one side of the pore in which it exists. If the root is initially at a distance from the soil surface, which is less than the diameter of a nearby pore, then the 'break' may be expected to arise at the weaker point, which will be the coarse pore (Fig. 7). The reinforcement of the root-soil contact by mucigel adsorption will of course help to maintain contact adjacent to the root (Weatherley, 1976) so that the root-soil gap develops away from the root. However the 'gap' may be just as significant in limiting mass flow and diffusion to the root even if it is a few microns away from the root. At present there is no adequate way of determining whether 'gaps' in the soil water continuum at the root soil interface, or divorced from it, are the more important.

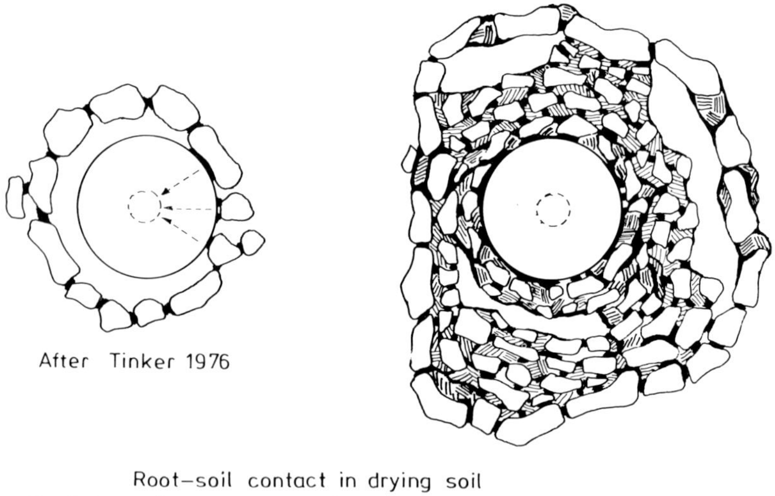

Fig. 7. Schematic representation of the development of air gaps at or near the root soil interface during drying of soils (a) at the interface (Tinker, 1976), (b) divorced from the interface.

A possible involvement of the mucigel in relation to ion transport was suggested by Jenny and Grossenbacher (1963). This is perhaps more speculative than other discussion relating to the root-soil interface, but is nevertheless a possibility. It has also been suggested that the mucigel can act as a lubricant, allowing the root to move into and through pores which are a tight fit (Barley and Greacen, 1967). If the mucigel is in fact similar to many polysaccharide gums, it will strengthen attachment between root and adjacent soil particles. The alternative suggestion of Barley and Greacen (1967) that the sloughing of the surface cells of the root cap, together with the mucilage, allows easier movement of the root is more convincing. The sloughed material would be left with the soil, and the root move through a gel which it had deposited.

Other Considerations

In this short paper only a few aspects of the physics and chemistry of the root-soil interface have been discussed. Much of this discussion has been speculative. There are few direct measurements of the properties of the interface. It is known that it is an area of intense microbiological activity. Also gas exchange proceeds actively at the interface, as well as ionic movements which result in changes in pH and other factors. These aspects have not been discussed. Obviously to characterize the interface adequately will require a great deal of careful and detailed work. Progress in understanding water and nutrient absorption may well come most rapidly by using as a working hypothesis the proposition that the interface has negligible effects on uptake, and then studying in detail those soil and plant systems where this hypothesis is shown to be inadequate.

References

Aylmore, L.A.G. and Quirk, J.P. (1967). The micropore size distribution of clay mineral systems. *J. Soil Sci.*, 18, 1.

Barley, K.P. and Greacen, E.L. (1967). Mechanical resistance as a soil factor influencing the growth of roots and underground shoots. *Adv. Agron.*, 19, 1.

Bradfield, R. (1937). Soil conservation from the viewpoint of soil physics. *J. Amer. Soc. Agron.*, 29, 85.

Brewer, R. (1960). Cutans: their definition, recognition and interpretation. *J. Soil Sci.*, 11, 280.

Carr, C.E. and Greenland, D.J. (1975). Potential application of polyvinyl acetate and polyvinyl alcohol in the structural improvement of sodic soils. *In* "Soil Conditioners", special publication, p.47. Soil Sci. Soc. Amer.

Currie, J.A. (1965). Diffusion within soil microstructure: a structural parameter for soils. *J. Soil Sci.*, 16, 279.

Eriksson, J., Håkansson, I. and Danfors, B. (1974). "The Effect of Soil Compaction on Soil Structure and Crop Yields". Bull. 354, p.17. Swedish Institute of Agricultural Engineering.

Foster, R.C. and Rovira, A.D. (1976). Ultrastructure of wheat rhizosphere. *New Phytol.*, 76, 343.

Gerber, T.D., Wilding, L.P. and Franklin, R.E. (1974). Ion diffusion across cutans: a methodology study. *In* "Soil Microscopy" (Ed. G.K. Rutherford), pp.730-746. Limestone Press, Kingston, Ontario.

Jenny, H. and Grossenbacher, K.A. (1963). Root-soil boundary zones as seen in the electron microscope. *Soil Sci. Soc. Amer. Proc.*, 27, 273.

Kemper, W.D., Maasland, D.E.L. and Porter, L,K. (1964). Mobility of water adjacent to mineral surfaces. *Soil Sci. Soc. Amer. Proc.*, 28, 164.

Lawrence, G.P. (1978). Measurement of pore sizes in fine-textured soils: a review of existing techniques. *J. Soil Sci.*, 28, 527.

Lüttge, U. and Pitman, M.G. (1976). "Transport in Plants. II. Tissues and Organs. Part B". Springer Verlag, Berlin.

Nye, P.H. and Tinker, P.B. (1977). "Solute Movement in the Soil Root System". Blackwells, Oxford.

Oades, J.M. (1978). Mucilages at the root surface. *J. Soil Sci.*, 29, 1.

Parfitt, R.L. and Greenland, D.J. (1970). Adsorption of polysaccharides by montmorillonite. *Soil Sci. Soc. Amer. Proc.*, 34, 862.

Quirk, J.P. (1957). Significance of surface areas calculated from water vapour sorption isotherms by use of the B.E.T. equation. *Soil Sci.*, 80, 423.

Russell, R.S. (1977). "Plant Root Systems: their Function and Interaction with the Soil". McGraw-Hill, London.

Russell, R.S. and Goss, M.J. (1974). Physical aspects of soil fertility - the response of roots to mechanical impedance. *Neth. J. agric. Sci.*, 22, 305.

Tinker, P.H.B. (1976). Transport of water to plant roots in soil. *Phil. Trans. Roy. Soc., Lond., B*, 273, 445.

Turchenek, L.W. and Oades, J.M. (1978). Organo-clay particles in soils. *In* "Modification of Soil Structure" (Eds. W.W. Emerson and A.R. Dexter), pp.137-144. Wiley, Chichester.

USDA. (1975). "Soil Taxonomy". USDA Soil Conservation Service, Agricultural Handbook No. 463. Government Printing Office, Washington.

Weatherley, P.E. (1976). Introduction: water movement through plants. *Phil. Trans. Roy. Soc., Lond., B*, 273, 435.

WATER POTENTIAL GRADIENTS AND RESISTANCES OF A SOIL-ROOT SYSTEM MEASURED WITH THE ROOT AND SOIL PSYCHROMETER

H.B. SO

Department of Agronomy and Soil Science, University of New England, Armidale, N.S.W., Australia

Summary

An improved version of the Fiscus root psychrometer was developed to measure the vertical distribution of root xylem water potentials in an actively transpiring soybean plant and soil psychrometers were used for measuring soil water potentials at various depths. The results confirmed the validity of the root psychrometer measurements and enabled partitioning of the resistances of root xylem, root cortex (radial resistance), soil rhizosphere and soil pararhizal regions under contrasting soil moisture conditions. The data showed that an interaction between the various resistances determined the pattern of water uptake and that xylem resistance was significant.

Introduction

Investigations into plant water relations, have disclosed a number of uncertainties associated with the root system and the soil immediately around it, the rhizosphere (Tinker, 1976), e.g. the relative magnitudes of plant (R_p) and soil rhizosphere resistances (R_s) and their influence on water uptake. Earlier work conveniently assumed a negligible plant resistance and hence a constant water potential at the root surface throughout the root system

(Gardner, 1960; Cowan, 1965). This led to the conclusion that the soil resistance is dominant and exerts its control at relatively high soil water potentials which agrees with experiments carried out in small pots (Reicosky and Ritchie, 1975; Faiz and Weatherley, 1977) and on small sections of a single root (So, Aylmore and Quirk, 1978). Other experiments however, mainly on plants grown in the field, have shown that plant resistance remains dominant until soil water potentials are relatively low and close to -15 bars (Newman, 1969a, b; Reicosky and Ritchie, 1975). One difference between these two cases is that a small pot is equivalent to a small section only of the root zone of a plant growing in the field and is depleted of water uniformly. In the field, the root zone is rarely depleted of water in a uniform manner, the region of greatest water uptake shifting downwards with time (Ogata, Richards and Gardner, 1962). Possible factors affecting this downward shift are evaporation from the soil surface, root distribution and root xylem resistance. It has been shown that the last, usually assumed to be negligible, can be significant (Wind, 1955; Passioura, 1972).

This paper briefly discusses the construction and use of a root psychrometer and experiments designed to partition the various resistances of the water flow pathway.

Theory

The pathway of water movement through the soil-plant system can be approximated by an analogue of Ohm's law (Van den Honert, 1948):

$$F = \frac{\psi_1 - \psi_2}{R_{12}} = \frac{\psi_2 - \psi_3}{R_{23}} \text{ etc.} \qquad (1)$$

where F is flux of water across a given cross-sectional area of the system; $\psi_1 - \psi_2$, $\psi_2 - \psi_3$ are potential differences across successive components of the system and R_{12}, R_{23} are corresponding resistances to water flow across these components.

It is assumed that the root system consists of an array of parallel roots which can be divided into j equal vertical sections. Following Cowan (1965), we express the water pathway in the soil-plant system as an arrangement of series and parallel resistances as shown in Figure 1. From equation (1) we can readily obtain the following relation-

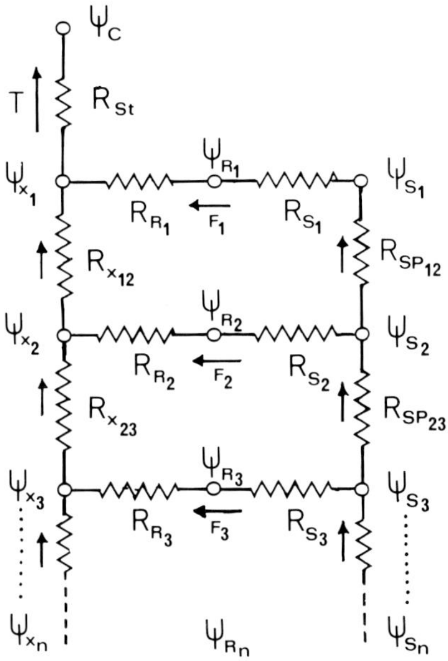

Fig. 1. Pathways of water transport in a soil-plant system. ψ_c, ψ_x, ψ_r and ψ_s are potentials of the canopy, root xylem, root surface and soil respectively. R_{st}, R_x, R_r, R_s and R_{sp} are resistances of stem, xylem, radial root tissues, soil rhizosphere and pararhizal soil respectively. T is the transpiration rate and F represents local fluxes. Numbers denote the horizons and arrows the direction of water flow.

ship for the transpiration rate (T) of the "whole soil-plant system",

$$T = \frac{\psi_c - \psi_{x_1}}{R_{st}}, \tag{2}$$

the water flux through the nth section of the root system (F_n),

$$F_n = \frac{\psi_{x_n} - \psi_{s_n}}{R_{r_n} + R_{s_n}} \qquad (3)$$

or when expressed per unit length of root for that section,

$$F_n{}^* = \frac{\psi_{x_n} - \psi_{s_n}}{R^*_{r_n} + R^*_{s_n}} \qquad (3a)$$

and the vertical flux of water through the xylem between the nth and $(n + 1)^{th}$ section:

$$T - \sum_{i=n}^{j} F_n = \frac{\psi_{x_n} - \psi_{x(n+1)}}{R_{x(n,n+1)}} \qquad (4)$$

where ψ_c, ψ_x, ψ_s are the water potentials of the canopy (average), xylem and soil respectively; R_{st}, R_r, R_s, R_x are the resistances of the stem (xylem), root (radial), soil rhizosphere and root xylem and n denotes the number of the sections from the soil surface.

The soil rhizosphere resistance per unit length of root (R^*_s) can be estimated from the equation for water flow to a single root derived by Gardner (1960):

$$\psi_r - \psi_s = \frac{q}{4\pi K_\psi} \ell n \; (b^2/a^2) \qquad (5)$$

where ψ_r is the water potential at the root surface, q is the water uptake rate per unit length of root, K_ψ the hydraulic conductivity of the soil as a function of ψ and b and a are the radius of soil rhizosphere and root respectively; b is equal to $(\pi \; Lv)^{-\frac{1}{2}}$ where Lv is the root density (root length per unit soil volume). q can be estimated from the change in moisture content (θ) as $(d\theta/dt)/Lv$. The soil rhizosphere resistance of a single root then becomes:

$$R^*_s = \frac{\ell n(1/\pi \; Lv \; a^2)}{4\pi K_\psi}$$

R^*_r can be calculated from equation (3a). Since ψ_r is not known, the value of $K\psi$ used is that at the associated ψ_s which will tend to underestimate R^*_s. Since all resistances

R^*_r and R^*_s in a unit layer of soil are arranged in parallel the relative magnitudes remain constant when pooled as R_r and R_s.

The pararhizal resistance (R_{sp}) between two soil sections which can be estimated from the vertical flow (V_z) of water through the soil, is

$$V_{z(n,n+1)} = \frac{\psi_{sn} - \psi_{s(n+1)}}{R_{sp(n,n+1)}} = \pi r^2 K_\psi \frac{(\psi_{sn} - \psi_{s(n+1)})}{d} \quad (7)$$

and

$$R_{sp(n,n+1)} = \frac{d}{\pi r^2 K\psi} \quad (8)$$

where r = radius of the soil column and d is the length of each soil section.

Methods and Materials

The root psychrometer

Construction of the root psychrometer which is similar to that of Fiscus (1972), is shown in Figure 2. In place of the single junction used by Fiscus, we used a commercial EMCO* three-wire psychrometer to improve its stability against rapid temperature fluctuations (Chow and de Vries, 1973). Details on the construction, theory and use of this psychrometer are published elsewhere (So, 1977).

Water uptake by soybeans

Four soybean seeds (*Glycine* max L. (Merr.) var. 'Wayne') were planted in each of three cylindrical clear acrylic columns, 100 cm long and 20 cm diameter, packed to a uniform bulk density of 1.51 g cm^{-3} with Sparta fine sand from near Ames, Iowa, USA. The acrylic column was split lengthwise into two equal sections and taped together for ease of sampling at the end of the experiment. Two holes (2 cm diameter) and two windows of 5 x 15 cm were provided

* Trade names are included for the reader's benefit and do not imply endorsement by the author.

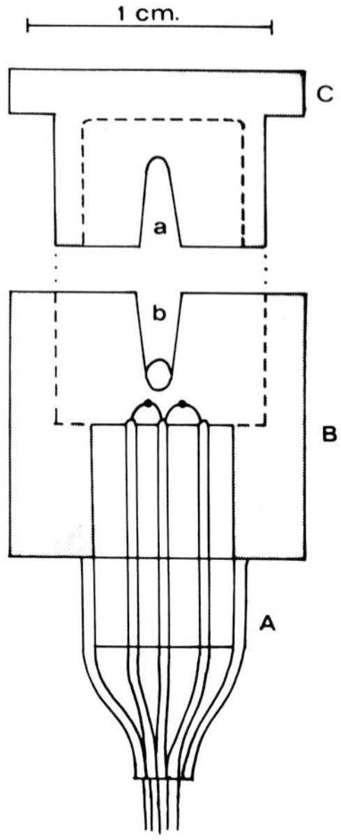

Fig. 2. Schematic diagram of the root psychrometer. A. Emco psychrometer, B and C, Teflon housing. When the slits a and b are aligned they form a hole for the root.

for inserting soil and root psychrometers, respectively, at depths of 5, 10, 20, 40, 60 and 90 cm from the soil surface. Two weeks after germination the plants were thinned to one plant per column. Until the plants were ready for the experiment, they were grown in a glass-house at Iowa State University. Day and night temperatures were maintained at 24° and 20°C respectively and winter sunlight was supplemented by a bank of high intensity fluorescent tubes (20 cm above the canopy), which provided a quantum flux of 800 $\mu E\ m^{-2}\ s^{-1}$ at the top of the canopy for 14 hours per day.

At six weeks, the roots had reached the bottom of the columns and these were then transferred to a constant temperature growth room where the experiment was conducted. Light was provided by a bank of four high intensity fluorescent tubes and eight incandescent spot lights providing a quantum flux of 250-800 $\mu E\ m^{-2}s^{-1}$ throughout the canopy for

14 hours per day. Room temperature was normally maintained at 21°C; however, during the daytime the heat of the light raised the temperature to 24°C which resulted in a very high rate of transpiration of approximately 4 cm/day. Two root and two soil psychrometers were inserted at each of the designated depths. Soil was removed through the window until a suitable root was found ($\geq$ 0.4 mm). The psychrometer was then attached to the root, which had been cleaned with a wash bottle, the soil was repacked and the window firmly taped into place.

The whole column was then insulated in a polystyrene foam box and the gaps packed with chips of the same material. The top was covered with aluminium foil to minimize heating through radiation. This insulation reduced the daily temperature fluctuations inside the soil column to less than 1°C and was also important in reducing temperature gradients throughout the column which might have caused large errors in measurements. The whole arrangement was then placed on a load cell with a pressure transducer connected to a recorder to monitor transpirational losses.

Measurements of soil and root water potentials were made at least three times daily throughout 4 drying cycles (28 January - 12 February, 1977). Leaf canopy water potentials were coninuously monitored via stem diameter changes which were calibrated against the pressure chamber method (So, 1978). Resistances of the soil-plant system were calculated using equations (2) to (8).

Other measurements included soil moisture characteristic curves using the filter paper method of Fawcett and Collis-George (1965) and soil hydraulic conductivity by the evaporation method of Passioura (1976). Root distribution was measured in duplicate cylinders at the beginning of the experiment and at the end of the third drying cycle; at the end of the fourth drying cycle, the experimental column was used. Roots were washed and cleaned of debris and then counted on a Rowse and Philips root counting machine modified to include a laser light source (J.F. Andrews, Iowa State University, personal communication).

Results and Discussion

Assuming no appreciable longitudinal movement of water in the root cortex, we believe that at equilibrium, the psychrometer measures the water potential of the root xylem (So, Aylmore and Quirk, 1976).

However, under transient conditions, the potentials measured by the psychrometer lag behind the potential of the xylem by an amount dependent on (1) the resistance to vapour diffusion between root surface and psychrometer chamber and (2) the resistance to liquid water flow across the root tissues. The time lag due to the vapour diffusive resistance is equal to the instrument response time which is estimated at approximately 15 minutes for the psychrometers used (So, 1977). The amount of water that the root has to supply or absorb to produce a relative humidity change of 2% (98% to 100%) which corresponds to the range of potentials encountered by the roots, is so small (10^{-9} cm^3) that even with the lowest rate of water uptake recorded at 10^{-4} cm^3h^{-1} per cm root (So *et al.*, 1976), the plant needs at most a few seconds to complete the process. Therefore, the resistance to liquid water flow is insignificant in relation to the diffusive resistance and the psychrometer readings when corrected for errors due to instrument response time, essentially correspond to the xylem water potentials and when soil moisture is not limiting, the pattern of changes in the latter must follow closely that of the leaves. One such observation is shown in Figure 3 which gives the diurnal changes of canopy and uncorrected xylem water potential of two roots at 10 cm depth. Corrections were not performed since in this case they were indeed very small. The soil during this period was close to or wetter than field capacity and its potential cannot be measured by the psychrometer. The time lag

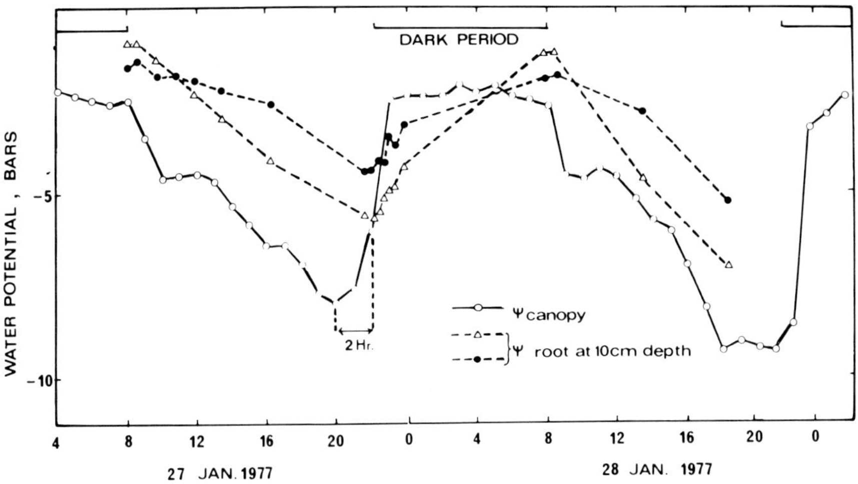

Fig. 3. Diurnal pattern of canopy (o) and xylem water potential of two soybean roots at 10 cm depth (Δ and •). Soil was close to field capacity.

between the two patterns was approximately ½ hour in the morning, increasing to approximately 2 hours at the end of the day, and was largely due to the resistance of the xylem. The potential profiles at various stages of the second drying cycle are shown in Figure 4. The increase in root water potential with depth is consistent with an upward flow of water and supports the validity of the measurements with the root psychrometer. No calibration against other instruments has been carried out as yet.

It is interesting to note that an inversion of the potential gradient between root and soil can occur in the upper layers of soil when the water potential is low (2-2-77 in Fig. 4); in fact, water uptake still continues at a low rate against this gradient (Fig. 6). Such an inversion has been reported previously and it is not clear what caused such apparently active water uptake; it may be associated with a pumping mechanism in relation to ion uptake (Taerum, 1973).

An analysis of the data was carried out for 29 January, 1977, three days after the whole column was re-watered and

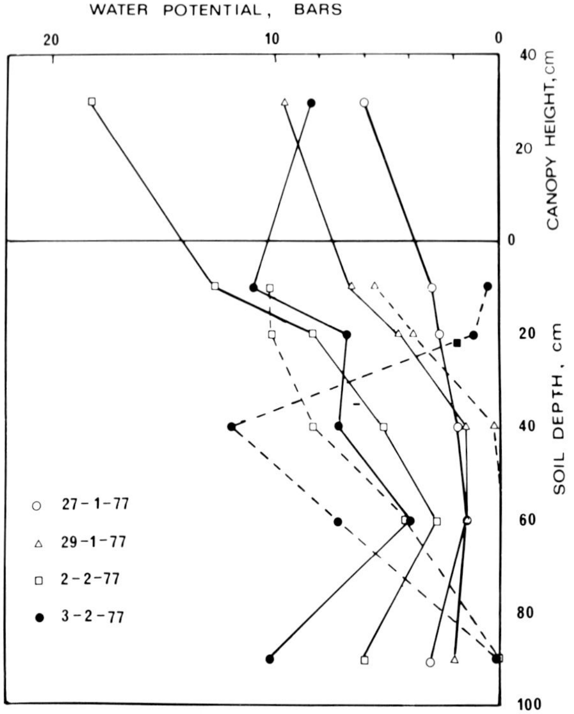

Fig. 4. Potential profiles of plant (——) and soil (----) at 1500 h on four different days. The soil was watered with 1500 ml on the morning of 3 February, 1977 and wetted the top 35 cm of soil. Soil water potentials on 27.1.77 are not shown, but were close to zero (field capacity).

soil moisture was not limiting (Fig. 5), also for 2 February, 1977, when the column was nearly depleted and the plants were severely wilted during the day (Fig. 6). Partitioning of the resistances involved is shown in both figures. Pathways with very high resistances which contribute very little to the transpiration stream and those that are not yet in operation are shown as dotted lines. The solid lines denote pathways that contribute to transpiration and the total resistances of the respective sections are given. Resistances in brackets are estimates based on a potential difference of 100 mb across the section since actual gradients are too small to be measured.

As the regions of greatest water uptake shift downwards with time, so do the contributing pathways. One can immediately see from the two figures that a simple partitioning between plant and rhizosphere resistance is misleading. In the regions of highest water uptake, soil rhizosphere resistances are very small (given as zero resistance since the actual magnitude is much less than 1 mb h cm^{-3}) and root radial (R_r) and xylem resistances (R_x) control the water uptake rates in combination with soil pararhizal resistance (R_{sp}) which controls the vertical flux of water through the soil. R_{sp} increases very rapidly as one moves away from the region of highest uptake, whereas the soil rhizosphere resistance (R_s) increases at a much slower rate. Even then, away from regions of high water uptake, R_s becomes of similar magnitude as R_r at relatively high soil water potentials of -2 to -3 bars (Fig. 4). The data show that the xylem resistance of the root system is such that it certainly cannot be ignored. Calculations were carried out only for the top 60 cm on 29 January as below this depth, soil or root water potentials were close to zero and were not measurable. It was assumed that so flow occurred below this depth. The total water uptake calculated for that day (1292 cm^3) agrees well with the water loss measured by the load cell (1105 cm^3) which gives us confidence in assuming a no flow condition below 60 cm. Therefore, although roots exist below this depth, they do not contribute significantly to water uptake if the soil near the surface can satisfy the plant's demand. However, when the upper soil layer is depleted of available water, uptake occurs almost exclusively from the lower layer (Fig. 6). Calculations of resistances cannot be carried out as water uptake occurs against the potential gradient.

One kind of resistance that could not be partitioned is the root-soil interface resistance (Tinker, 1976) which in this paper is included in the radial root resistance R_r and could possibly contribute significantly to the vari-

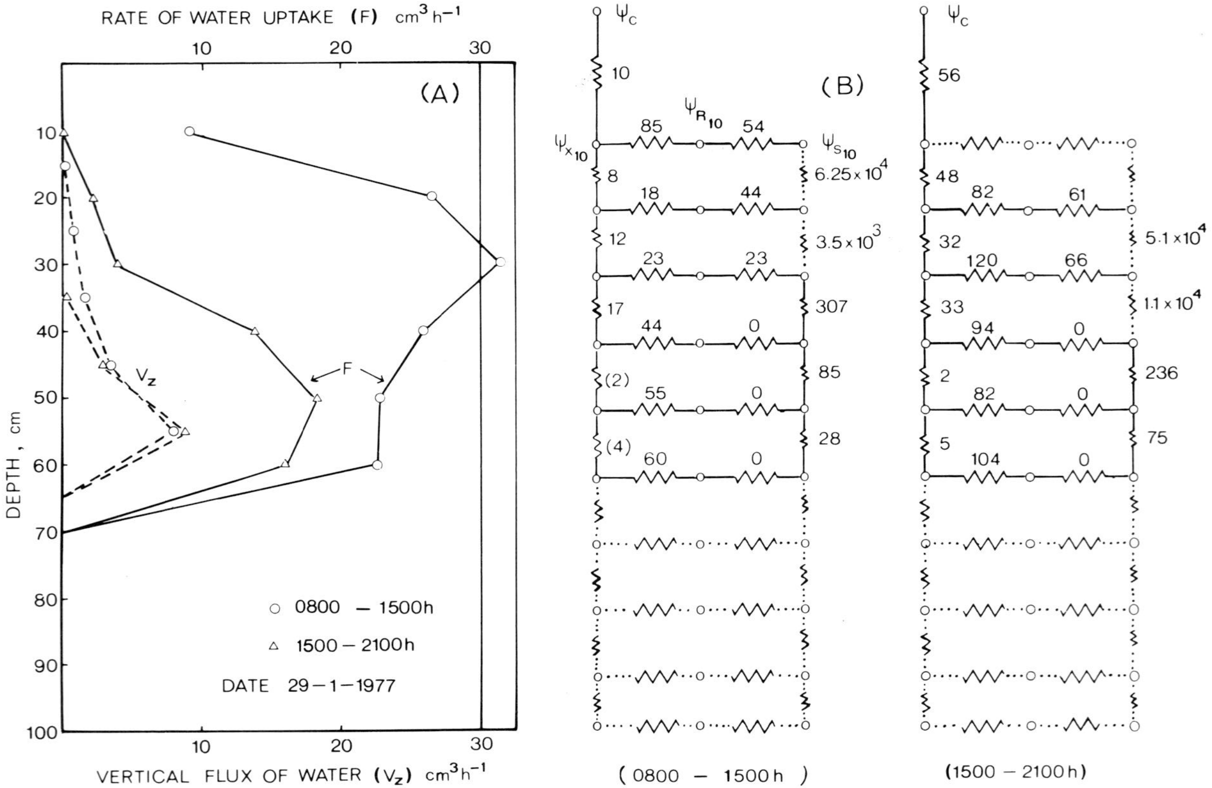

RATE OF WATER UPTAKE (F) cm³h⁻¹
(A)
DEPTH, cm
VERTICAL FLUX OF WATER (Vz) cm³h⁻¹
F
Vz
○ 0800 — 1500h
△ 1500 — 2100h
DATE 29—1—1977
(B)
ψc
ψR10
ψx10
ψs10
6.25 x 10⁴
3.5 x 10³
5.1 x 10⁴
1.1 x 10⁴
(0800 — 1500h)
(1500 — 2100h)

ability of R_r values. Such a resistance is probably the reason why R_r at 10 cm depth was very high in a dry soil such as occurred on 1 February, 1977.

No error analysis has been carried out on the values of the resistances and at this stage they should be viewed as approximate values. However, this should not change the conclusions materially.

Acknowledgements

The author wishes to thank Dr. Howard M. Taylor of A.R.S., USDA, Ames, Iowa for his help with equipment and valuable advice and Iowa State University for the opportunity to carry out this work while on sabbatical leave. The Iowa Soybean Promotion Board has provided financial support for this work.

References

Chow, T.L. and de Vries, J. (1973). Dynamic measurements of soil and leaf water potential with a double loop Peltier type thermocouple psychrometer. *Proc. Soil Sci. Soc. Amer.*, 37, 181.

Cowan, I.R. (1965). Transport of water in the soil-plant-atmosphere system. *J. Appl. Ecol.*, 2, 221.

Fawcett, R.G. and Collis-George, N. (1967). A filter paper method for determining the moisture characteristics of a soil. *Aust. J. Exp. Agric. & An. Husb.*, 7, 162.

Fiscus, E.L. (1972). *In situ* measurement of root water potentials. *Plant Physiol.*, 50, 191.

Gardner, W.R. (1960). Dynamic aspects of water availability to plants. *Soil Sci.*, 89, 63.

Newman, E.I. (1969a). Resistance to water flow in soil and plant. I. Soil resistance in relation to amounts of root: Theoretical estimates. *J. Appl. Ecol.*, 6, 1.

Newman, E.I. (1969b). Resistance to water flow in soil and plant. II. A review of experimental evidence on the rhizosphere resistance. *J. Appl. Ecol.*, 6, 261.

Ogata, G., Richards, L.A. and Gardner, W.R. (1960). Transpiration of alfalfa determined from soil water content changes. *Soil Sci.*, 89, 179.

Fig. 5. (opposite)
(A) Rate of water uptake (F) and vertical fluxes (V_z) from each 10 cm section of the column plotted against depth on 29 Jan. 1977 when soil water was non-limiting. (B) The resistances of the soil-plant system early (0800 - 1500 h) and late (1500 - 2100 h) in the day.

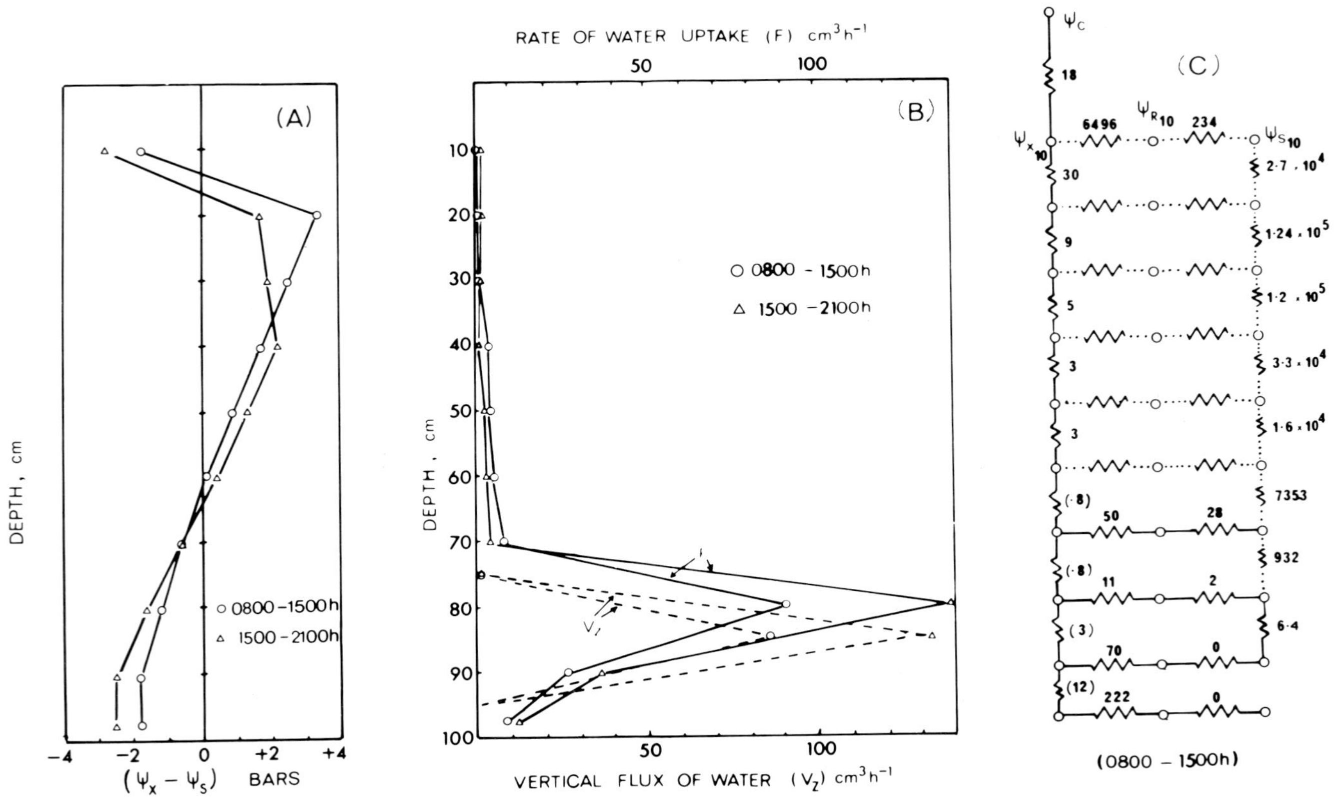
(A)
DEPTH, cm
0800–1500h
1500–2100h
$(\Psi_x - \Psi_s)$ BARS
(B)
RATE OF WATER UPTAKE (F) cm^3h^{-1}
DEPTH, cm
0800 - 1500h
1500 -2100h
F
V_z
VERTICAL FLUX OF WATER (V_z) cm^3h^{-1}
(C)
Ψ_C
18
$\Psi_{x_{10}}$
6496
$\Psi_{R_{10}}$
234
$\Psi_{S_{10}}$
30
$2{\cdot}7 \times 10^4$
9
$1{\cdot}24 \times 10^5$
5
$1{\cdot}2 \times 10^5$
3
$3{\cdot}3 \times 10^4$
3
$1{\cdot}6 \times 10^4$
(·8)
7353
50
28
(·8)
932
11
2
(3)
6·4
70
0
(12)
222
0
(0800 – 1500h)

Passioura, J.B. (1977). Determining soil water difffusivities from one-step outflow experiments. *Aust. J. Soil Res.*, **15**, 1.

Reicosky, D.C. and Ritchie, J.T. (1976). Relative importance of soil resistance and plant resistance in root water absorption. *Proc. Soil Sci. Soc. Amer.*, **40**, 293.

So, H.B. (1977). *In situ* measurement of root water potentials. *Aust. Soil Sci. Soc. Inc.* ACT Branch Conference, Canberra. May 19-21.

So, H.B. (1978). An analysis of the relationship between stem diameter and leaf water potentials. *Agron J.* (in press).

So, H.B., Aylmore, L.A.G. and Quirk, J.P. (1976). Measurement of water fluxes and potentials in a single root-soil system. I. The Tensiometer-potometer system. *Plant and Soil*, **45**, 577.

So, H.B., Aylmore, L.A.G. and Quirk, J.P. (1978). Measurement of water fluxes and potentials in a single root-soil system. II. Applications. *Plant and Soil* (in press).

Taerum, R. (1973). Occurrence of inverted water potential gradients between soil and bean roots. *Physiol. Plant.*, **28**, 471.

Tinker, P.B. (1976). Roots and water - Transport of water to plant roots in soil. *Phil. Trans. R. Soc. Lond. B.*, **273**, 445.

Van den Honert, T.H. (1948). Water transport as a catenary process. *Disc. Faraday Soc.*, **3**, 146.

Fig. 6. (opposite)
(A) Potential difference (ψ_r - ψ_s) versus depth on 1 Feb. 1977. Positive values indicate that soil water potential is lower than root xylem water potential. (B) Rate of water uptake (F) from each 10 cm section of the column and the associated vertical fluxes (V_z) versus depth. Water uptake is still occurring although root potentials are higher than the soil water potentials at depth of 20-60 cm.
(C) The resistances of the soil-plant system in the early part of the day (0800 - 1500 h).

THE EFFECT OF ORGANIC ROOT PRODUCTS ON THE AVAILABILITY OF PHOSPHORUS TO PLANTS

D.G. KEPERT, A.D. ROBSON and A.M. POSNER

Department of Soil Science and Plant Nutrition, University of Western Australia, Australia

Summary

The evidence for the involvement of organic materials derived from plant roots growing in the soil in phosphate nutrition from insoluble phosphate sources is briefly reviewed and found to be inconclusive. An experiment is described in which wheat plants are grown in media of increasing abrasiveness and hydroxyapatite is used as an insoluble source of phosphate. Growth appears to be greatest in the most abrasive medium, carborundum.

Introduction

The role of plant roots in modifying phosphorus availability in the rhizosphere is not completely understood. Indirect evidence that plant roots interact with soil components to influence phosphorus availability comes from several studies. Roots of a number of plants (squash, soybeans, barley and wheat) absorbed many times more phosphate when in close contact with Virginia fluor-apatite than when they were in the supernatant (Johnston and Olsen, 1972). Swaby and Sperber (1959) also related the growth responses of lupins and serradella to apatites in acid soils to the ability of some of the rhizosphere microflora to solubilize hydroxy-apatite.

Further indirect evidence comes from a study of desorption isotherms from some of the Hoosfield soils on the long term rotations at Rothamsted (Posner, unpublished data). Desorption was measured by determining the phosphate concentration in 0.01M $CaCl_2$ as a function of the soil/solution ratio. The desorption isotherms were compared with "plant desorption isotherms" obtained from results of Webber (1966) and Webber and Mattingly (1970) who measured phosphate uptake by rye grass from the corresponding soils. Once the plant was removed from the soil the null point equilibrium phosphate concentration in 0.01M $CaCl_2$ was determined after a two week equilibration. The amount of phosphate that was desorbed in the presence of the plant was considerably greater than that removed by the simple desorption process in 0.01M $CaCl_2$ (Fig. 1). This result

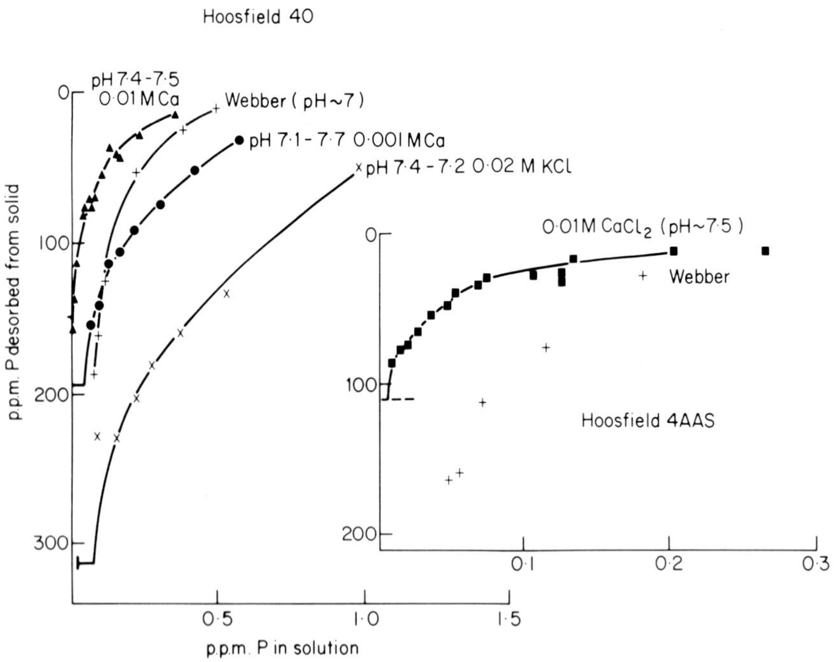

Fig. 1. The relationships between desorption and solution concentration for two Rothamsted soils. These are compared with rye-grass uptake curves of Webber (1966) and Webber and Mattingly (1970). (Posner, unpublished.)

may have been due to microbial activity during the equilibration period after plant growth. The desorption of phosphate was also strongly affected by cation, calcium ion concentration and ionic strength (Fig. 1). Thus changes in cation concentration and ionic strength around a growing root may be of importance in determining uptake from a soil.

Bhat, Nye and Baldwin (1976) have shown that it was not possible to account for phosphorus uptake by rape by considering diffusion alone. They suggested that there was some induced desorption from the soil which was not due to pH changes. By contrast the work of Sanders and Tinker (1971) and Bhat and Nye (1974) suggests that in the case of phosphate uptake by onion, it is not necessary to ascribe any role to root exudates to account for uptake.

Recently it has been demonstrated that rhizosphere products from wheat seedling roots can increase the dissolution of calcium phosphates but not iron and aluminium phosphates (Moghimi, Lewis and Oades, 1978, in press). In subsequent studies 2-keto-gluconic acid was demonstrated to be particularly important in dissolution of calcium phosphates (Moghimi, Tate and Oades, 1978, in press).

There are several mechanisms by which the plant root may modify the availability of phosphorus. Firstly, the release of phosphate from insoluble phosphates (Ca, Al, Fe) may be increased by the formation of soluble complexes between metal ions and metabolites either of microbial or plant origin (Barber, 1968). Secondly, organic acids found in root exudates that are capable of forming adsorption complexes and more particularly chelates are very effective competitors for phosphate adsorbed on kaolinite and aluminium oxides but less so on goethite (Nagarajah, Posner and Quirk, 1970). Chelating agents produced by plant roots may also be involved in the utilization of calcined rock phosphates. Harrison and Posner (unpublished) demonstrated that in 1N neutral ammonium citrate the solubilities of A grade Christmas Island rock phosphate and C grade rock phosphate either untreated or calcined at 400 and 900°C were in no way related to their solubilities in 0.01M $CaCl_2$ at pH 4.0 (Table 1). Plant yields appear to be related to solubilities in the citrate reagent (Doak *et al.*, 1965).

TABLE 1

Solubility of Christmas Island Rock Phosphates
Harrison and Posner (unpublished)

	% Released at pH 4 after 17 days (0.01M $CaCl_2$) (1g/1)	% P soluble in neutral ammonium citrate (1g/1)	N_2 surface area m^2/g
A grade rock P	16	3.9	23
C grade rock P	4	2.6	41
C grade calcined 400°C	0.8	82	31
C grade calcined 900°C	1.3	12.5	4

A third mechanism by which plant roots may modify phosphate availability is by affecting the rhizosphere pH. When cation absorption exceeds anion absorption the rhizosphere pH will fall. Such changes in acidity could markedly increase or decrease phosphate availability. McLachlan (1976) ascribed differences among species in ability to obtain phosphorus to different abilities to acidify the rooting media. However, buck wheat which made the rhizosphere most acid and which obtained the most phosphate had also the largest rooting system. Differences among species in ability to acidify the rooting media may result from differences in plant growth.

Factors affecting Rhizosphere Secretions

Martin (1977a) has proposed that all materials released from actively growing roots should not be termed root exudates but should be subdivided into (i) root exudates (compounds released into nutrient solution from sterile seedlings), (ii) mucigel (an actively secreted substance) and (iii) root lysates (compounds released by autolysis through degeneration of root hairs, epidermal cells and cortical cells). In recent studies using labelled carbon it has been demonstrated that substantial amounts of carbon (as much as 40% of root material) can be deposited in soil during growth (Shamoot, McDonald and Bartholomew, 1968; Martin, 1977a). The amount of material deposited is increased by moisture stress, micro-organisms, low temperature and abrasiveness (Barber and Martin, 1976; Martin, 1977a). Most of the organic loss from wheat roots appears to be due to autolysis of cortical tissue (Martin, 1977a, b) which is the first stage of root decomposition (Holden, 1975).

Most early studies of root exudates measured water soluble material released by seedlings grown in aseptic solution culture thereby greatly underestimating the quantity of material released from the soil under natural conditions (Barber and Gunn, 1974). Recently Nye and Tinker (1977) calculated that if 20% of the plant weight was lost as lysates then these compounds could be important in modifying nutrient availability. Hence we designed an experiment to examine the effect of organic products lost from the root on the dissolution of hydroxy-apatite as measured by phosphorus uptake by wheat plants. Rooting media of increasing abrasiveness were used to vary the amount of organic material lost from the growing root. A microbial culture was added so as to include secondary compounds produced by microbial metabolism.

The design of the experiment was a factorial combination of (i) three phosphate treatments (two rates of synthetic hydroxy-apatite (HA) and a control with adequate soluble phosphate applied daily) and (ii) three rooting media (carborundum (SiC), sand and solution culture) (Table 2). The growth media were selected for their chemical inert-

TABLE 2

Levels of applied phosphate and calcium, and expected phosphate and calcium concentrations in the media solutions

Treatment Designation			SiC			Sand			Solution Culture	
			$(Ca+P)_{sol}$	$Ca+HA_1$	$Ca+HA_{10}$	$(Ca+P)_{sol}$	$Ca+HA_1$	$Ca+HA_{10}$	$(Ca+P)_{sol}$	$Ca+HA_1$
Total Nutrient per pot (μ moles)	P	HA	-	480	4,800	-	480	4,800	-	480
		Soluble	500	-	-	500	-	-	500	-
	Ca	HA	-	800	8,000	-	800	8,000	-	800
		Soluble	835	1,600*	1,600*	835	1,140*	1,140*	835	5,000*
Expected Nutrient Concentrations (μM)	P		0-30†	2	2	0-45†	2	2	0-10†	2
	Ca		0-50#	5,000	5,000	0-75#	5,000	5,000	0-17#	5,000

(* † #) Variations are due to the different field capacities of the rooting media

ness and differing degrees of abrasiveness.

For the HA treatments calcium in the solution was maintained at 5,000μM using $CaSO_4$ to suppress the phosphate concentration to a level limiting for plant growth. For the soluble phosphate treatment calcium was maintained at a level adequate for plant requirements, but low enough to prevent precipitation of the phosphate. All other essential nutrients were provided. Initially nitrate was used as a source of nitrogen to prevent dissolution of HA by H^+ released by the root as a result of cation uptake exceeding anion uptake. At 3-4 weeks the pH values of the solution cultures rose from their previously steady values close to the pK_2 of phosphate, namely pH = 6.8-7.2, to pH = 8. Thereafter pH values were maintained close to pH = 7 by using NH_4NO_3 or potassium nitrate.

At harvest, 56 days, most growth had been made in the three rooting media by the soluble phosphate control (Fig. 2; Table 3). For the carborundum but not for the sand,

Fig. 2. The effect of rooting medium on the growth of wheat at day 44.
1 SiC:$(Ca+P)_{sol}$, 2 SiC:Ca+HA_{10}, 3 SiC:Ca+HA_1
4 Sand:$(Ca+P)_{sol}$, 5 Sand:Ca+HA_{10}, 6 Sand:Ca+HA_1
7 Solution Culture:$(Ca+P)_{sol}$, 8 Solution Culture:Ca+HA_1

growth at the high rate of HA approached that for the soluble phosphate treatment. At the lower rate of HA, growth was markedly depressed in all three media, with the depression for solution culture > sand > SiC. These growth differences between the three media were not due to decreased rooting power as the ratio of roots/tops was in the same order, namely solution culture > sand > SiC (Table 3).

TABLE 3

The effect of rooting medium on the growth of wheat at two levels of HA compared with the soluble phosphate control

	Treatment	SiC	Sand	Solution Culture
Fresh Weight Tops (g/5 plants)*	$(Ca+P)_{sol}$	19.94 ± 0.04	19.14 ± 0.03	16.70 ± 0.38
	$Ca+HA_{10}$	18.60 ± 0.06	14.92 ± 0.26	-
	$Ca+HA_{1}$	11.95 ± 0.15	7.94 ± 0.10	5.37 ± 0.79
Fresh Weight Roots (g/5 plants)*	$(Ca+P)_{sol}$	24.35 ± 0.09	26.10 ± 0.03	25.05 ± 0.33
	$Ca+HA_{10}$	23.48 ± 1.66	21.98 ± 0.46	-
	$Ca+HA_{1}$	11.71 ± 0.36	10.98 ± 1.56	11.77 ± 0.59
Ratio Fresh Weight Roots/Tops*	$(Ca+P)_{sol}$	1.22 ± 0.01	1.36 ± 0.00	1.50 ± 0.06
	$Ca+HA_{10}$	1.26 ± 0.09	1.47 ± 0.04	-
	$Ca+HA_{1}$	0.98 ± 0.04	1.38 ± 0.20	2.28 ± 0.34

* Data are expressed as means with standard errors.

Plants from all HA treatments had dark green foliage compared with the normal green colour of the controls, suggesting that the growth depression was due to phosphate deficiency.

Roots grown in SiC, in addition to appearing thinner and less healthy than those from the corresponding sand treatment, retained more rooting media on their surface after washing.

These preliminary results are consistent with the idea that abrasive media that increase the amount of organic material lost from the root will increase the dissolution of HA.

Residual HA, calcium and phosphate concentrations of the soil solutions, calcium and phosphate concentrations of the plant tops and roots, and amounts and phosphate concentrations of residual organic material are being determined to give a balance sheet of HA utilization by the plants. The roots are to be examined by the scanning electron microscope.

References

Barber, D.A. (1968). Microorganisms and the inorganic nutrition of higher plants. *Ann. Rev. Plant Physiol.*, 19, 71.

Barber, D.A. and Gunn, K.B. (1974). The effect of mechanical forces on the exudation of organic substances by the roots of cereal plants grown under sterile conditions. *New Phytol.*, 73, 39.

Barber, D.A. and Martin, J.K. (1976). The release of organic substances by cereal roots into the soil. *New Phytol.*, 76, 69.

Bhat, K.K.S. and Nye, P.H. (1974). Diffusion of phosphate to plant roots. III. Depletion around onion roots without root hairs. *Plant and Soil*, 41, 383.

Bhat, K.K.S., Nye, P.H. and Baldwin, J.P. (1976). Diffusion of phosphate to plant roots in soil. IV. The concentration distance profile in the rhizosphere of roots with root hairs in a low-P soil. *Plant and Soil*, 44, 63.

Doak, B.W., Gallaher, P.J., Evans, L. and Muller, F.B. (1965). Low-temperature calcination of "C" grade phosphate from Christmas Island. *N.Z. J. Ag. Res.*, 8, 15.

Holden, J. (1975). Use of nuclear staining to assess rates of cell death in cortices of cereal roots. *Soil Biol. Biochem.*, 7, 333.

Johnston, W.B. and Olsen, R.A. (1972). Dissolution of fluorapatite by plant roots. *Soil Sci.*, 114, 29.

McLachlan, K.D. (1976). Comparative phosphorus responses in plants to a range of available phosphorus situations. *Aust. J. Agric. Res.*, 27, 323.

Martin, J.K. (1977a). Factors influencing the loss of organic carbon from wheat roots. *Soil Biol. Biochem.*, 9, 1.

Martin, J.K. (1977b). Effect of soil moisture on the release of organic carbon from wheat roots. *Soil Biol. Biochem.*, 9, 303.

Nagarajah, S., Posner, A.M. and Quirk, J.P. (1970). Competitive adsorption of phosphate with polygalacturonate and organic anions on kaolinite and oxide surfaces. *Nature*, London, 228, 83.

Nye, P.H. and Tinker, P.B. (1977). "Solute Movement in the Soil Root System". Blackwell, Oxford.

Sanders, F.E. and Tinker, P.B. (1971). Mechanism of absorption of phosphate from soil by endogone mycorrhizas. *Nature*, London, 232, 278.

Shamoot, S., McDonald, L. and Bartholomew, W.V. (1968). Rhizo-deposition of organic debris in soil. *Soil Sci. Soc. Amer. Proc.*, 32, 817.

Swaby, R.J. and Sperber, J. (1959). Phosphate dissolving microorganisms in the rhizosphere of legumes. *In* "Nutrition of Legumes" (Ed. E.G. Hallsworth), p.289. Butterworths Scientific Publications, London.

Webber, M.D. (1966). "Studies on Inorganic Phosphate". Ph.D. Thesis. University of London, London.

Webber, M.D. and Mattingly, G.E.G. (1970). Inorganic soil phosphorus. I. Changes in monocalcium phosphate potentials on cropping. *J. Soil Sci.*, 21, 111.

MODIFICATIONS TO MECHANICAL AND MINERALOGICAL COMPOSITION OF SOIL WITHIN THE RHIZOSPHERE

A.N. SARKAR, D.A. JENKINS and R.G. WYN JONES

Department of Biochemistry and Soil Science,
University College of North Wales, Bangor, UK

Summary

During an experimental study of rhizosphere development around French Bean roots, mechanical and mineralogical modifications were recorded. The former entailed a relative decrease in material 63 - 6.3μm e.s.d. in size, probably due to disaggregation of polymineralic shale particles. The latter was most obvious in the finest (< 0.063 μm) fraction and involved a decrease in both regularly and irregularly interstratified 10/14Å material: "amorphous" (extractable) Al, Fe and C accumulated in the rhizosphere, whilst Si was depleted. These effects are interpreted as being due to intensified "weathering" of soil materials in the vicinity of the plant root.

Introduction

The "rhizosphere" may be defined as the volume of interaction between a plant's rooting system and its immediate soil environment. Its validity in terms of microbiological and chemical gradients has been demonstrated convincingly (e.g. Rovira and Davey, 1974; Tinker, 1969); so far, however, little attention has been paid to the mechanical or mineralogical differentiation of the rhizosphere from the bulk soil. As an incidental part of experimental study of the influence of various treatments on rhizosphere development (Sarkar, 1977), modifications to both particle size distribution and composition of the clay fraction became apparent. Modification of solid phases by such relatively short-term plant growth was considered to be of sufficient interest to merit comment.

Materials and Methods

French Bean (*Phaseolus vulgare* L) was grown on a homogenized < 6 mm fraction of the Lowland Phase of the Denbigh Series from the U.C.N.W. Farm at Aber, Gwynedd. This is a silt loam whose clay mineralogy is dominated by hydrous mica and chlorite, derived from the parental Ordovician shales, together with minor amounts of vermiculite and interstratified material. The growth experiments were run in 150 mm diameter pots containing approximately 2 kg of soil. One germinated seed was planted centrally per pot. Three different treatments, replicated six times, were applied: (a) no additions, (b) choline chloride, or (c) calcium nitrate at levels calculated to provide 500 ppm nitrogen. Phosphorus was added to all pots at 1000 ppm as potassium dihydrogen phosphate, and the pH of the soil was adjusted to an initial value of 7.0 with $Ca(OH)_2$. After six weeks "rhizosphere" samples were taken by the excision of a soil core, circa 15 mm in diameter, centred on the main shoot/root axis. This was carefully separated from the rooting system to yield some 10g soil material per plant, the product of six replicates being bulked. "Non-rhizosphere" samples were taken at approximately 30 mm from the main root axis.

The samples were dispersed ultrasonically in the presence of "Calgon" and fractionated into seven particle size classes by the normal processes of wet-sieving (>63μm), sedimentation (63-20μm, 20-6.3μm and 6.3-2μm e.s.d.) and centrifugation (2-0.63μm, 0.63-0.2μm, 0.2-0.063μm and <0.063μm e.s.d.). In addition, a turbidometric method was developed for estimation of particle size distribution on 100 mg samples of pre-sieved (<63μm) soil.

For clay mineral analysis by X-ray diffraction, subsamples were treated to remove organic matter (with NaOCl; Anderson, 1963) and "free iron oxides and amorphous silica and alumina" (Mehra and Jackson, 1960; Hashimoto and Jackson, 1960), and were then fractionated by centrifugation to obtain the four subsamples 2-0.63μm, 0.63-0.2μm, 0.2-0.063 μm and <0.063μm. For X.R.D.A. samples were prepared by pipetting ~50 mg in suspension onto glass slides following a range of standard pretreatments (i.e. K^+ saturation; heating to 500°C; Mg^{+2} saturation: treatment with ethylene glycol, dimethyl sulphoxide etc.), although such treatments were precluded in some cases by the small amounts available of some of the finest fractions; slides of untreated material were also prepared. The samples were analysed on Philips equipment (PW.1011/1050) employing Ni-filtered CuK_α radiation generated at 40kV/20mA, scanning at 1/4° 2θ/min with a slit system of 1/4°/0.1mm/1/4° from 2-5°2θ, and at 1° 2θ/min, and with a slit system of 1°/ 0.1mm/1° over the range 5-15°2θ. The "amorphous" components of the clay fraction, extracted successively with pyrophosphate (Bascomb, 1968), dithionite (Mehra and Jackson, 1960) and NaOH (Hashimoto and Jackson, 1960) were estimated by atomic absorption spectroscopy (Fe and Al) and colorimetric methods (Si).

Results

The experiments were designed primarily to enable those soil properties which differed significantly (P = 0.05 - 0.01) between rhizosphere and non-rhizosphere soil to be recognized, namely extractable phosphate, calcium and potassium, organic carbon and total nitrogen. Of relevance to this discussion, however, are the changes in particle size distribution and clay fraction composition: the significance of the other properties will be reported sep-

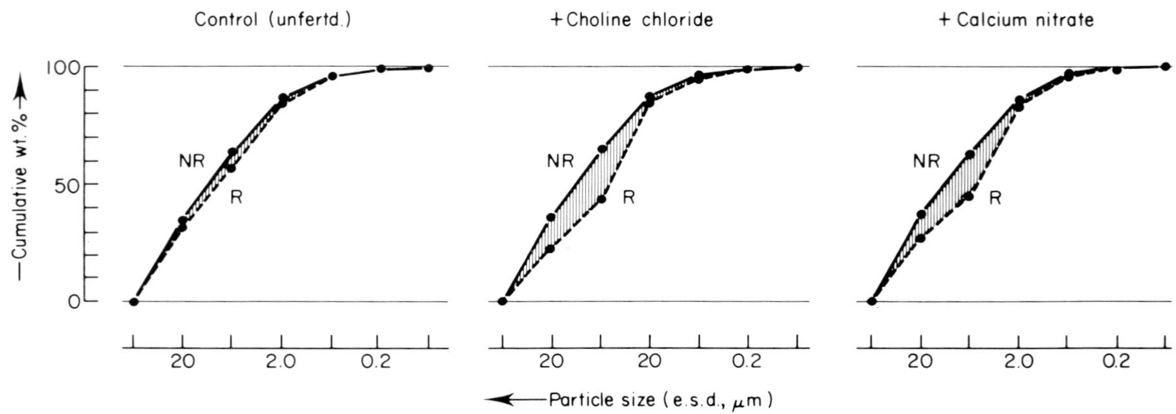

Fig. 1. Mechanical analysis of rhizosphere (R) and non-rhizosphere (NR) samples (<63μm)

arately, as will the micromorphological and chemical changes within the rhizosphere which were investigated by thin-section microscopy, micro-autoradiography and electron-probe microanalysis.

From Fig. 1 it is evident that, irrespective of the treatment involved, there is a consistent shift in the mechanical composition of the $<63\mu m$ fraction between the rhizosphere and non-rhizosphere soil samples. This involves a relative increase in the finer fractions ($<6.3\mu m$) of 25-35%, and of the fine silt (6.3 - 2.0μm) in particular, at the expense of the medium and coarse silt fractions. Explanations which could be offered for these observations include the concentration of organic debris within this "size" (settling velocity) range, and, in particular, the comminution of polymineralic material - i.e. the disaggregation of compound shale particles through intensified "weathering" in the root zone (c.f. Stewart, Adams and Abdullah, 1970). The accentuated effects in the two experiments receiving fertilizer treatment correlate with their consistently lower values for rhizosphere pH and higher levels of K^+, Na^+ and Mg^{2+}: they are also associated with a marked stimulation of both plant growth and microbial activity within those rhizospheres where various sources of nitrogen had been added (Sarkar, 1977).

The clay mineral composition of the bulk soil material - the starting point for the clay mineral analyses - is presented in Figure 2. From this it can be seen that the coarser fractions are dominated by a hydrous mica and an Fe-rich chlorite (002:001 ~ 5) with minor vermiculite and traces of regularly, and possibly irregularly, interstratified vermiculite/chlorite (heat unstable) and vermiculite/hydrous mica. Traces of kaolinite are detectable in the 0.6 - 0.2μm fraction though not, surprisingly, in the coarser fraction: this suggests some contamination of the local shale-derived drift with material of northern origin. With decreasing particle size, the proportions of hydrous

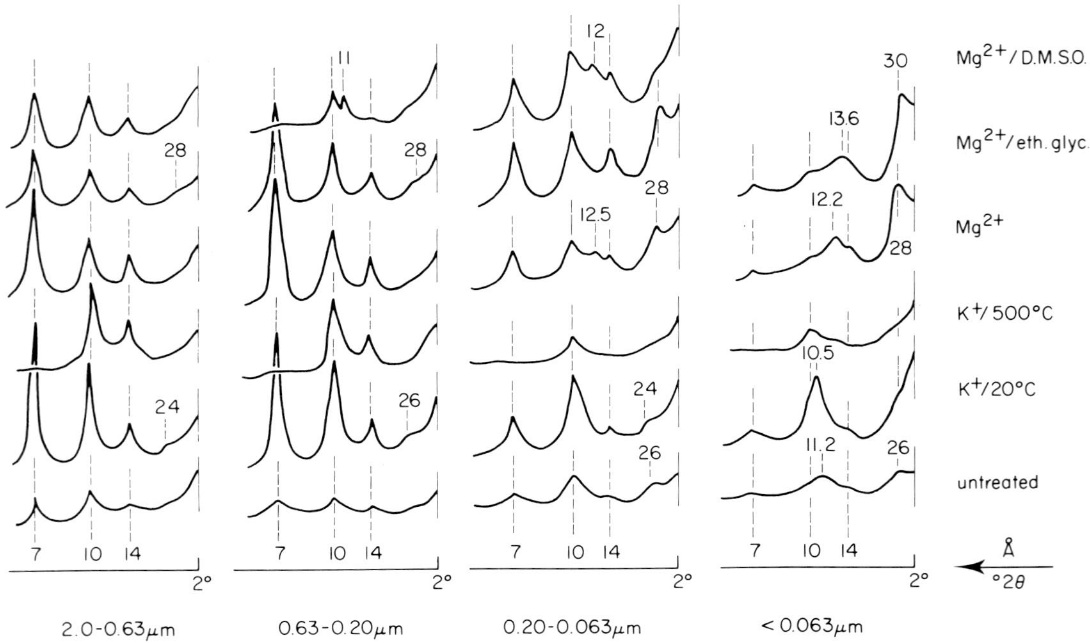

Fig. 2. X.R.D.A. traces for different size fractions of the initial soil

mica and chlorite decrease, the latter showing signs of thermal instability, whilst those of vermiculite and interstratified material increase progressively to become dominant in the fine fractions (<0.2μm): traces of smectite also appear in the 0.2 - 0.06μm fraction. In the finest fraction (<0.063μm) the interstratified material, both regular and irregular, now includes a swelling component (i.e. smectite/hydrous mica?). From this it may be inferred that the hydrous mica and/or chlorite are undergoing progressive alteration within the finer fractions to vermiculite and interstratified material with the ultimate development of swelling layers. Such trends have been suggested in comparable soil systems (e.g. Droste, Bhattacharya and Sandeman, 1962; Evans and Adams, 1975; Bain, 1977) although it has been shown in Silurian shale-derived soils from North Wales that the chlorite undergoes direct dissolution without intermediary products (Rezk, 1976).

Following the growth experiments, analyses of non-rhizosphere (NR) and rhizosphere (R) material were again carried out. The former differed little from the initial material, but the rhizosphere samples showed a consistent divergence in mineralogical composition (Fig. 3). There is a distinct reduction in height of those peaks ascribable to irregularly interstratified material (smectite/hydrous mica?) with a concomitant relative increase in their swelling component: at the same time the peaks due to vermiculite and hydrous mica are unmasked and possibly also reduced. Reduction in peak height of the regularly interstratified minerals containing swelling components is even more pronounced. These modifications tend to be more marked in the control soil rhizosphere samples than in those from fertilized soils. Such low-angle X.R.D.A. results are always vulnerable to variations in sample preparation and unfortunately the small amounts of <0.063μm material obtained from rhizosphere samples precluded replication. However, this pattern of divergence of rhizosphere mineral-

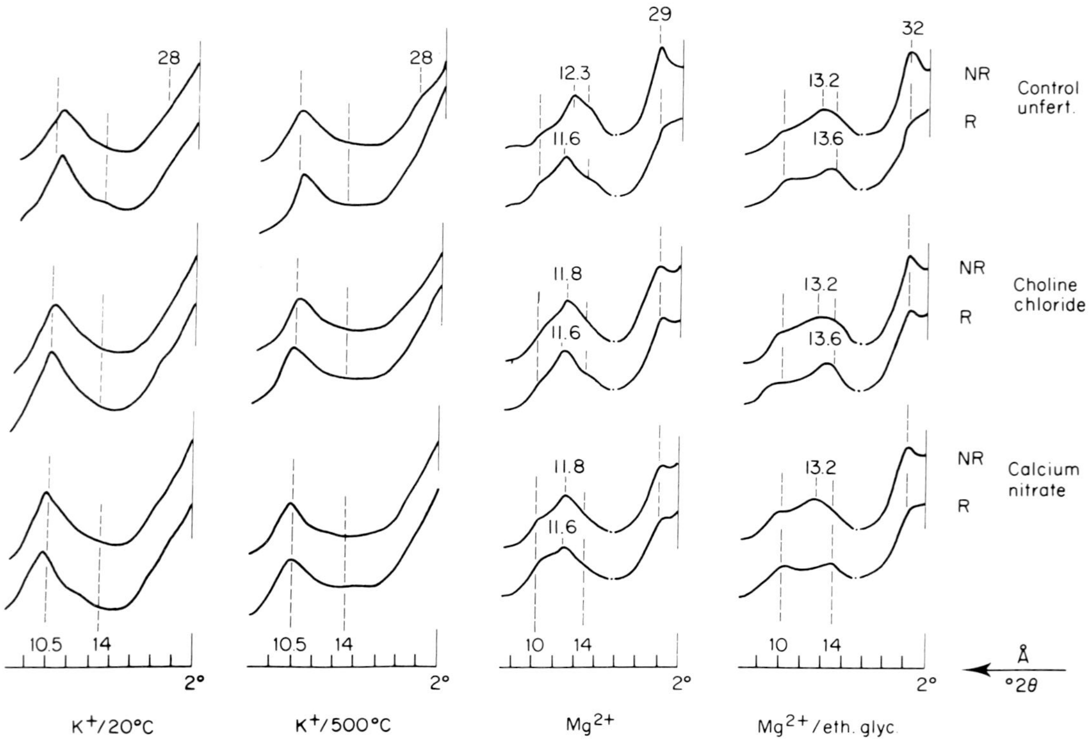

Fig. 3. X.R.D.A. traces of rhizosphere and non-rhizosphere samples (<0.063μm)

ogy was qualitatively similar for all the different treatments and also for the other plant species studied.

Representative results for the "amorphous" (i.e. extractable) constituents in the <0.063μm fraction are presented in Table 1. Sodium pyrophosphate and dithionite/citrate proved to be efficient extractants for Fe and Al and sodium hydroxide for Si and Al. Overall, larger amounts of Al and Fe were extracted from rhizosphere as compared to non-rhizosphere clay samples, whilst the reverse was true for Si. From the figures it may be shown that the $SiO_2:Al_2O_3$ for "amorphous" material in the rhizosphere (0.29-0.30) is lower than that in the bulk soil (0.36-0.37). Values for extractable Al and, in particular, Fe showed significant positive correlations with extractable carbon indicating a chemical association (i.e. complexing?) between these components as suggested by Greenland (1971).

The cation exchange capacities for the treated ("amorphous-free") clay fractions proved to be systematically higher (+ 6-8 me/100g) than those from untreated samples. This could be explicable in terms of the coatings of amorphous materials being above their isoelectric points and thus masking with their positive charges negative exchange sites on clay mineral particles. Cation exchange capacities were also systematically higher (+ 5-8 me/100g) for clay fractions from the rhizosphere as compared to those from the bulk soil: this would be consistent with the observed higher proportion of swelling components in the former.

Conclusions

These experiments have provided clear evidence of the mechanical and mineralogical differentiation by a growing plant of its rhizosphere. The change in particle size distribution is one of comminution and, in the particular soils studied, probably reflects an accelerated disaggregation of compound shale fragments into their component silt/

TABLE 1

Extractable ("amorphous") components in the clay fractions (wt.%) and C.E.C. (Na^+)

		Organic carbon			Fe_2O_3			Al_2O_3				SiO_2			(me/100g)	
		Pyr.	Dic.	Total	Pyr.	Dith.	Total	Pyr.	Dith.	NaOH	Total	Dith.	NaOH	Total	Tr.	Untr.
Control (unfert.)	R	1.20	1.80	3.00	1.32	8.89	11.2	0.40	0.74	1.83	2.97	0.21	0.62	0.83	47	52
	NR	0.90	1.80	2.70	1.23	9.42	10.7	0.37	0.74	1.59	2.70	0.18	0.81	0.99	41	47
Choline chloride	R	1.80	2.40	4.20	1.89	10.7	12.5	0.46	0.54	1.98	2.98	0.32	0.58	0.90	48	55
	NR	0.90	1.95	2.85	1.03	8.97	10.0	0.37	0.67	1.59	2.63	0.22	0.77	0.99	41	47
Calcium nitrate	R	1.50	2.25	3.75	1.13	10.8	11.9	0.53	0.57	2.40	3.50	0.21	0.93	1.14	46	54
	NR	0.80	1.80	2.60	0.94	8.49	9.4	0.32	0.73	1.68	2.73	0.17	0.81	0.98	40	46

N.B. negligible amounts of Fe were extracted by NaOH, and of Si by pyrophosphate.

Key: R - rhizosphere soil; NR - non-rhizosphere soil; Pyr. - pyrophosphate; Dith. - dithionite/citrate; Dic. - dichromate.

clay particles. The mineralogical changes are manifested most clearly in the finest clay fractions: this is reasonable considering that the modifications probably progress in from the edges of the phyllosilicate lattices involved and therefore become a more significant proportion in the finest particles. The mineralogy of this fraction is relatively obscure and complex to begin with and it is planned to carry out the more detailed analysis required to establish the reproducibility and full significance of the changes detected in the X-ray diffraction patterns. Destruction of interstratified material together with a relative increase in swelling components is paralleled by an increase in "amorphous" material (of higher $Al_2O_3:SiO_2$), of C.E.C., and probably of organically complexed Fe. All these effects illustrate the pedogenic significance of the rhizosphere as a zone of intensified alteration of the inorganic constituents of the soil and it is interesting to note that such effects were generally more marked in the absence of fertilizer treatments when the plant presumably made greater demands on its soil environment.

That such effects should be detectable within the mineral fraction of the soil is surprising in view of the relatively short-term nature of the experiments. Also, the sampling of the "rhizosphere" was of necessity crude in that it must have involved considerable dilution of any "rhizosphere effect" by initial unmodified soil: by extrapolation, the effects within close proximity of the roots must be of proportionately greater magnitude. However, this experimental approach is apparently sufficiently sensitive to monitor changes induced within the rhizosphere of a growing plant and thus offers a means of obtaining a clearer picture of the actual nature of the root environment.

References

Anderson, J.V. (1968). An improved pretreatment for mineralogical analysis of materials containing organic matter. *Clays Clay Miner.*, 10, 380.

Bain, D.C. (1977). The weathering of chloritic minerals in some Scottish soils. *J. Soil Sci.*, 28, 144.

Bascombe, C.L. (1968). Distribution of pyrophosphate-extractable iron and organic carbon in soils of various groups. *J. Soil Sci.*, 19, 251.

Droste, J.B., Bhattacharya, N. and Sanderman, J.A. (1962). Clay mineral alteration in some Indiana soils. *Clays Clay Miner.*, 9, 329.

Evans, L.J. and Adams, W.A. (1975). Chlorite and illite in some lower Palaeozoic mudstones of mid-Wales. *Clay Min.*, 10, 387.

Greenland, D. (1971). Interaction between humic and fulvic acids and clay. *Soil Sci.*, 111, 34.

Hashimoto, I. and Jackson, M.L. (1960). Rapid dissolution of allophane and kaolinite-halloysite after dehydration. *Clays Clay Miner.*, 7, 102.

Mehra, O.P. and Jackson, M.L. (1960). Iron oxide removal from soils and clays by a dithionite-citrate system buffered with sodium bicarbonate. *Clays Clay Miner.*, 7, 317.

Rezk, M.M.A.A. (1975). "Studies on chlorite in soils derived from Silurian Rocks in Denbighshire". Ph.D. Thesis, Univ. Wales.

Rovira, A.D. and Davey, C.B. (1974). Biology of the Rhizosphere. *In* "The Plant Root and its Environment" (Ed. E.W. Carson), pp.153-204. University Press of Virginia, Charlotteville, 1974.

Sarkar, A.N. (1977). "Biochemistry and Micromorphology of the Rhizosphere". Ph.D. Thesis, University of Wales.

Stewart, V.F., Adams, W.A. and Abdulla, J.H. (1970). Quantitative Pedological studies of soils derived from Silurian Mudstones. I. *J. Soil Sci.*, 21(2), 242.

Tinker, P.B. (1969). The transport of ions in the soil around plant roots. *In* "Ecological Aspects of the Mineral Nutrition of Plants" (Ed. I.H. Rorison), pp.135-147. Blackwell Scientific Publications, Oxford.

OVERESTIMATE OF N_2-FIXATION IN THE RHIZOSPHERE BY THE ACETYLENE REDUCTION METHOD

J.F. WITTY

Department of Soil Microbiology, Rothamsted Experimental Station, Harpenden, England

Summary

Much of the data on N_2-fixation in the rhizosphere is based on the reduction of C_2H_2 *in situ* or in soil cores. Experiments with $^{14}C_2H_2$ showed that C_2H_4 produced during the assay by *Phaseolus vulgaris*, *Nostoc ellipsosporum*, *Clostridium pasteurianum*, *Bacillus polymyxa* and active pasture grass cores came entirely from $^{14}C_2H_2$. This was not, however, the case with some less active cores. Although these cores produced only negligible quantities of C_2H_4 when incubated without C_2H_2, as much as half of C_2H_4 produced in the presence of $^{14}C_2H_2$ was unlabelled. This is explained by the oxidation of endogenously produced C_2H_4 which is inhibited by 10% C_2H_2. Cores incubated without C_2H_2 are thus valueless as controls for endogenous C_2H_4 production. The C_2H_2 assay can seriously overestimate N_2-fixation when fixation rates are low.

Introduction

Much of the quantitative data on N_2-fixation in the rhizosphere of non-leguminous plants is based on the acetylene reduction assay (see Knowles, 1977). This procedure is well established and has been extensively used to estimate fixation in a range of organisms and associations (Hardy, Burns and Holsten, 1973). All N_2-fixing systems so far tested, except those meeting their energy requirement by the oxidation of lower alkanes (de Bont and Mulder, 1976) will reduce C_2H_2 to C_2H_4 and nitrogenase is the only enzyme known to mediate this reduction. It is assumed in the procedure that C_2H_2 is the sole substrate for C_2H_2-dependent C_2H_4 production.

Ratios of C_2H_4 produced to N_2-fixed often depart from the theoretical ratio of 3:1, particularly in systems containing soil where ratios of 8:1 (Steyn and Delwiche, 1970) and exceptionally 25:1 (Brouzes, Mayfield and Knowles, 1971) are reported. Because of this anomaly, because of the considerable lag between the addition of C_2H_2 and the evolution of C_2H_4 and because of the possible production endogenous C_2H_4 over a protracted (commonly 24 h) incubation period, C_2H_2 reduction based on data on fixation by undisturbed roots either *in situ* (Balandreau and Dommergues, 1973) or in large cores (Day, Dart, Harris and Van Berkum, 1975) is difficult to interpret. The experiments with $^{14}C_2H_2$ here described were undertaken to distinguish between C_2H_4 derived from C_2H_2 and that arising from other processes in the soil.

Materials and Methods

Ampoules of C_2H_2 (250 μ Curies in 2 ml from the Radiochemical Centre, Amersham, Bucks) were diluted to 100 ml with unlabelled C_2H_2. Contaminating C_2H_4 was removed using a bath made by pouring liquid N_2 into n-propanol and stirring vigorously to give a 'slush' of alcohol crystals at -127°C. The gas sample in a 100 ml glass tube (Fig. 1) was immersed in the bath for *ca.* 30 min to freeze out C_2H_4 (M.P. - 81.8°C). Ethylene (B.P. - 103.9°C, M.P. - 169.4°C) volatilizes readily and was removed by several cycles of evacuation (of 1-2 seconds duration) and replacement with N_2. After final evacuation the tube was transferred to a Dewar flask of liquid N_2 and fully evacuated before returning to room temperature. Acetylene reverts rapidly to the gas phase when the boiling point is reached and the tube should not be loaded with more C_2H_2 than it will hold at s.t.p.

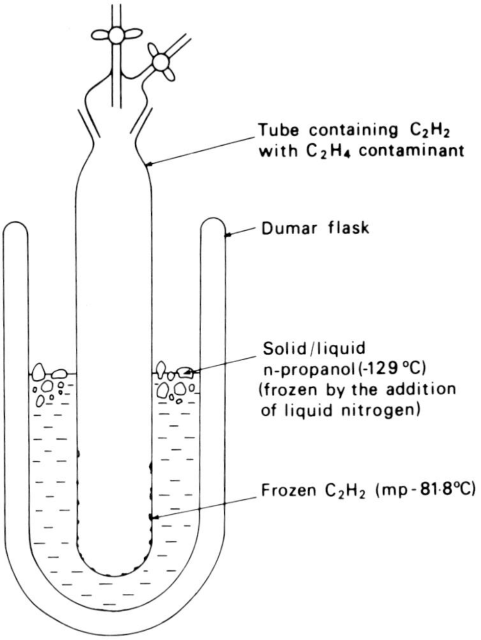

Fig. 1. Apparatus for the purification of C_2H_2. Contaminating C_2H_4 can be removed by gentle evacuation at -129°C.

Samples were incubated in the normal manner (Hardy, Burns and Holsten, 1973) in a gas phase containing 10% C_2H_2. Grass cores (predominantly *Lolium perenne* from long term pastures at Rothamsted Experimental Station) were taken in 11 cm dia. x 16 cm steel cylinders. The $^{14}C_2H_4$ produced was resolved by gas chromatography on a 4 m x 2 mm column of 80 - 100 Mesh Durapak (Phenyl isocyanate on porasil C) cooled to 0°C by circulated ethylene glycol. N_2 carrier gas at 15 ml min^{-1} gave good definition and wide, even spacing of the hydrocarbon peaks. The flame ionization detector was modified to minimize dead volume and sealed so that individual hydrocarbons, oxidized in the flame to CO_2 and flushed out of the detector in the gas stream, could be trapped in hyamine hydroxide (2 ml 15% hyamine hydroxide in scintillation fluid on 3 mm glass beads in a 10 ml burette). The trapped CO_2 was eluted from the column with 18 ml of scintillation fluid. Counts on $^{14}C_2H_2$ dilutions prepared to calibrate the technique were correlated with C_2H_2 peak height at $r > 0.996$ (12 values).

The predicted values (shown as solid lines in Figs. 2 and 3) are based on $^{14}C_2H_2$ calibration curves and represent the quantity of label which should occur in the C_2H_4 if it were derived entirely from C_2H_2. Allowance has been made for the difference in detector response to C_2H_2 and C_2H_4.

Ethylene oxidation was investigated using field cores of *Lolium perenne* and pure bacterial cultures (University of Warwick EU1 and EU5).

Results

Figures 2 and 3 show the relationship between radioactive count and chromatographically determined C_2H_4 concentration. The data from soil cores containing *Lolium perenne* was chosen as representative of results obtained with cores of relatively low (<100g $N.ha^{-1}d^{-1}$) and relatively high (>300g $N.ha^{-1}d^{-1}$) activity.

Values obtained from nodules of *Phaseolus vulgaris* from an active core of *Lolium perenne* with non-leguminous weeds (reducing 30.3 μ moles C_2H_2 d^{-1}, theoretically equivalent to 360g $N.ha^{-1}d^{-1}$), from *Anabeana cylindrica* and from anaerobic suspensions of *Bacillus polymyxa* and *Clostridium pasteurianum* are shown in Figure 2; these fit well to the predicted value.

Figure 3 shows the results obtained with a similar but less-active core which produced a total of 5.36 μ moles C_2H_2 d^{-1} in the gas phase. The quantity of label in the C_2H_4 was 57% lower than that predicted from the C_2H_2 calibration curve and from legume nodules; only 2.3 μ moles of C_2H_4 d^{-1} were derived from C_2H_2. Similar results were

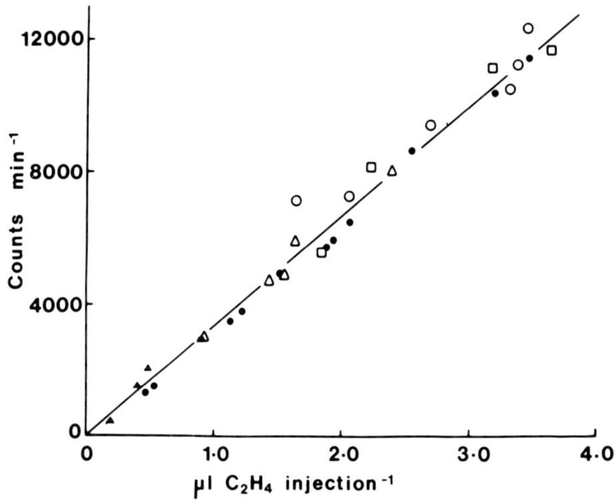

Fig. 2. Production of $^{14}C_2H_4$ by legume root nodules (○), *Bacillus polymyxa* (△), *Clostridium pasteurianum* (□), *Anabaena cylindrica* (▲) and a soil core with grass and non-leguminous weeds (●), estimated to fix 360g $N.ha^{-1}day^{-1}$. Predicted values, based on the count in the $^{14}C_2H_2$ substrate, are shown as a continuous line.

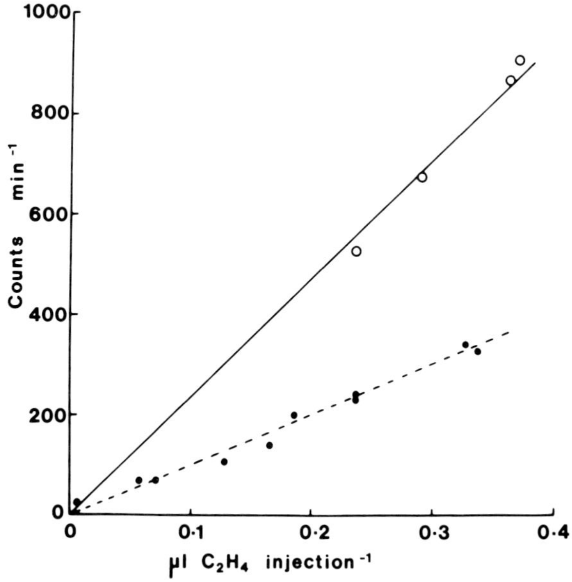

Fig. 3. Production of $^{14}C_2H_4$ by a temperate pasture grass core (●) estimated to fix 62g $N.ha^{-1}day^{-1}$. Comparison with the predicted value (solid line) and legume root nodules (○) shows that only 43% of the C_2H_4 is derived from C_2H_2.

obtained with two other less active cores (mean total C_2H_2 production 4.71 μ moles C_2H_4 d^{-1}) where, on average, only 53% of the C_2H_4 produced was derived from C_2H_2.

In none of these experiments were significant quantities of C_2H_4 detected in the gas phase of control samples incubated aerobically without C_2H_2. Rates of oxidation, sufficient to remove most of the endogenously produced C_2H_4, were, however, observed in many of the field cores, and this oxidation was investigated using bacterial cultures and soil cores. Figure 4 shows the effect of C_2H_2 on C_2H_4 oxidation by shaken bacterial suspensions (University of Warwick, EU1). The results are similar to those of de Bont (1976) with an immediate cessation of oxidation in response to added C_2H_2.

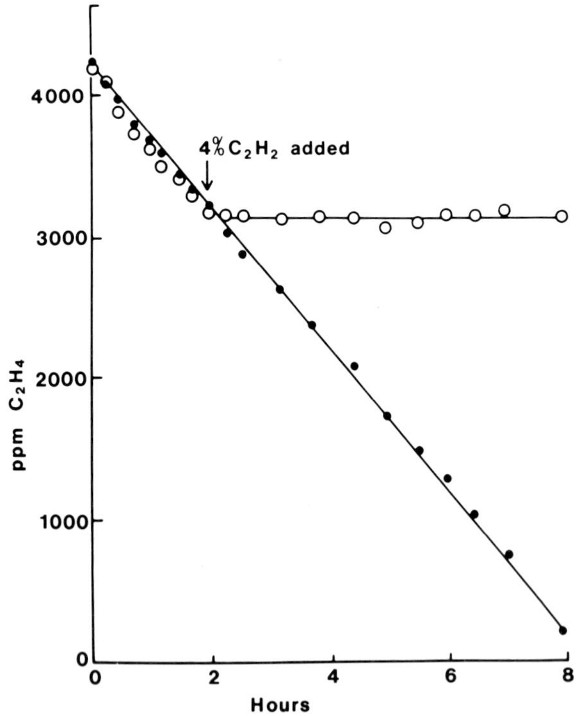

Fig. 4. C_2H_4 oxidation by bacterial cultures with (o) and without (●) added C_2H_2. The values plotted are the means of two replicates which varied by less than 10%.

Discussion

Cornforth (1975) notes that the potential rate of aerobic decomposition of C_2H_4 in soil exceeds, by about 50 fold, its rate of production at anaerobic microsites. The present study shows that it is only when C_2H_2, which blocks C_2H_4 oxidation, is added that total endogenous C_2H_4 production can be detected. Control samples incubated without C_2H_2 are thus valueless in determining endogenous C_2H_4 production. This leads to an overestimate of N_2-fixation, and may explain the anomalously high $C_2H_4:N_2$ ratio obtained in some soil systems (Steyn and Delwiche, 1970; Brouzes, Mayfield and Knowles, 1971). Extrapolation of the total C_2H_4 production from the less-active core (Fig. 3) gives an erroneous fixation rate of 62g $N.ha^{-1}d^{-1}$ compared with 27g $N.ha^{-1}d^{-1}$ based on the reduction of $^{14}C_2H_2$. Many soils will produce and oxidize C_2H_4 (Smith and Restal, 1971; Abeles, Craker, Forrence and Leather, 1971; Lynch and Harper, 1974) and although some indication of potential rates of C_2H_4 production might be obtained using anaerobic controls, the data would be of no quantitative significance because the increased numbers of anaerobic microsites would lead to increased C_2H_4 production. Moreover if C_2H_4 oxidation is to be entirely suppressed stringent anaerobiosis must be achieved since oxidation continues at O_2 concentrations of 0.2% (Cornforth, 1975).

The degree to which fixation is overestimated by the C_2H_2 reduction assay depends on the rate of endogenous C_2H_4 production which varies with soil type and planting regime and it is not therefore possible to define a universal lower limit to the use of the technique. With temperate grass cores errors can become significant when C_2H_2-dependent C_2H_4 production extrapolates to less than 100g $N.ha^{-1}d^{-1}$ and data from soil-containing systems fixing less than this (see Knowles, 1977) should be interpreted with caution unless $^{14}C_2H_2$ controls are used.

Acknowledgements

I am indebted to Dr. F.A. Skinner and Dr. S.B. Primrose for providing bacterial cultures. The work was supported by MOD Research Scheme R3041.

References

Abeles, F.B., Craker, L.E., Forrence, L.E. and Leather, G.R. (1971). Fate of air pollutants: Removal of ethylene, sulphur dioxide, and nitrogen dioxide by soil. *Science*, 173, 914.

Balandreau, J. and Dommergues, Y. (1973). Assaying nitrogenase (C_2H_2) activity in the field. *Bull. Ecol. Res. Comm.* (Stockholm), 17, 247.

Brouzes, R., Mayfield, C.I. and Knowles, R. (1971). Effect of oxygen partial pressure on nitrogen fixation and acetylene reduction in a sandy loam soil amended with glucose. *In* "Biological Nitrogen Fixation in Natural and Agricultural Habitats" (Ed. by T.A. Lie and E.G. Mulder), Pl. Soil (special volume), p.48.

Cornforth, I.S. (1975). The persistance of ethylene in aerobic soils. *Plant and Soil*, 42, 85.

Day, J.M., Harris, D., Dart, P.J. and Van Berkum, P. (1975). The Broadbalk Experiment: an investigation into gains from non-symbiotic nitrogen-fixation. *In* "Nitrogen Fixation by Free-living Microorganisms" (Ed. by W.D.P. Stewart), p.71. Cambridge University Press.

De Bont, J.A.M. (1976). Bacterial degradation of ethylene and the acetylene reduction test. *Can. J. Microbiol.*, 22, 1060.

Hardy, R.W.F., Burns, R.C. and Holsten, R.D. (1973). Applications of the acetylene-ethylene assay for measurement of nitrogen fixation. *Soil Biol. Biochem.*, 5, 47.

Knowles, R. (1977). The significance of asymbiotic Dinitrogen Fixation by Bacteria. *In* "A Treatise on Dinitrogen Fixation" (Ed. by R.W.F. Hardy and A.H. Gibson), Section IV. p.33. John Wiley and Sons, London.

Lynch, J.M. and Harper, S.H.T. (1974). Formation of ethylene by soil fungus. *J. gen. Microbiol.*, 80, 187.

Rice, W.A. and Paul, E.A. (1971). The acetylene reduction assay for measuring nitrogen fixation in waterlogged soil. *Can. J. Microbiol.*, 17, 1049.

Steyn, P.L. and Delwiche, C.C. (1970). Nitrogen fixation by non-symbiotic microorganisms in some southern California soils. *Environ. Sci. Technol.*, 4, 1122.

Smith, K.A. and Restall, S.W.F. (1971). The occurrence of ethylene in anaerobic soil. *J. Soil Sci.*, 22, 430.

BIOLOGY OF THE SOIL-ROOT INTERFACE

A.D. ROVIRA

C.S.I.R.O. Division of Soils, Glen Osmond, South Australia

Summary

A short account is given of research on subjects relevant to the understanding of the biology of the soil-root interface, in which recent developments have occurred.

Introduction

Since this subject was reviewed by Rovira and Davey (1974) there has been a considerable change of emphasis in studies of the rhizosphere on the soil-root interface and Bowen and Rovira (1976) drew attention to some of these changes from the traditional studies of the numbers and types of micro-organisms and the characterization of organic materials released from roots into solution under axenic conditions. This paper is not a comprehensive review and deals with areas of research which reflect new thinking of the biology of rhizosphere and the relevance of this approach to plants in the field.

These areas of research in the biology of the soil-root interface are:

(a) Release of organic materials from roots
(b) Microbial colonization of roots
(c) Application of light microscopy to the root surface
(d) Application of electron microscopy
(e) Modelling of the populations in the rhizosphere

(f) Root-associated nitrogen fixation
(g) Mycorrhizas
(h) Interactions between root pathogens and root function

Release of Organic Materials from Roots

For many years estimates of the organic compounds released from roots were based upon results with plants grown under axenic conditions in nutrient solution into which 0.1 to 0.4% of the photosynthate was "exuded" (Rovira and McDougall, 1965). Subsequent studies by Barber and Gunn (1974) showed that roots growing in glass ballotini released from the roots sugars and amino acids equivalent to 5 to 9% of the total weight.

Such a finding brought a new dimension in rhizosphere studies and was supported by Barber and Martin (1976), Martin (1977) and Sauerbeck and Johnen (1977) all of whom reported that carbon released from roots growing in soil amounted to some 20% of the total plant dry matter. Barber and Martin (1976) found that micro-organisms increased the percentage of photosynthetically fixed carbon released into soil by roots from 5-10% under axenic conditions to 12 to 18%, most of the added carbon released as CO_2. Martin (1977) reported that up to 39% of the carbon translocated to the roots was released into the soil and he attributed this to the lysis of root cell walls rather than actual exudation through intact cell walls. This was supported by the E.M. studies of Foster and Rovira (1976) which showed considerable penetration by bacteria of the epidermal and cortical cells in field-grown wheat at flowering.

Microbial Colonization of Roots

To understand the dynamics of colonization of roots by micro-organisms there has been a departure from the plate counting methods by which workers estimated the R:S ratio.

(viz. ratio of the population in the rhizosphere to that in the soil), to the use of model systems in which growth rates on or around the roots are measured. Bowen and Rovira (1976) stressed the need to consider the "generation time" of bacteria on roots as the basis for studying the dynamics of colonization. Barber and Lynch (1977) found that with barley plants growing in solution culture the "doubling time" for the mixed species of bacteria was 23.9, 29.7 and 103.4 hours for the growth periods 0-4, 4-7 and 7-16 days respectively.

Bowen and Foster (1977) eliminated the problem of different colonization times on parts of roots of different ages by planting axenically-grown roots into soil and counting the bacteria on selective media at intervals. They found that pseudomonads had a generation time on roots of 5.2 h compared with 39h for *Bacillus* spp. while in soil *Pseudomonas* spp. had a generation time of 77 h and *Bacillus* spp. over 100 h. Bowen and Foster noted large differences in generation time between species of *Pseudomonas*. This concept of "generation" or "doubling" time on roots or in the rhizosphere is considered the best way to characterize the colonizing ability of different organisms and is an approach which will help establish the basic principles behind the colonization of roots by micro-organisms e.g. growth rates, prior occupation of sites, competition for nutrients, antagonism, etc.

Application of Light Microscopy to the Root Surface

Direct light microscopy of the micro-organisms on roots has shown that only 5 to 10% of the root surface is covered by micro-organisms (Bowen and Theodorou, 1973; Rovira *et al.*, 1976). These studies, in which sound quantitative techniques have been used to assess the populations densities, should dispel the myth that roots were covered by a mantle of micro-organisms - such a myth comes from the astronomical counts expressed per gramme of root or soil. The application of pattern analysis to the microbial microflora on roots has shown an irregular distribution with the bacteria aggregating on two scales viz. - small clumps and also larger aggregations (Newman and Bowen, 1974). Other work has shown that aggregation is associated with cell junctions (Rovira, 1956; Malajczuk and Bowen, unpublished) and with moribund epidermal cells (Bowen and Foster, 1977).

Fluorescence microscopy using either fluorochromes on roots (Schippers and van Vuurde, 1977) or fluorescent antisera on slide impressions of roots colonized by different strains of *Rhizobium* (Trinick, 1970) has revealed the distributions of bacteria and the competition for colonization between strains of *Rhizobium*.

Application of Electron Microscopy

Our concept of the spatial distribution of the various components of the soil/root interface has been helped greatly by the application of both transmission and scanning electron microscopy. (T.E.M. - Dart and Mercer, 1964; Greaves and Darbyshire, 1972; Old and Nicolson, 1975, 1977; Foster and Rovira, 1976, 1977; S.E.M. - Dart (1971); Campbell and Rovira, 1973; Rovira and Campbell, 1974, 1975.)

Scanning electron micrographs in Figures 1 and 2 show the surface of roots of wheat (1) a healthy root with mucilage spreading over the surface enveloping bacteria and clay particles, (2) a root infected with Take-all fungus in which the mucigel is no longer present and the root is heavily colonized with bacteria. The T.E.M. view of a clover root in Figure 3 shows the dense packing of bacteria for 10 to 15 μm between the root and soil with many types of bacteria, some forming small colonies. The bacteria some distance from the soil-root interface occur either singly or as pairs completely enclosed in clay by comparison with the formation of colonies near the root.

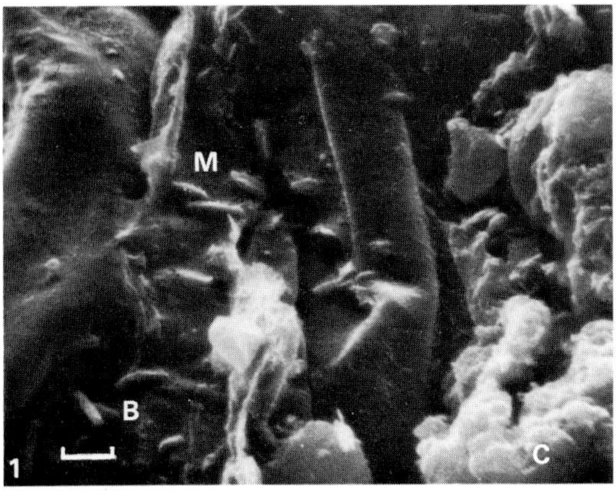

Fig. 1. Scanning electron micrograph of wheat root, M = Mucigel, B = Bacteria, C = Clay. Bar = 5 μm.

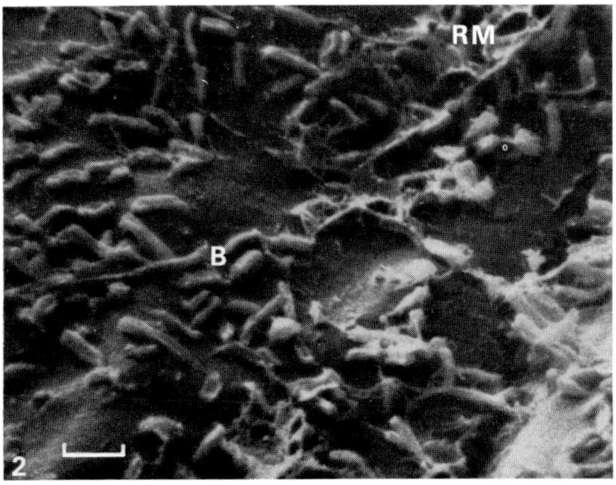

Fig. 2. Scanning electron micrograph of wheat root infected with Take-all fungus. B = Bacteria, RM = Remnants of mucigel. Bar = 5 μm.

The rhizosphere of *Paspalum* grown in permanent pasture under irrigation shows a thick layer of mucigel being dissolved by *Spirillium*-like bacteria then a sharp boundary to the clay-organic matter material 10 to 30 μm from the root; the colony of bacteria developed in this clay-organic material was most probably nourished by soluble material from the root (Fig. 4).

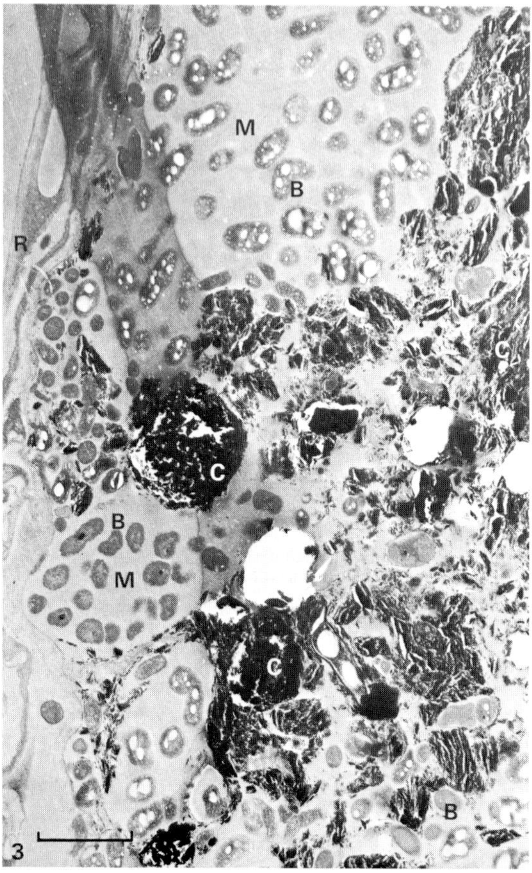

Fig. 3. Transmission electron micrograph of root of subterranean clover (*Tr. subterraneum*). R = Root, B = Bacteria, C = Clay, M = Mucilage - probably bacterial in origin. Bar = 5 μm.

Modelling of the Populations in the Rhizosphere

The quantification of the energy material lost by roots into soil and of the microbial population on roots led Bowen and Rovira (1973) to ask "Are modelling approaches useful in rhizosphere biology?". They concluded that there was then still insufficient information to produce such a model. However, in 1977, Newman and Watson developed a computer model to predict the effects of changes in soil water, root density, root exudation, initial microbial populations and in response of micro-organisms to substrate on the concentrations of substrates and micro-organisms at different distances from the root. Their prediction of a sharp reduction of the population away from the root surface has been supported by the T.E.M. studies. The modelling concept will be useful to predict the changes in microbial populations with changes in conditions at the soil-root interface and will help workers ask critical questions to guide future research.

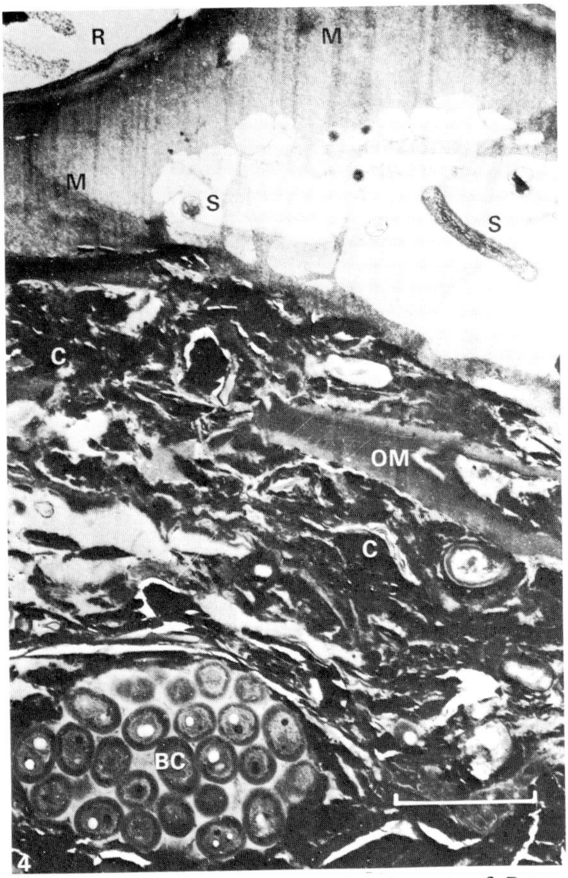

Fig. 4. Transmission electron micrograph of root of *Paspalum notatum*. R = Root, M = Mucigel, S = *Spirillum*-like bacterium, C = Clay, O.M. = Organic Matter, B.C. = Bacterial Colony. Bar = 5 μm.

Root-associated Nitrogen Fixation

Reports of agronomically significant amounts of nitrogen being fixed in the rhizosphere of grasses in the cooler regions of Western Australia came from Parker (1957) and later from Dobereiner and Campelo (1971) with the association between the roots of *Paspalum* and *Azotobacter paspali*. Later reports that linked nitrogen fixation under tropical grasses with photosynthesis (Dobereiner and Day, 1975) and associated the fixation with *Azospirillum (Spirillum lipoforum)* living inside the cortical cells of the roots (Dobereiner, 1977) indicate that significant amounts of nitrogen are fixed when the soil is at high temperature and wet. The manipulation of this system for agronomic benefit may be more difficult than managing the legume-*Rhizobium* symbiosis but considerable research is being undertaken in the hope that such fixed nitrogen may replace fertilizer nitrogen.

Mycorrhizas

The biology and function of the ectomycorrhizal fungi have been understood for a long time (Marks and Kozlowski, 1973) but the appreciation of the importance of the vescicular-arbuscular mycorrhizas (VAM) is a recent phenomenon. The tremendous growth responses of plants in soils low in phosphate (Mosse, 1973; Tinker, 1975) and the improvement in nitrogen fixation by legumes when infected with V/A mycorrhiza (Crush, 1974; Daft and El-Gaimi, 1975) has aroused interest in countries with extensive areas of phosphate-deficient soil. This is an important area of research in the biology of the soil-root interface and, like the root-associated nitrogen fixation, justifies more research with agronomic aims (Bowen and Bevege, 1976).

Interactions between Root Pathogens and Root Function

The presence of root pathogens in the soil-root interface may be more common than their absence. Hence it is relevant to present some results on the effects of pathogens on root function.

One of the major root pathogens of cereals in Southern Australia is cereal cyst nematode (*Heterodera avenae*) (Meagher, 1972; Rovira, 1977). This pathogen affects the shoot to root ratio, response of plant to phosphate (Table 1), response of root growth to soil temperature (Fig. 5) and results in the formation of thick layers of mucigel colonized by fungi and bacteria (Figs. 6 and 7).

TABLE 1

Interactions between cereal cyst nematode (H.avenae), superphosphate and growth of wheat*

Nematicide**	Superphosphate	Seminal Roots/Plant		Tops/Plant		Ratio:- Top Wt./Root Wt.
		Length (cm)	Dry Wt. (mg)	Leaf area (cm^2)	Dry Wt. (mg)	
-	-	434	63	7	37	0.59
+	-	859	60	16	103	1.70
-	+	664(53)#	76(19)	13(94)	73(84)	0.96
+	+	892(4)	59(0)	19(19)	119(16)	2.02

* 580 eggs/100 g soil
** Aldicarb (Temik, Union Carbide) for complete control at 9 mg a.i./kg soil (equiv. 9 kg a.i./ha)
# ( ) Percent increase with superphosphate at 200 mg/kg

Pot experiment: Soil kept moist at 15°C for 4 weeks before seeding to stimulate hatching of nematodes, wheat grown at a root temperature of 10°C for 4 weeks. Values are the means per plant calculated from five replicate pots with 5 plants/pot containing 400 g soil.

The losses of cereal yields due to infection of roots by Take-all fungus (*Gauemannomyces graminis*) make it second in importance to stem rust on a world-wide basis. One of its symptoms is the premature ripening or "haying-off" of crops, and we have shown that infection of roots of wheat at both the seedling and flowering stages increases the water potential of leaves and the osmotic potential of the sap (Table 2). These changes might be due to the effects of the *G.graminis* on the translocation of assimilate from leaves to roots (Clarkson *et al.*, 1975) and the blocking of the xylem vessels by hyphae and gum associated with the hyphae (Fig. 8).

The influence of these two soil-borne root diseases on the mineral composition of the leaves of wheat is shown in Table 3.

The interactions between micro-organisms at the soil-root interface are important in determining the balance of micro-organisms and, in the case of root diseases, the extent of biological control. Rovira and Campbell (1975) demonstrated with S.E.M. that in soils in which Take-all was suppressed the hyphae of *G.graminis* were colonized by rod-shaped bacteria (Fig. 9) and at a later stage the

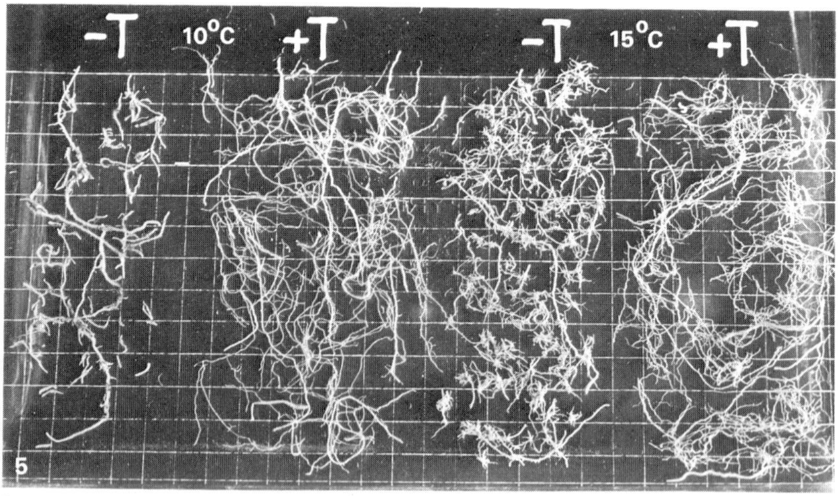

Fig. 5. The effect of soil temperature on growth of wheat roots with (-T) and without (+T) cereal cyst nematode (CCN). The same soil, containing CCN, was incubated at 15°C for 4 weeks to promote hatching, planted to wheat grown in the glasshouse with 10° and 15°C root temperatures for 4 weeks. -T = Superphosphate only. +T = Superphosphate plus the nematicide "Temik". 1 cm grid in background. Cm root/plant at 10°: with CCN 133, without CCN 460. Cm root/plant at 15°: with CCN 400, without CCN 616.

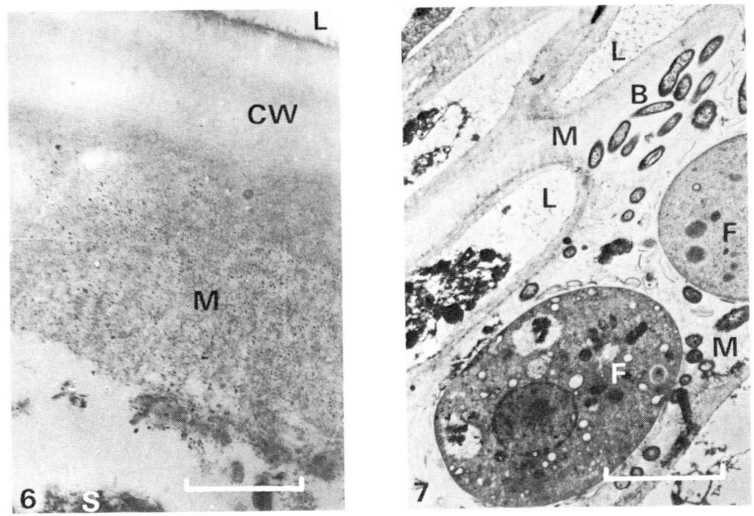

Figs. 6 and 7. Transmission electron micrograph showing the formation of mucilage in and around root infested with CCN. L = Lumen of plant cell, C.W. = Plant cell wall, M = Mucilage, B = Bacteria, F = Fungus, S = Soil. Bar = 5 μm.

TABLE 2

Influence of Take-all fungus (G. graminis) on the osmotic potential of sap and water potential+ of leaves of wheat

Stage of Growth:	Three leaf		Flag leaf at Flowering	
	Healthy	Diseased	Healthy	Diseased
Leaf Water Potential	-554 ± 80	-1286 ± 203	-1640 ± 56	-1640 ± 53
Osmotic Potential	N.M.*	N.M.	-1370 ± 89	-1780 ± 83

± Standard error of the mean
* N.M. = not measured
\+ Water potential, kPa

TABLE 3

The effect of cereal cyst nematode (H. avenae) and Take-all fungus (G. graminis) on the mineral content of the tops of wheat

(a) Cereal cyst nematode

Soil Treatment	Top Dry Wt. g/m row	Mineral Content (%)**						
		P	S	K	Ca	Si	Cl	Mg
Nil	3*	0.46	0.55	4.5	0.92	1.6	0.86	0.19
Nil	8*	0.46	0.43	5.1	0.55	1.6	0.97	0.16
Nil	22#	0.45	0.41	6.2	0.28	1.3	1.6	0.16
Nematicide***	32	0.49	0.40	5.8	0.31	1.4	1.6	0.16

(b) Take-all

Disease Rating+	Top Dry Wt. (g/plant)	Mineral Content (%)**				
		P	S	K	Ca	Si
5	0.2	0.44	0.28	2.2	0.28	1.3
4	0.5	0.48	0.35	5.4	0.32	1.3
1 to 3	0.6	0.61	0.43	6.7	0.30	1.6
0	0.8	0.72	0.45	6.9	0.36	2.3

Superphosphate applied at 200 kg/ha.
* Plants severely stunted.
** XRF analyses conducted by Mr. J.T. Hutton on plants sampled 8 weeks after seeding at Turretfield Research Centre.
*** Aldicarb ("Temik" from Union Carbide Ltd.) at 2.4 kg a.i./ha gave complete control.
\# Roots with light infestation.
\+ Disease rating 0 = no disease; 1 to 3 = "light" infection confined to lesions on the seminal roots; 4,5 = "heavy" infection extending from seminal roots to crown.

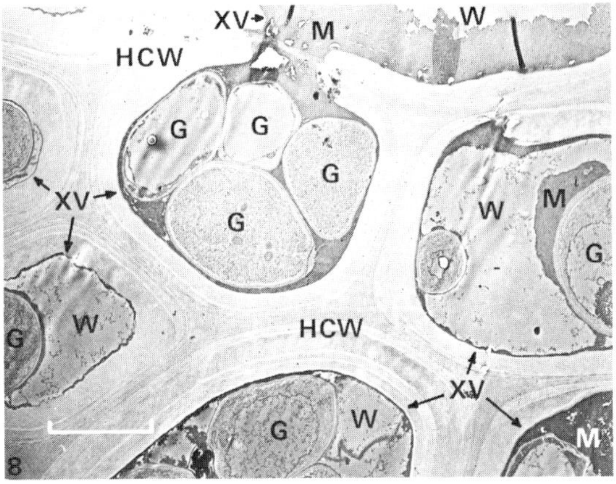

Fig. 8. Micrograph of cross section of wheat root infested with Take-all fungus (*G.graminis*). G = *G.graminis*, M = Mucilage, H.C.W. = Host-cell walls, XV = Xylem vessels, W = Water filled space. Bar = 5 μm.

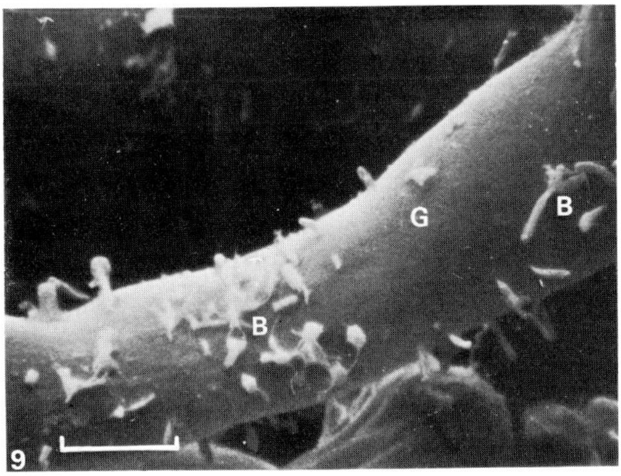

Fig. 9. Scanning electron micrograph of hyphae of *G.graminis* (G) colonized by bacteria (B). Bar = 5 μm.

hyphae collapsed and showed holes most probably caused by the bacteria (Fig. 10).

Conclusions

The biology of the soil-root interface is extremely complex but by the development of new techniques we now have a far better understanding of the release of organic materials from roots, the dynamics of root colonization

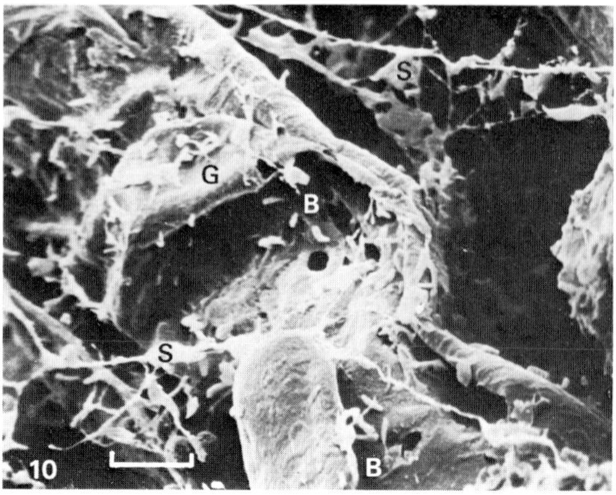

Fig. 10. Scanning electron micrograph of hyphae of *G. graminis* (G) showing the lysis by bacteria (B). Remnants of the hyphae and associated mucilage appear as strands (S). Bar = 5 μm.

and the ultra-structure of the soil-root interface. The application of mathematical models to this zone of soil with validation by experiments will help define the critical questions to be answered.

The ultimate agricultural goal in studies of the biology of the soil-root interface must be the manipulation of micro-organisms in this zone to increase plant growth. Information is required for such manipulation, apart from the Rhizobium-legume association, on endotrophic mycorrhizas, nitrogen fixation with grasses, biological control of root diseases, and the introduction of micro-organisms which stimulate plant growth.

References

Barber, D.A. and Gunn, K.B. (1974). The effect of mechanical forces on the exudation of organic substances by the roots of cereal plants grown under sterile conditions. *New Phytol.*, 73, 39.

Barber, D.A. and Lynch, J.M. (1977). Microbial growth in the rhizosphere. *Soil Biol. Biochem.*, 9, 305.

Barber, D.A. and Martin, J.K. (1976). The release of organic substances by cereal roots into soil. *New Phytol.*, 76, 69.

Bowen, G.D. and Bevege, D.I. (1976). Micro-organisms, the third dimension in phosphate nutrition and ecology. *Rev. in* Rural Science III "Prospects for Improving Efficiency of Phosphorus

Utilization". (Ed. G.J. Blair), p.103. University of New England, Armidale, N.S.W.
Bowen, G.D. and Foster, R.C. (1978). Dynamics of microbial colonization of plant roots. *In* Proc. Symposium on: "Soil Microbiology and Plant Nutrition (SOMIPLAN)" (Ed. W.J. Broughton and C.K. John), University Malaysia Press (in press).
Bowen, G.D. and Rovira, A.D. (1973). Are modelling approaches useful in rhizosphere biology? *Bull. Ecol. Res. Comm. Stockholm*, 17, 443.
Bowen, G.D. and Rovira, A.D. (1976). Microbial colonization of plant roots. *Ann. Rev. Phytopath.*, 14, 121.
Bowen, G.D. and Theodorou, C. (1973). Growth of ectomycorrhizal fungi around seeds and roots. *In* "Ectomycorrhizae: Their Ecology and Physiology" (Eds. O.C. Marks and T.T. Kozlowski), p. 107. Academic Press, New York.
Campbell, R. and Rovira, A.D. (1973). A study of the rhizosphere by scanning electron microscopy. *Soil Biol. Biochem.*, 5, 747.
Clarkson, D.T., Drew, M.C., Ferguson, I.B. and Sanderson, J. (1975). The effect of the Take-all fungus, *Gaeumannomyces graminis*, on the transport of ions by wheat plants. *Physiol. Pl. Path.*, 6, 75.
Crush, J.R. (1974). Plant growth responses to vesicular-arbuscular mycorrhiza. VII. Growth and nodulation of some herbage legumes. *New Phytol.*, 73, 743.
Daft, M.J. and El-Ghamhi, A.A. (1975). Effects of *Glomus* infection on three legumes. *In* "Endomycorrhizas" (Eds. F.E. Sanders, B. Mosse and P.B. Tinker), p.580. Academic Press, New York.
Dart, P.J. (1971). Scanning electron microscopy of roots. *J. Exp. Bot.*, 22, 163.
Dart, P.J. and Mercer, F.V. (1964). The legume rhizosphere. *Arch. Mikrobiol.*, 44, 344.
Döbereiner, J. (1977). Nitrogen fixation in grass-bacteria associations in the tropics. Proc. Advis. Comm. Meeting on Nitrogen Int. Atom. Energy Agency. Vienna (in press).
Döbereiner, J. and Campelo, A.B. (1971). Non-symbiotic nitrogen fixing bacteria in tropical soils. *Pl. Soil*, Spec. Vol., 457.
Döbereiner, J. and Day, J.M. (1975). Dinitrogen fixation in the rhizosphere of tropical grasses. *In* "Nitrogen Fixation by Free-living Micro-organisms" (Ed. W.D.P. Stewart), p.39. Cambridge University Press.
Foster, R.C. and Rovira, A.D. (1976). Ultrastructure of wheat rhizosphere. *New Phytol.*, 76, 343.
Foster, R.C. and Rovira, A.D. (1978). The ultrastructure of the rhizosphere of *Trifolium subterraneum* L. Proc. Int. Symp. on Microbial Ecology, Dunedin, New Zealand, August 1977. Springer Verlag (in press).
Greaves, M.P. and Darbyshire, J.F. (1972). The ultrastructure of the mucilaginous layer on plant roots. *Soil Biol. Biochem.*, 4, 443.
Marks, G.C. and Kozlowski, T.T. (1973). (Eds.) "Ectomycorrhizae - Their Ecology and Physiology", p.444. Academic Press, New York.
Martin, J.K. (1977). Factors influencing the loss of organic carbon from wheat roots. *Soil Biol. Biochem.*, 9, 1.
Meagher, J.W. (1972). Cereal cyst nematode (*Heterodera avenae* Woll.). Studies on ecology and control in Victoria. *Dept. Agric. Vic. Tech. Bull.*, 24, 1.

Mosse, B. (1973). Advances in the study of vesicular-arbuscular mycorrhizas. *Ann. Rev. Phytopath.*, 11, 171.
Newman, E.I. and Bowen, H.J. (1977). Patterns of distribution of bacteria on root surfaces. *Soil Biol. Biochem.*, 6, 205.
Newman, E.I. and Watson, A. (1977). Microbial abundance in the rhizosphere: A computer model. *Pl. Soil*, 48, 17.
Old, K.M. and Nicolson, T.H. (1975). Electron microscopical studies of the microflora of roots of sand dune grasses. *New Phytol.*, 74, 51.
Old, K.M. and Nicolson, T.H. (1977). A new concept of the rhizosphere/root cortex micro-environment. Proc. Int. Symp. on Microbial Ecology, Dunedin, New Zealand, August 1977. Springer Verlag (in press).
Parker, C.A. (1957). Non-symbiotic nitrogen fixing bacteria in soil. III. Total nitrogen changes in a field soil. *J. Soil Sci.*, 8, 48.
Rovira, A.D. (1956). A study of the development of the root surface microflora during the initial stages of plant growth. *J. appl. Bact.*, 19, 72.
Rovira, A.D. (1977). Assessment of yield losses in wheat due to cereal cyst nematode (*Heterodera avenae*). Aust. Pl. Path. Soc. Workshop on "Epidemiology and Crop Loss Assessment". Lincoln College, New Zealand, August 1977.
Rovira, A.D. and Campbell, R. (1974). Scanning electron microscopy of micro-organisms on the roots of wheat. *Microbiol. Ecol.*, 1, 15.
Rovira, A.D. and Campbell, R. (1975). A scanning electron microscope study of interactions between micro-organisms and *Gaeumannomyces graminis* (Syn. *Ophiobolus graminis*) on wheat roots. *Microbiol. Ecol.*, 3, 177.
Rovira, A.D. and Davey, C.B. (1974). "Biology of the Rhizosphere in the Plant Root and its Environment" (Ed. E.W. Carson), p.153. University Press of Virginia, Charlottesville.
Rovira, A.D. and McDougall, B. (1967). Microbiological and biochemical aspects of the rhizosphere. *In* "Soil Biochemistry" (Eds. A.D. McLaren and G.H. Peterson), p. 417. Marcel Dekker.
Rovira, A.D., Newman, E.I., Bowen, H.J. and Campbell, R. (1974). Quantitative assessment of the rhizosphere microflora by direct microscopy. *Soil Biol. Biochem.*, 6, 211.
Sauerbeck, D.R. and Johnen, B.G. (1977). Root formation and decomposition during plant growth. *From* "Soil Organic Matter Studies" - Int. Symp. on Soil Organic Matter Studies, Brunswick, F.D.A., Sept. 1976.
Schippers, B. and van Vuurde, J.W.L. (1977). Studies of microbial colonization of wheat roots and the manipulation of rhizosphere microflora. "Proc. Int. Symp. on Microbial Ecology" (Ed. M.W. Loutit). Springer Verlag (in press).
Tinker, P.B. (1975). Soil chemistry of phosphorus and mycorrhizal effects on plant growth. *In* "Endomycorrhizas" (Eds. F.E. Sanders, B. Mosse and P.B. Tinker), p.353. Academic Press, New York.
Trinick, M.J. (1969). "*Rhizobium* Interactions with Soil Microorganisms". Ph.D. Thesis, University of Western Australia.

ROOT MICROORGANISMS IN MIXTURES AND MONOCULTURES OF GRASSLAND PLANTS

E.I. NEWMAN, R. CAMPBELL, P. CHRISTIE,* ALISON J. HEAP and R.A. LAWLEY

Botany Department, University of Bristol, UK

Summary

The abundance of root surface fungi and bacteria and of internal mycorrhizal infection was determined on four species common in British lowland grassland, *Anthoxanthum odoratum*, *Lolium perenne*, *Plantago lanceolata* and *Trifolium repens*, grown alone and in two-species mixtures. Differences between species and experiments in abundance of bacteria and fungi were correlated with differences in plant nitrogen concentration. Abundance of all three microbial groups was often significantly affected by the presence of another species of plant, both in glasshouse pot experiments and in the field. Three possible mechanisms for such neighbour effects seem likely. (1) In some, but not all, two-species mixtures fungal abundance was intermediate between the two monoculture values, suggesting that the fungi were using substrates from both plants. (2) Some results could be explained if one plant exuded a substance which inhibited growth of mycorrhizal fungi of nearby plants. (3) Changes in bacterial abundance were consistently correlated with changes in nitrogen concentration in the host plant, suggesting that plant competition, by altering plant nutrient status, influenced exudation and

* *British Antarctic Survey, Madingley Road, Cambridge.*

hence bacterial growth.

Introduction

Much of the world's land surface is covered by vegetation consisting of several or many species growing together. In Britain monocultures of crop plants occupy less than half of the land area, and a large proportion of the remainder is permanent grassland, most of which consists of several or many species intimately intermingled both above and below ground. This raises the question of whether the root microorganisms of one plant can be influenced by another plant close by. In this paper we present information on the abundance of mycorrhizas and root-surface saprophytes on a few common grassland plants, showing that microbial abundance can vary between these plant species and can also be influenced by the presence of a neighbouring species. We discuss possible mechanisms by which these changes in microbial abundance are brought about.

Methods

Most of the research involved two grasses, *Anthoxanthum odoratum* L. and *Lolium perenne* L. and two dicotyledonous herbs, *Plantago lanceolata* L. and *Trifolium repens* L. All four commonly grow together in permanent pasture. In glasshouse experiments four plants, either all of the same species or two of one species plus two of another, were grown per pot, in unsterilized grassland soil without added fertilizer. In a permanent pasture near Bristol (the field from which the soil for the glasshouse experiments was collected) *P. lanceolata* plants growing with and without competition from other species were compared; *Plantago* on its own was obtained by weeding out surrounding plants and preventing lateral root spread with a plastic cylinder. The percentage of the root surface covered by bacteria and the length of fungal mycelium per unit of root surface area were determined by direct microscopic examination of stained segments from healthy roots. Internal infection of vesicular-arbuscular mycorrhiza was assessed by staining other root segments by the method of Phillips and Hayman (1970) and then estimating the intensity of infection within numerous small squares, on a scale 0-4. For further details of methods see Rovira *et al.* (1974) and Christie, Newman and Campbell (1978).

TABLE 1

Abundance of root-surface microorganisms and internal mycorrhizal infection on plants growing in monoculture in four pot experiments

(a)

Experiment	Bacteria (cover %) A	B	C	Fungi (mm/mm^2) A	B	C
Lolium perenne	7.7	4.3	n.d.	12.1	0.7	8.4
Plantago lanceolata	6.3	5.6		14.3	1.8	8.8
Statistical significance of difference*	$P < 0.05$	ns		ns	ns	ns

(b)

Experiment D.	Bacteria (cover %)	Fungi (mm/mm^2)	Mycorrhiza+
Anthoxanthum odoratum	0.64b	0.8a	0.39a
Lolium perenne	0.31a	1.4a	0.60b
Trifolium repens	0.48ab	9.2b	1.07c

* ns = not significant
+ intensity of infection, scale 0-4
n.d. = not determined
In part (b) any two figures within a column which are not followed by the same letter are significantly different ($P < 0.05$).
Experiment A is Experiment 2 of Rovira *et al.* (1974).

Results

Table 1 shows the microbial abundance on species growing separately in pots. All four experiments were conducted in the same glasshouse, though at different times, and used soil from the same field. Part (a) shows that in each of the first three experiments there was little difference in abundance of bacteria or fungi between *Lolium* and *Plantago*, but between the experiments there was a marked difference in the abundance of fungi. Table 1 part (b) shows that within a single experiment there can be significant differences between species in bacteria, fungi and mycorrhiza. *Anthoxanthum* was highest in bacterial abundance but lowest in fungi and mycorrhiza. Figure 1 suggests a possible cause for part of the variation in microbial abundance between species and experiments. The vertical axis is the volume of root surface fungi and bacteria (which is presumably closely related to the biomass), the horizontal axis is percentage nitrogen in the shoot.

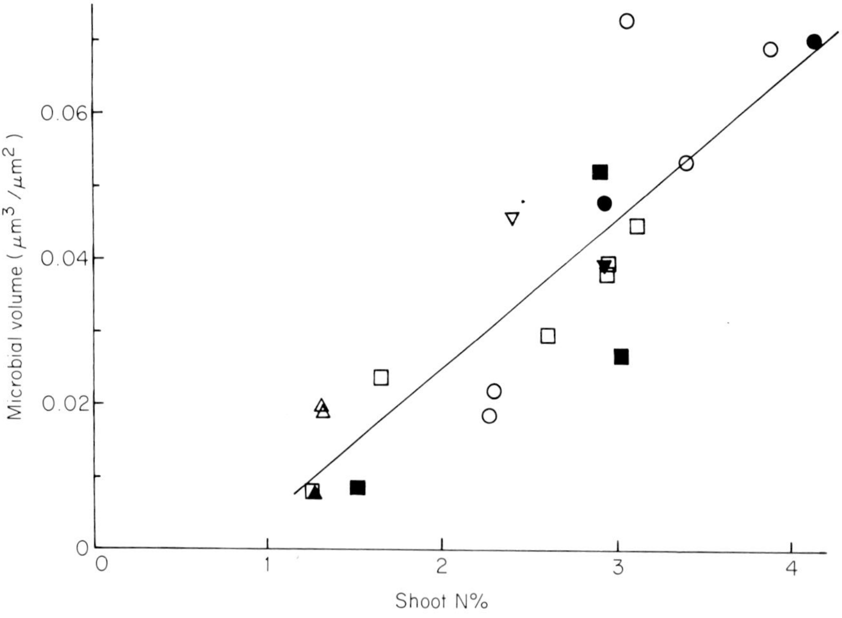

Fig. 1. Relationship between shoot nitrogen concentration and combined volume of bacteria and fungi per unit root surface area, in three pot experiments. Black symbols, plants in monoculture; open symbols, in mixture; △▲ *Anthoxanthum*, □■ *Lolium*, ▽▼ *Plantago*, □■ *Trifolium*.

The data are from three pot experiments, using the same soil but conducted at different times of the year, involving four species with or without other species present.

The correlation coefficient is 0.873 ($P < 0.001$). Such a correlation could in theory be caused by the microorganisms increasing the plant's nitrogen uptake, but a more likely explanation is that increased nitrogen concentration in the plant caused increased microbial growth, perhaps through increased exudation.

Influence of Neighbouring Species

Figure 2 shows, as an example, the effect of neighbouring grass plants on the microbial abundance on *Plantago* roots. In the glasshouse pots, pasture and garden plots (swards) the effect of grass was to reduce mycorrhizal abundance in *Plantago* roots. Root surface fungi, in contrast, were increased by the presence of grass in the pots and pasture, and bacteria were also increased in the pasture. In the plots, where the soil was different, both fungi and bacteria were sparse on the root surface and not significantly influenced by the partner species. This example shows that bacteria, fungi and mycorrhiza can all be affected by a neighbouring species. In the various species combinations we have tested, fungi and bacteria have often been affected by neighbouring species, sometimes to a greater extent than in Figure 2. Usually the amount of fungi and bacteria changed in the same direction, but fungi changed more. Mycorrhiza changed least, and not necessarily in the same direction as the other groups (as shown in Fig. 2).

Possible Mechanisms for Neighbour Effects

We now consider, in turn, three possible mechanisms by which one plant could influence another's root microorganisms, and see how far our results would fit in with each of them.

Combined substrate supply

Since in a mixture the roots of the plants are closely

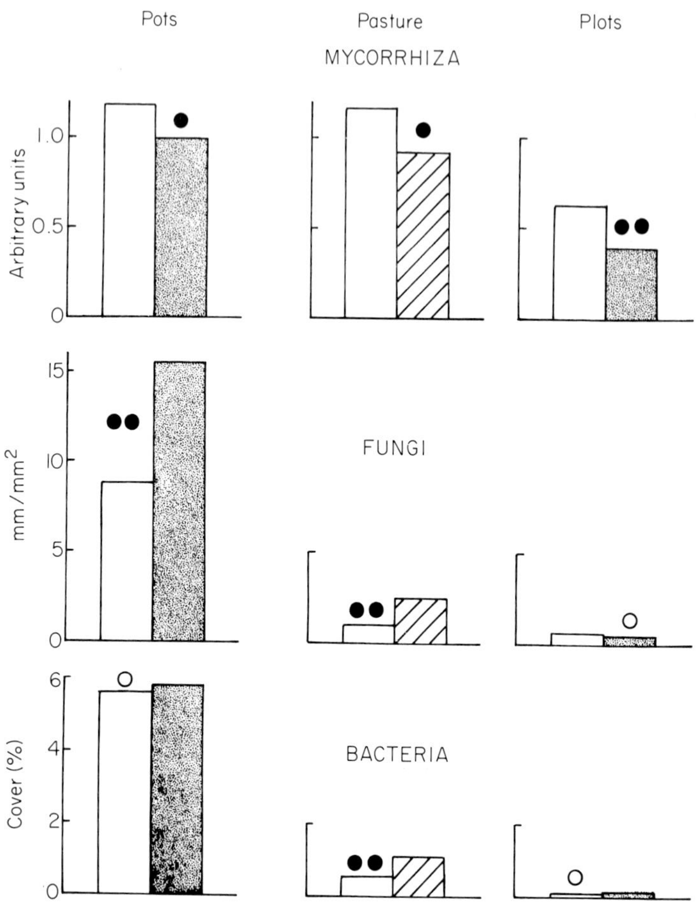

Fig. 2. Abundance of root microorganisms on *Plantago* roots in presence or absence of other species. Open columns, *Plantago* growing alone; stippled columns, *Plantago* growing with *Lolium*; hatched columns, *Plantago* growing with other pasture species, mainly grasses. Statistical significance of differences: ●● $P < 0.001$, ● $P < 0.05$, ○ not significant.

intermingled it is possible that the microorganisms could be obtaining substrate from the exudate of both species. If this were so the microbial abundance in the mixture would probably be intermediate between the values in the two monocultures. Table 2 shows that *Trifolium* had much higher fungal abundance than the two grasses *Lolium* and

Anthoxanthum, when they were grown separately; but in a mixture the abundance on the two species was virtually the same and about intermediate between the monoculture values. This would fit the hypothesis of a combined substrate supply for the fungus. On the other hand in other species combinations the fungal abundance in the mixture did not fit this pattern, as shown in the lower part of Table 2.

TABLE 2

Fungal abundance (mm/mm^2) on root surfaces of plants in pots, growing either separately or in two-species mixtures

	Lolium	*Trifolium*	*Anthoxanthum*	*Trifolium*
Separate	1.4a	9.2c	0.8a	9.2c
Together	4.1b	4.2b	3.1b	3.4b
	Lolium	*Plantago*	*Lolium*	*Anthoxanthum*
Separate	8.4a	8.8a	1.4a	0.8a
Together	11.3a	15.5b	1.4a	2.7b

Within any group of four (two species, separate/together), any two figures not followed by the same letter are significantly different ($P < 0.05$).

We have attempted to find out whether the species composition of fungi on the roots of *Lolium* and *Plantago* is altered when they grow together. Fungal abundance was determined by placing numerous washed 4-mm root segments on agar and identifying the fungi which grew out. Most of the fungal species which grew were profusely sporing species, e.g. *Penicillium* spp and phycomycetes, which could have grown from spores rather than mycelium on the root, and no significant difference was found between *Lolium* and *Plantago*, or between mixture and monoculture, in the abundance of any of these fungi. However one non-sporing grey fungus, a *Rhizoctonia* species, occurs commonly on *Lolium* roots, though less often on *Plantago* and two other species examined at the pasture site (Table 3(a)). If this fungus used substrate from both plant species its abundance in mixture should be intermediate between the monoculture values, but this was not true in either of two pot experiments (Table 3(b)).

TABLE 3

Abundance of Rhizoctonia isolate L3 on root surfaces expressed as percentage of 4-mm root segments infected

(a) Plants in permanent pasture				
Lolium perenne	11.3c			
Holcus lanatus	6.0b			
Plantago lanceolata	3.3ab			
Trifolium repens	0 a			
Any two figures not followed by the same letter are significantly different ($P < 0.05$)				
(b) Pot experiments	Experiment 1		Experiment 2	
	Lolium	*Plantago*	*Lolium*	*Plantago*
Separate	8.0	2.0	22.4	4.0
Together	11.3	6.0	21.6	4.4
Statistical significance:				
Lolium v. *Plantago*	$P < 0.05$		$P < 0.001$	
Separate v. Together	not sig.		not sig.	

Examples can be found where mycorrhizal abundance in mixtures tends towards the average of the two monocultures, though the two species in mixture never achieve exactly equal values as in the upper part of Table 2. None of our bacterial results show this sort of convergence in mixture.

In conclusion this hypothesis fits most convincingly for fungi in grass-clover mixture (Table 2), but clearly it cannot be the main mechanism involved in neighbour effects.

Exudation of inhibitory substances

The roots of one species might exude a substance inhibitory to the growth of some microorganisms, which reached the roots of other plants and influenced their microbial population. There is evidence that living roots of *Calluna vulgaris* exude a substance which inhibits growth of ecto-mycorrhizal fungi of some trees (Handley, 1963; Robinson, 1972). An experiment by Hayman, Johnson and Ruddlesdin (1975) also suggested an exudate effect: swede, which is not mycorrhizal, reduced mycorrhizal infection of onion when the two grew together. In our results it is mycorrhiza which fit best with the hypothesis of an exudate effect. Figure 3 shows that *Anthoxanthum* consistently reduced the mycorrhizal abundance in neighbouring plants, including other *Anthoxanthum*. By analysis of variance the partner effect is significant ($P < 0.01$) whereas the host x partner interaction is not ($P > 0.25$). These results, combined with the low mycorrhizal abundance in *Anthoxanthum* itself, would be in agreement with the hypothesis that *Anthoxanthum* roots exude a substance inhibitory to mycorrhizal fungi.

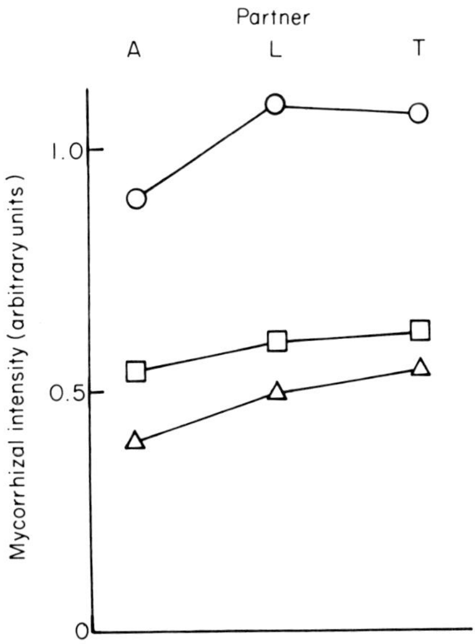

Fig. 3. Intensity of mycorrhizal infection in *Anthoxanthum* (△), *Lolium* (□) and *Trifolium* (○) when grown with one of the same three species (A, L or T respectively) in pots.

Effect via the host plant

If plants are competing this could influence their internal chemical composition, and thus the amount and composition of their root exudate, so this is an indirect means by which one plant could influence another's microbes. For example, if plant A shades plant B this might reduce B's exudation rate and so reduce the abundance of saprophytes on its roots. Direct evidence that plants can influence each other's root microbes even when the root systems are not intermingling was provided by an experiment in which two *Lolium* plants and two *Trifolium* plants were grown per pot, with their roots kept separate by a partition, and their shoots either allowed to intermingle or not depending on the orientation of an above-ground partition. Fungal mycelium length on *Lolium* roots was 9.0 mm/mm^2 when the two species were completely separated, but was significantly reduced, to 4.5 mm/mm^2, when they could intermingle above ground only. Fungal abundance on *Trifolium* differed little between the two treatments.

To study the effect of neighbouring plants on each other's nutrient status the N, P and K concentrations of the shoots have been determined. It was not possible to determine the concentration in the roots, but this would probably vary in a parallel way. The ratio (shoot N% of species A when growing with species B)/(shoot N% of A when alone) can be used to express the effect of species B on A's nitrogen status. When expressed in this way there is a consistent relationship between changes in nitrogen concentration and bacterial abundance. Figure 4(a) is based on two experiments involving four species in various combinations; Figure 4(b) is from the *Lolium-Trifolium* partition experiment mentioned above, but now including treatments where the two species could intermingle below ground only, above ground only, or both above and below ground. In both parts of Figure 4 there is a significant correlation, showing that when the presence of a neighbouring plant resulted in an increased nitrogen concentration in the 'host' plant, its bacterial abundance increased also. Correlation with the ratios of P and K percentage was less strong. These results would fit the hypothesis that plants were competing for nitrogen and that a changed nitrogen status affected the composition or amount of root exudate and hence the bacterial abundance on the root surface. Fungi and mycorrhizas, analysed in the same way, show some correlations with nutrient percentage, but they are less consistent, suggesting that for them the mineral nutrient status of the host does not have such crucial importance.

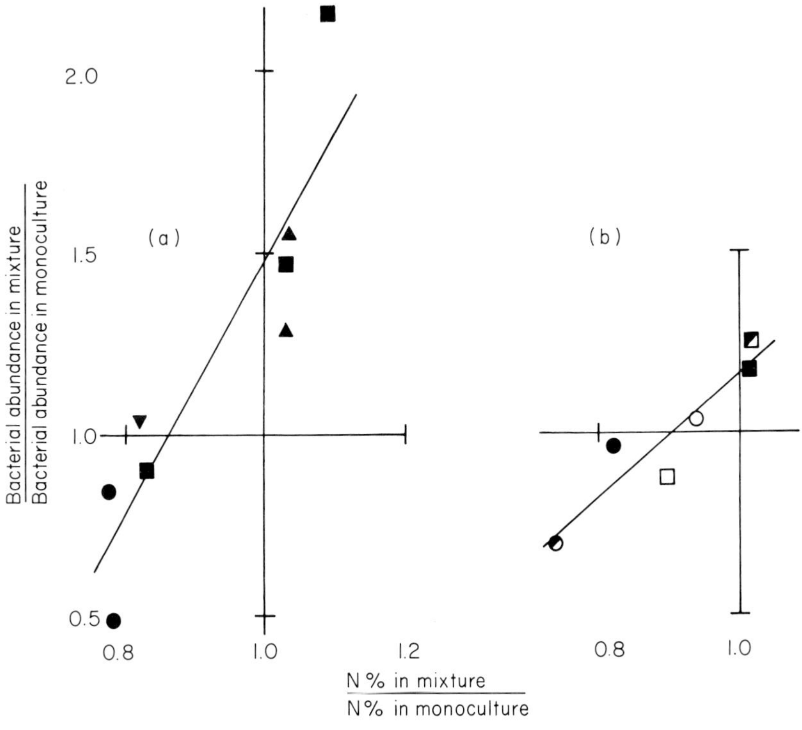

Fig. 4. Relationship between effect of neighbouring plants on shoot nitrogen percentage and on bacterial abundance on root surface. 'Host species' ▲ *Anthoxanthum*, ▼ *Plantago*, ■ *Lolium*, ● *Trifolium*. Open symbols shoot association only, pied symbols root association only, black symbols root and shoot association. Correlation coefficients: (a) $r = 0.902$, $P < 0.01$; (b) $r = 0.928$, $P < 0.01$.

Discussion

The primary conclusion from this research is that one plant species can influence the abundance of root surface bacteria and fungi and of mycorrhizas on another. This could provide an indirect means by which one plant influences another. If the root microorganisms of plant A are influencing its growth, then any alteration of their abundance caused by plant B will result in an indirect effect of B on A's growth.

We should emphasize that bacteria and fungi were studied on the root surface only. The microorganisms in the rhizosphere proper may behave differently.

Conclusions about the mechanisms by which one plant affects another's microbes must at present be tentative, but the suggestions made here agree with other known facts. The clearest evidence is that bacteria are strongly influenced by the nitrogen status of their host, and other plants influence the bacteria via the host plant. This fits with the evidence that plants supplied with extra mineral nitrogen exude more amino acids (Bowen, 1969) and that many rhizosphere bacteria are amino acid users (Gray and Williams, 1971). The computer model of Newman and Watson (1977) predicts that most of the soluble root exudate will be taken up by microorganisms near the root surface so that bacteria on one root would be expected to receive little substrate from another root unless the two roots were almost touching. For root surface fungi, on the other hand, our results (Table 2) show that when the abundance on two plant species in monoculture was very different the values in mixture were intermediate, perhaps because the fungi grew from one root to another and obtained substrate from both. This would fit with the known lack of specificity of many saprophytic root fungi and their ability to translocate organic materials. However, in other species combinations (lower part of Table 2) some other mechanism was apparently operating, perhaps because the fungal abundance in monoculture differed less markedly. Our experiments were not well designed for testing the hypothesis that toxic exudates from one plant inhibit microorganisms on another. Nevertheless the reduction of mycorrhizal abundance when *Anthoxanthum* is present could be due to this mechanism, since *Anthoxanthum* contains coumarin (Gibbs, 1974), and coumarins are fungitoxic (Uritani and Akazawa, 1969).

Acknowledgement

This research was supported by a grant from the Natural Environment Research Council.

References

Bowen, G.D. (1969). Nutrient status effects on loss of amides and amino acids from pine roots. *Plant and Soil*, **30**, 139.

Christie, P., Newman, E.I. and Campbell, R. (1978). The influence of neighbouring grassland plants on each others' endomycorrhizas and root-surface microorganisms. *Soil Biol. Biochem.* (in press).

Gibbs, R.D. (1974). "Chemotaxonomy of Flowering Plants". McGill-Queen's University Press, Montreal.

Gray, T.R.G. and Williams, S.T. (1971). "Soil Micro-organisms". Oliver and Boyd, Edinburgh.

Handley, W.R.C. (1963). Mycorrhizal associations and *Calluna* heathland afforestation. *Forestry Commission Bulletin* 36.

Hayman, D.S., Johnson, A.M. and Ruddlesdin, I. (1975). The influence of phosphate and crop species on *Endogone* spores and vesicular-arbuscular mycorrhiza under field conditions. *Plant and Soil*, **43**, 489.

Newman, E.I. and Watson, A. (1977). Microbial abundance in the rhizosphere: a computer model. *Plant and Soil*, **48**, 17.

Phillips, J.M. and Hayman, D.S. (1970). Improved procedures for clearing roots and staining parasitic and vesicular-arbuscular mycorrhizal fungi for rapid assessment of infection. *Trans. Brit. Mycol. Soc.*, **55**, 158.

Robinson, R.K. (1972). The production by roots of *Calluna vulgaris* of a factor inhibitory to growth of some mycorrhizal fungi. *J. Ecol.*, **60**, 219.

Rovira, A.D., Newman, E.I., Bowen, H.J. and Campbell, R. (1974). Quantitative assessment of the rhizoplane microflora by direct microscopy. *Soil Biol. Biochem.*, **6**, 211.

Uritani, I. and Akazawa, T. (1959). Alteration of the respiratory pattern in infected plants. *In* "Plant Pathology" (Ed. J.G. Horsfall and A.E. Dimond), p.349. Academic Press, New York.

RESPONSE OF NONLEGUMINOUS PLANTS TO ROOT INOCULATION WITH FREE LIVING DIAZOTROPHIC BACTERIA

MURRAY H. GASKINS* and DAVID H. HUBBELL**

**Agronomy Department, IFAS, University of Florida, USA*

Summary

Pearl millet, *Pennisetum americanum* (L) Shum cv Gahi 3, plants achieved higher rates of growth after inoculation with a diazotrophic bacterial culture, *Azospirillum brasilense* (formerly *Spirillum lipoferum*) strain 13t. However nitrogenase activity increased only when competing bacteria were excluded. Cytokinins and other plant growth substances were identified as metabolic products in bacterial cultures, and treating plants with kinetin or gibberellic acid induced dry weight increases similar to those obtained from inoculation treatments. These results suggest that their production of growth-modifying phytohormones, rather than, or in addition to their fixation of dinitrogen, explains the plant responses to inoculation with *Azospirillum*.

Introduction

Within recent years convincing evidence has accumulated of the existence of loose associations of various diazotrophic bacteria with the roots of certain tropical C-4 grasses (Mulder and Brotonegoro, 1974; Döbereiner, 1974;

*** Soil Science Department, IFAS, University of Florida, Gainesville, Florida, USA*

Döbereiner and Day, 1975). Efforts to utilize these associations to replace partially the heavy chemical nitrogen requirements of cereal grains and forage grasses have received considerable attention.

We report here investigations of plant responses to inoculation with a strain of *Azospirillum brasilense*, under controlled laboratory conditions.

Materials and Methods

Experiments were conducted with pearl millet, *Pennisetum americanum* (L) Shum cv Gahi 3, as the host plant and *Azospirillum brasilense* (formerly *Spirillum lipoferum* strain 13t) as the inoculant.

Plants used for collection of root exudate were grown from surface sterilized seeds under sterile conditions in culture tubes (50 x 300 mm) containing 50 ml of dilute (¼ strength) Hoagland's mineral nutrient solution. Stainless steel screens supported the seeds in the tubes and permitted the roots of the plants to grow into the solutions underneath. Filtered air pumped through the tubes maintained the CO_2 concentration at ambient level. A controlled environmental chamber provided daytime illumination for 14 h at a PAR level of $600\mu E \cdot sec \cdot cm^{-2}$. Temperature was controlled at 30°C in the day and 25°C at night.

After 20 days, organic materials released into the culture solutions were assayed. Carbohydrates were determined by the phenolsulfuric acid method, proteins by the biuret method, and organic acids by titration with 0.01 N NaOH. An aliquot of the cell culture was extracted with cold 80% methanol after homogenization with a sonic disintegrator, evaporated to the aqueous phase, and partitioned with ethyl acetate as described by Shindy and Smith (1975). Attempts were made to identify plant growth hormones in the extracts, by chromatography on a reverse phase HPLC column.

Another group of plants was used to determine whether

organic compounds produced by plant roots supported dinitrogen fixation by *Azospirillum*. After plants had been maintained for 20 days, the root solutions were inoculated with 1 ml of water containing 10^8 cells of *Azospirillum*. After 24 h at 30°C all the tubes were flushed with a 1:9 acetylene:air mixture. Half of the tubes were then flushed with a 1:9 acetylene:nitrogen mixture until they contained 1-2 per cent oxygen. In earlier tests of various bacterial strains, in defined culture media, this procedure had been found to give near optimum rates of acetylene reduction. After 8 h incubation at 37°C, ethylene concentrations were determined by gas chromatography.

To determine effects of inoculation on growth rates of plants in soil, pearl millet seedlings were grown in 15 cm pots, each containing 1 kg of loamy sand which had been steam sterilized one month before and stored in closed but not sterile containers. In some of the pots, the organic matter content was increased by addition of 1/8 or 1/4 by weight of peat.

Bacterial cells were grown in trypticase soy broth (Difco), harvested by centrifugation and resuspended in water at a cell concentration of $10^8 \cdot ml^{-1}$ for inoculation of the plants. The cells in one half of the inoculum were killed by autoclaving, and 20 ml per pot of the live or killed cell suspensions were applied to the soil five days after plants emerged.

Treatments were arranged as a split plot experiment in three replicates with organic matter levels as whole plots and live or killed cells as sub-plots. After six weeks, plant tops were removed for dry weight determination. The root systems, intact in the pots in which they grew, were enclosed in polyethylene bags and the bags flushed with 1:9 acetylene:air mixture. After an incubation period of 8 h, ethylene was determined by gas chromatography.

To test plant responses to mixtures of plant growth hormones such as those which might be produced by rhizosphere organisms, kinetin, gibberellic acid, and indole acetic acid were added to soil in which pearl millet seedlings had been established. The plants were treated five days after emergence and three times thereafter at 3 day intervals. Plants also were treated by inoculation with live or killed bacteria, as in the earlier test. The plants were harvested after 32 days, and dry weights of tops and roots were determined.

Results and Discussion

Quantities of organic matter found in the root solutions were low (Table 1). Most of these materials were carbohydrates. Organic acids, which support *Azospirillum* growth better than sugars, were present only in very small amounts.

TABLE 1

Organic materials released from the root tissue of pearl millet

	Weight per mg of root tissue			
	Total Carbohydrates	Protein	Organic Acids	Free Amino Acids
	(μg)	(μg)	(μg)	
In solution[a]	40.5	13.7	1.1	Trace
Loosely bound[b]	9.3	7.4	0.4	-
Firmly bound[c]	9.0	14.1	-	-

[a] Plants were grown aseptically in tubes for 30 days as described in text, and checked for sterility before analysis.
[bc] Determined as the amount added to the solution after 5 and 30 sec sonication, respectively.

We have noted that when plants are maintained for several days after inoculation sediment accumulates in the tubes. This may indicate accelerated production of organic materials by the roots, in response to inoculation (Umali-Garcia, Hubbell, and Gaskins, 1978). It will be necessary to grow plants and bacteria together for longer periods of time, to evaluate fully interactions between the species which affect root exudation. When Barber and Martin (1976) and Martin (1977) used ^{14}C labelling to follow carbon loss from roots in the soil, they found much higher values than those from our experiments. They found also that carbon movement into the soil was depressed in sterile systems. This does not necessarily indicate however that acetylene reduction rates would have increased greatly in the experiments reported here had there been more organic material in the root solutions, since adding succinic acid, a carbon substrate readily metabolized by *Azospirillum*, produced only a modest increase in acetylene reduction activity.

The results do show that the bacteria utilize organic substances produced by the plants, and fix dinitrogen, in the absence of competition from other organisms. Data from the other experiments suggest however that this may occur only in axenic systems.

Rates of acetylene reduction were low in all tests (Table 2). Adding succinic acid to the root solution and incubating in an oxygen depleted atmosphere increased activity. Removing the roots from the solution reduced activity substantially, presumably because most of the bacteria were bound to the root surfaces.

TABLE 2

Acetylene reduction by Azospirillum *cells in plant root solutions*

Material tested[a]	Assay conditions and C_2H_4 produced/h: +Succinate[b] Incubated in: Air	+Succinate[b] Incubated in: Air/N_2	-Succinate Incubated in: Air/N_2
	(μM)	(μM)	(μM)
Plant and root solution	1.6	3.1	1.5
Plant only	1.2	1.5	0.8
Root solution only	0.7	0.8	0.4
Control - Succinate only	-	0.9	-

[a] 10^8 bacterial cells were added to the tubes 24 h before C_2H_4 was determined.

[b] Added to final concentration of $4 \cdot 10^{-5}$ M, as succinic acid buffered to pH 7 with KOH.

When the broth extracts were chromatographed on a reverse-phase liquid chromatograph column, peaks tentatively identified as indole acetic acid and a cytokinin were found. Since various rhizosphere organisms have been shown to produce these materials (Brown, 1972; Phillips and Torrey, 1972) these results were not surprising.

Pot studies

Plant growth was affected by both soil organic matter levels and by inoculation (Table 3). However, since both killed and live cells increased growth, the response of the plants to inoculation could not be attributed solely to dinitrogen fixation by the bacteria. Acetylene reduction rates were not increased by inoculation with *Azospirillum*.

TABLE 3

Effect of Azospirillum *inoculation and soil organic matter content on dry weight yield of pearl millet plants*

Azospirillum inoculation*	Soil organic matter content and dry weight of plant tops at 6 wks		
	Low (g)	Intermediate (g)	High (g)
Live cells	3.66	3.09	5.12
Killed cells	3.58	3.15	4.68
None	3.12	2.76	4.79

* Cells were washed in water and collected by centrifugation to remove culture broth. Plants were inoculated 3 times at 1-week intervals, with one ml water containing 10^8 cells. The levels of soil organic matter were approximately 3, 10, and 15%.

The experiment comparing effects of growth regulating compounds with *Azospirillum* inoculation indicated that short term responses are similar (Table 4). While these observations do not resolve the question whether *Azospirillum* inoculation increases dinitrogen fixation in the rhizosphere, they demonstrate clearly that increased fixation is not invariably associated with observed growth responses.

TABLE 4

Effect of Azospirillum *inoculation and growth hormone treatments on growth of pearl millet plants grown in soil*

	Plant Dry Weight	
Treatment	Tops (mg)	Roots (mg)
Kinetin[1]	190	41
GA_3[1]	222	46
IAA[2]	92	34
AZ 13[3]	250	55
Control	137	38

[1] Total applied was $4 \cdot 10^{-5}$ M
[2] Total IAA applied was $4 \cdot 10^{-6}$ M
[3] Cells were washed in water and collected by centrifugation. Plants were inoculated twice with one ml water containing 10^8 cells.

Conclusions

This work confirms our earlier report that energy substrates readily metabolized by *Azospirillum* are not produced in abundance by aseptically grown plants (Gaskins, Napoli and Hubbell, 1976). It is probable that the relatively slow plant growth imposed by the artificial conditions used in our experiments limited availability of organic substrates to the bacteria, and restricted nitrogen fixation. Achieving maximum rates was not the objective of these experiments.

Our failure to find increased acetylene reduction rates in inoculated pots confirms results of earlier experiments not reported here, and supports our hypothesis that inoculation with *Azospirillum* can increase plant growth without increasing nitrogen fixation in the soil.

The tentative evidence for plant growth hormone production by the bacteria, and the similarity of plant responses to *Azospirillum* inoculation and to treatment of the soil with growth substances, leads us to conclude that these bacterial metabolites are a major determinant of the plant responses to *Azospirillum* inoculation.

References

Barber, D.A. and Martin, J.J. (1976). The release of organic substances by cereal roots into the soil. *New Phytol.*, 76, 69.

Brown, M.E. (1972). Plant growth substances produced by microorganisms of soil and rhizosphere. *J. appl. Bact.*, 35, 443.

Döbereiner, J. (1974). Nitrogen-fixing bacteria in the rhizosphere. *In* "The Biology of Nitrogen Fixation" (Ed. A. Quispel), 760pp. American Elsevier Publishing Company, New York.

Döbereiner, J. and Day, J.M. (1975). Nitrogen fixation in the rhizosphere of tropical grasses. *In* "Nitrogen Fixation by Free Living Microorganisms" (Ed. W.D.P. Stewart), 471pp. Cambridge University Press, London.

Gaskins, M.H., Napoli, Carolyn and Hubbell, D.H. (1976). Root exudates in relation to associative nitrogen fixation in grasses. *Agron. Abstr. (U.S.)*, 1976, 71.

Martin, J.K. (1977). Factors influencing the loss of organic carbon from wheat roots. *Soil Biol. Biochem.*, 9, 1.

Mulder, E.G. and Brotonegoro, S. (1974). Free-living heterotrophic nitrogen-fixing bacteria. *In* "The Biology of Nitrogen Fixation" (Ed. A. Quispel), 769pp. American Elsevier Publishing Company, Inc., New York.

Phillips, D.A. and Torrey, J.G. (1972). Studies on cytokinin production by *Rhizobium*. *Plant Physiol.*, 49, 11.

Shindy, W.D. and Smith, O.E. (1975). Identification of plant hormones from cotton ovules. *Plant Physiol.*, 55, 550.

Umali-Garcia, Mercedes, Hubbell, D.H. and Gaskins, M.H. (1978). Infection of *Panicum maximum* roots by *Spirillum lipoferum*. *In* "Environmental Role of Nitrogen-fixing Blue-green Algae and Asymbiotic Bacteria" (Ed. U. Granhall). Swedish Natural Science Research Council, Stockholm

ESTIMATING CARBON TRANSFERS IN THE PLANT RHIZOSPHERE

F.R. WAREMBOURG and G. BILLES

C.E.P.E. Louis Emberger, B.P. 5051, 34033 Montpellier Cedex, France

Summary

Short term labelling of plants with $^{14}CO_2$ made it possible to separate root and microbial activity in the rhizosphere.

Measurements on wheat plants grown in the presence of root microflora and exposed to an atmosphere containing $^{14}CO_2$ showed a succession of two phases of $^{14}CO_2$ evolution around the roots when only one phase occurred with plants grown under axenic conditions.

A series of experiments was then undertaken in order to test the hypothesis of a time sequence for root CO_2 and microbial CO_2 resulting from utilization of exudates by the rhizosphere microflora.

Quantitative analysis of labelled root exudates found in the soluble phase around the roots showed large differences between plants grown under axenic or non-axenic conditions.

In both cases, labelling of the plants induced a liberation of labelled substances in the root solution with a maximum after a few hours. In the presence of microorganisms, this was followed by a second phase of liberation. The time sequence was approximately that of $^{14}CO_2$ evolution.

A qualitative study of the substances liberated in both kinds of cultures has been made using gel filtration techniques. It showed that the nature and composition of the compounds liberated during the second phase were different from those measured during the first phase and in axenic cultures.

A bacterial strain producing polysaccharides of known molecular weight and isolated from the rhizosphere of a graminaceae was used to inoculate axenic cultures of wheat. The progressive evolution of the soluble substances towards one type of compound, the bacterial polysaccharide, demonstrated the existence of microbial activity on root exudates, the maximum being observed during the second phase of $^{14}CO_2$ evolution. These results demonstrated a succession of root and microbial activities in the rhizosphere and the validity of respirometric methods as a means to quantify them.

Introduction

The functioning of rhizosphere microflora is mainly supported by carbonaceous compounds liberated by intact and healthy roots. Present methods of estimating amounts of root-derived substances in natural environments fail to give proper information without unduly disturbing the system.

The use of $^{14}CO_2$ has already permitted us to distinguish the total plant effect (roots + rhizosphere microflora respiration) in the overall soil respiration (Warembourg and Paul, 1977). A previous study (Warembourg, 1975) utilizing short term labelling of plants grown on soil or nutrient solution, showed a succession of two phases of $^{14}CO_2$ evolution around the roots colonized by microorganisms, against only one phase with plants grown under axenic conditions. This led to the hypothesis of a time sequence of root and microbial respiration in the rhizosphere, based on a time delay for the appearance and utilization of root and microbial respiratory substrates (labelled assimilates and labelled root exudates respectively).

The present study was conducted in order to demonstrate

the reality of this concept, by comparing the time course of $^{14}CO_2$ evolution and the liberation and dynamics of labelled substances around the roots of plants labelled with $^{14}CO_2$ during very short periods of time. Because of the lack of techniques to estimate the real amount of root exudates, mainly due to the difficulties of separating compounds like root mucigel which are insoluble or remain attached to the surface, this study was set up to compare soluble substances liberated by plants in axenic and non-axenic cultures.

This consisted of a series of experiments initiated with wheat plants to evaluate 1) quantitative liberation of labelled substances and 2) qualitative evolution of these substances with time following labelling of the plants. The final objective was to detect the role of root micro-flora in the overall activity measured in the rhizosphere, and in the short term carbon transfers in the root-soil interface.

Materials and Methods

Quantitative and qualitative determination of soluble substances liberated in the rhizosphere are described in experiments A and B respectively.

Culture and labelling of plants

Seeds of wheat (*Triticum* sp. var. Florence Aurore) were surface sterilized in 30% H_2O_2 for 20 minutes after wetting *in vacuo*. They were then germinated and transferred into sterilized culture units (3 per unit) consisting of a photosynthesis chamber and a root compartment containing plant nutrient solution.

For experiment A, this is illustrated in Figure 1 and was provided with a sampling tube closed by a rubber septum. This permitted successive sampling and replacing aliquots of the solution. A nylon net was used to support the seedlings above the liquid.

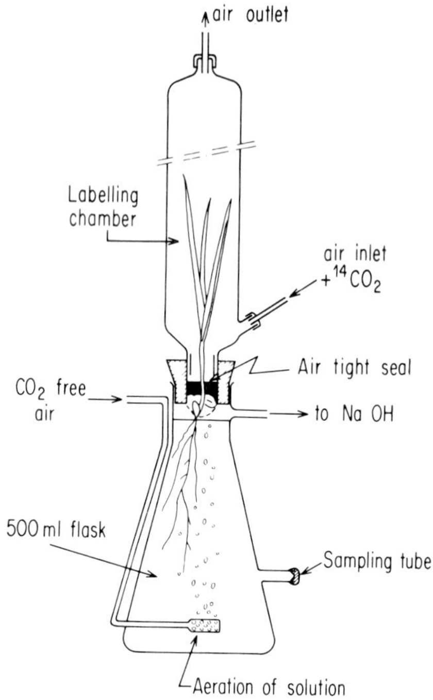

Fig. 1. Schematic diagram of the apparatus used for growing and labelling the plants with $^{14}CO_2$.

For experiment B, the root compartment was smaller and contained glass beads supported by a plate of porous glass. Nutrient solution and air were circulated through the beads by a peristaltic pump. The total amount of solution was collected and replaced at each sampling time by means of a tube located at the bottom of the root container and under the porous glass plate. In this case, the seedlings were placed directly on top of the glass beads.

The air lines of both kinds of units were provided with bacteriological filters.

After one week, half of the units were inoculated either with 0.1 ml of a 2 per cent suspension of soil collected in rhizosphere of growing wheat (experiments A and B), or with a specific bacterial strain called S_2 (experiment B). These bacteria isolated from the rhizosphere of a grass,

Brachypodium ramosum, by Cortez and Billes (1975), are capable of producing a soluble polysaccharide of molecular weight (MW) 51,000. In the present study, they were used to trace microbial activity.

Axenic and non-axenic cultures were then brought to the 5-6 leaf stage with a succession of 9 h night at 18°C and 15 h light at 28°C, with a light intensity of 30,000 lux. At that point, and for experiment A only, fiberglass wool was placed around the stems and the hole was filled with RTV silicone rubber to ensure a tight partition between the roots and the upper portion of the plant's atmosphere.

$^{14}CO_2$ labelling was then done according to a technique similar to the pulse labelling described by McDougall and Rovira (1965). Each unit received a $^{14}CO_2$ air mixture with a calculated radioactivity of 25 μCi and a CO_2 concentration to provide 0.03% in the chamber. Plant exposure to $^{14}CO_2$ was maintained for 1.5 hours.

Respiration measurements

A gas flow technique was used for measurements of CO_2 output from the whole root system as described earlier (Warembourg, 1975). Carbon and carbon-14 fluxes were recorded during 3 days by replacement, counting and titration of NaOH solution placed on the root aeration train. The results are expressed as rates of C and ^{14}C and as specific activities (SA) of the evolved carbon.

Collection of the root solution and measurements

Twenty ml aliquots of the nutrient solution in experiment A, and the total volume in experiment B were removed and replaced aseptically at each sampling time. Volatile compounds and dissolved $^{14}CO_2$ were removed by adding one drop of NHcl and by stirring the solutions under vacuum. They were then centrifuged in order to eliminate microorganisms and cell debris. For experiment B, the latter were retained by the plate of porous glass.

Quantitative measurements of radioactivity content of the supernatant were made by scintillation counting of two 1 ml aliquots. The residue of centrifugation was counted after addition of liquid scintillant. For experiment A, corrections were made for successive samplings.

The results were expressed in dpm per unit and per hour. For experiment B the ^{14}C content of non-soluble substances mostly associated with microbial bodies was also measured.

Qualitative analysis was based on molecular weight determination using gel filtration methods. Concentrated aliquots of each sample were fractionated using gels of different size and nature: Sephadex G 10, Biogel P 30 and P 60 which separate respectively at 700, 25,000 and 50,000 MW. Eluates of each column were counted in the liquid scintillation counter in order to determine the ^{14}C distribution between the various MW classes.

Results

Experiment A: Quantitative analysis of labelled soluble substances liberated in the rhizosphere of wheat, and comparison with $^{14}CO_2$ evolution after labelling of the plants with $^{14}CO_2$

Figure 2 summarizes the results obtained with axenic and non-axenic cultures. The middle curves which represent the continuous recording of $^{14}CO_2$ fluxes (expressed in radioactivity, dpm, per hour) show fluctuations that follow a diurnal pattern, with peaks occurring around noon, when the total respiratory activity is at its maximum, and a decreasing intensity with days after labelling.

The upper curves that record the variations in specific activity (SA) of the evolved carbon show a fundamental difference between the two types of cultures. In both cases, short term labelling of the plants induces an increase of SA that reaches a maximum a few hours after cessation of exposure to $^{14}CO_2$. There is then a gradual decrease; but in the presence of root microflora only the curve shows a second maximum that occurs approximately 12

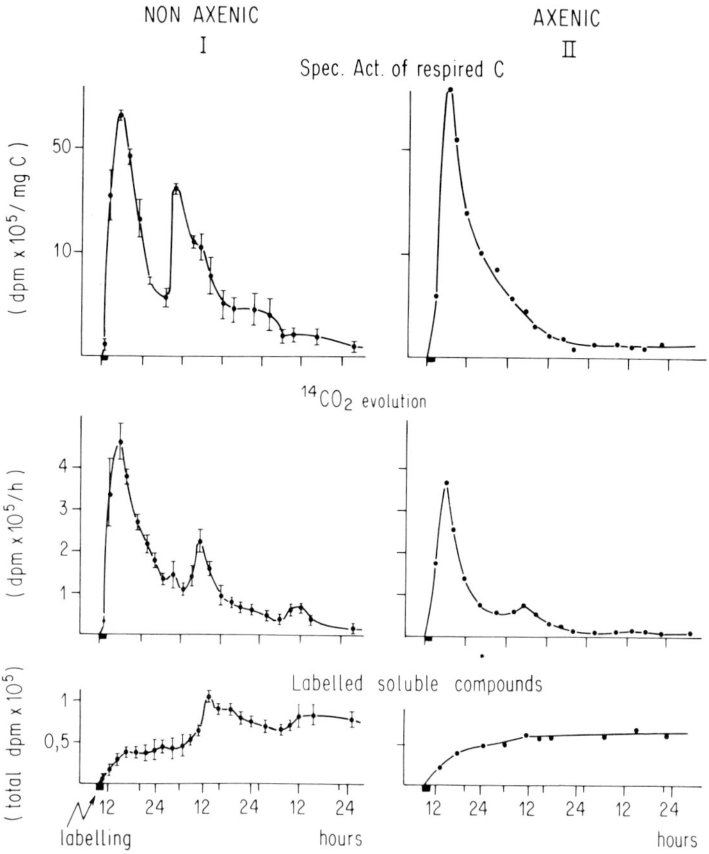

Fig. 2. Simultaneous variations of specific activity of CO_2, $^{14}CO_2$ fluxes and labelled soluble substances liberated in the axenic and non-axenic rhizosphere of wheat labelled with $^{14}CO_2$. (Plants grown in nutrient solution with 9 h night at 18°C and 15 h light at 28°C with a light intensity of 30,000 lux.)

hours after the first one. This corroborates the results described earlier (Warembourg, 1975), i.e. a succession of two phases of activity in the presence of microorganisms against only one in axenic cultures. As indicated by the two sets of curves, the specific activity, which expresses indirectly the relative proportion of carbon originated by the labelling process, varies independently of the diurnal rhythm. This is attributed to the succession of root and then microbial respiration.

The lower curves illustrate the simultaneous variations of labelled substances in the nutrient solution surrounding the roots. Comparison between axenic and non-axenic cultures is of interest, since in both cases plant exposure to $^{14}CO_2$ induces a liberation of labelled compounds as already demonstrated by McDougall (1968). This is illustrated on the curves by a regular increase that follows the same timing as $^{14}CO_2$ evolution. In axenic conditions, the amount of radioactivity measured in the nutrient solution has a tendency to stabilize, and the curve ends in a plateau. In the presence of microorganisms, there is a second rise which occurs on the second day around noon and is quantitatively more important than the first one. The third day is also marked by an increase that happened at the same time of the day. These later phases follow a diurnal pattern and can be attributed to temperature fluctuations. Because of the very small amounts of these substances, it has not been possible to measure the carbon content. Therefore data on specific activity, which could be compared to specific activity of respired CO_2 in the upper curve, are not available. However, the presence of root microflora is characterized by the following:

1. The microorganisms induce an extended period of liberation of substances that is not observed in axenic conditions.

2. They utilize the soluble compounds and this is reflected on the curve by a decrease in the amount of radioactivity in solution. Therefore, the results are only a picture of an equilibrium between production and breakdown, and they underestimate the real amount of soluble substances liberated in the rhizosphere of non-axenic plants. If there is a succession of root and microbial activities, the nature of the compounds must change with time from plant to microbial materials and it is necessary to make a qualitative analysis of the soluble substances in the root solution.

Experiment B: Qualitative characterization of the liberated substances, and their evolution with time after labelling of the plants with $^{14}CO_2$

The use of gel filtration techniques to give a qualitative picture of the various substances liberated around the roots has been dictated by the previous work of Cortez and Billes (1975), i.e. the isolation of a rhizosphere bacteria, called S_2, producing a specific polysaccharide of known molecular weight. Our experimental approach was (1) to compare the MW of the compounds liberated in axenic and non-axenic cultures and the quantitative distribution of ^{14}C in the various fractions after labelling of the plants with $^{14}CO_2$; (2) to detect through time the appearance of the microbial metabolite (the polysaccharide) in cultures inoculated with S_2.

Preliminary results have been obtained with simultaneous labelling of axenic and non-axenic plants inoculated with the bacterial strain S_2.

In axenic cultures, the ^{14}C distribution between the different classes of MW was the following: The compounds recovered in the < 25,000 fraction represented 98 to 100% of the radioactivity contained in the nutrient solution. Of that, about 90% was included in the < 700 class. The results were confirmed for two cultures and for nutrient solution samples collected at different periods following labelling of the plants. This indicated that soluble substances exuded by axenic roots and originated from a substrate photosynthesized at a given time, are of the same composition during the exudation period. They are almost entirely composed of light molecules such as simple sugars and amino-acids. This corroborates the findings of other authors (Vancura, 1964; McDougall, 1968).

In the presence of root microflora, the ^{14}C recovered in the root solution is illustrated in Figure 3.

The upper diagram (Fig. 3a) represents the quantitative and qualitative variations of the soluble substances lib-

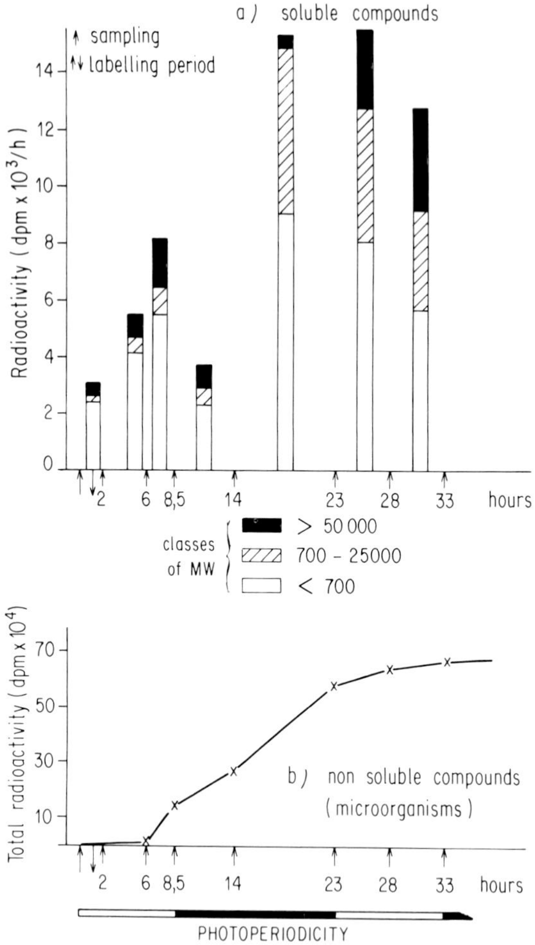

Fig. 3. Quantitative and qualitative variations with time of labelled substances liberated in the rhizosphere of non-axenic wheat inoculated with a bacterial strain producing a polysaccharide of 51,000 MW and labelled with $^{14}CO_2$. a) quantitative and qualitative variations of soluble substances. b) quantitative variations of non-soluble substances. (MW = molecular weight.)

erated in the rhizosphere of non-axenic wheat inoculated with S_2. The vertical columns record the rate of production (dpm/h). The different samplings are indicated by arrows on the time axis. The ^{14}C content of the various MW fractions, represented by different patterns has been calculated from the elution curves obtained with the analytical data of each chromatogram.

The figure indicates the two phases of liberation of soluble substances described previously. The ^{14}C distribution was different from that described for axenic cultures and changed markedly with time. After 2 hours, the solution already contained high MW compounds of more than 50,000 and the proportion increased to reach more than 20% after 8 and 14 hours. The second phase was characterized by a sudden increase of light and intermediate products: 58% < 700, 39% between 700 and 25,000, with only 3% in the higher MW fraction. Thereafter greater proportions of higher MW compounds were released as in the first phase.

We know that the bacteria S_2 synthetizes a polysaccharide of MW 51,000. There is therefore evidence that ^{14}C detected in the higher MW fraction may be mostly associated with this polysaccharide and can be attributed to microbial activity.

In addition, the relative proportion of compounds of MW between 700 and 25,000 shows clearly the existence of intermediate substances liberated in high amounts during the second phase, which were not produced in such high concentrations in the beginning of the first phase or in axenic conditions. This other significant change can also be attributed to the influence of the root microflora.

The Figure 3b represents for the same culture a record of the radioactivity in microorganisms, estimated by measuring the ^{14}C content of insoluble compounds. The cummulative curve indicates the more intense activity towards the end of the first phase of liberation of soluble substances, when the production of microbial polysaccharides reached its maximum and in the beginning of the second phase. This has not been recorded in axenic cultures which showed very small amounts of ^{14}C associated with non-soluble substances and non-significant changes with time. Everything seems to indicate that microbial activity which depends on the release of substances exuded from roots is continually increasing with time, and reaches its maximum at the end of the period of exudation of ^{14}C material, causing the appearance of the second phase of liberation of soluble compounds.

Discussion

Comparison between the results concerning respiration, production and evolution of soluble substances in the axenic and non-axenic rhizosphere of wheat helps to elucidate the dynamics of biological activity measured at the root-soil interface. We suggest the formulation of a model describing the various carbon transfers and utilization in the rhizosphere. This is illustrated by the diagram of Figure 4.

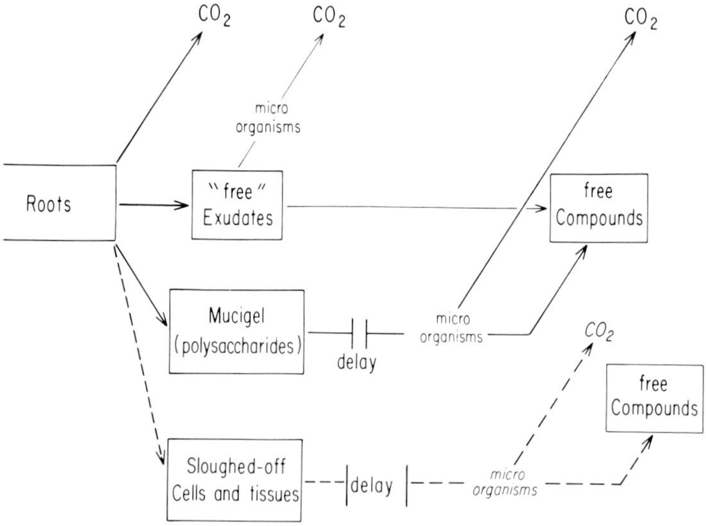

Fig. 4. Word model describing carbon fluxes in the rhizosphere of plants.

Newly photosynthesized carbonaceous compounds translocated to the roots can be used for biosynthesis of new structures or for respiratory substrates causing a release of CO_2, or liberation of exudates. Some of the exuded substances are in a soluble form and diffuse away from the roots; some are non-diffusible and stay in the outer part of the roots (Rovira, 1969). This latter fraction includes the root mucigel recently observed using electron microscopy. It can be assumed that these various processes occur simultaneously for a known translocated substrate.

If microorganisms are present in the rhizosphere, the most available substances, such as soluble compounds (mainly sugars and amino-acids), will be utilized and incorporated in microbial tissues. This will cause a release of CO_2 and a qualitative change from root exudates to microbial products. During the same period, increasing amounts of non-diffusible compounds around the roots will be slowly utilized by the microflora colonizing the roots. Long chain substances like polygalacturonic acids of the root mucigel (Leppard, 1974) will be broken down to smaller units that will be released into the root solution. The microbial activity will be marked by an intensive evolution of $^{14}CO_2$, together with the production of synthetic products of microbial origin; this could explain the second phase of activity.

Works of Campbell and Rovira (1973), Old and Nicholson (1975) and Foster and Rovira (1976) confirm the progressive disappearance of non-diffusible compounds such as root mucigel along the root and away from the root tip. First protected from microbial attack for unknown reasons, these compounds are progressively colonized by root microflora. The authors also mentioned the penetration of bacteria in the cortical cells. Together with utilization of sloughed-off cells and tissues, this phenomenon will cause a delayed evolution of $^{14}CO_2$. As indicated in the diagram, it is assumed that the measurement period used in this study was not long enough to detect $^{14}CO_2$ coming from this source.

In conclusion, the different results presented show that the "pulse labelling" technique can give a new insight into the very closed and intertwined processes that occur at the root-soil interface. Further developments of this research should give a means to utilize this method to estimate root and microbial activity, and therefore carbon budgets in the rhizosphere, without requiring axenic cultures and therefore with little disturbance of the system.

Acknowledgements

We thank Mrs. F. Lafont for competent technical assistance.

References

Campbell, R. and Rovira, A.D. (1973). The study of the rhizosphere by scanning electron microscopy. *Soil Biol. Biochem.*, 6, 747.

Cortez, J. and Billes, G. (1975). Etude des polysaccharides synthétisés par deux souches bactériennes rhizosphériques. *Soc. Bot. Fr., Coll. Rhizosphère*, **122**, 47.

Foster, R.C. and Rovira, A.D. (1976). Ultrastructure of wheat rhizosphere. *New Phytol.*, 76, 343.

Leppard, G.G. (1974). Rhizoplane fibrils in wheat. Demonstration and derivation. *Science*, **185**, 1066.

McDougall, B. (1968). The exudation of C^{14} labelled substances from roots of wheat seedlings. 9th I.C.S.S. Adelaide, Australia. 647.

McDougall, B. and Rovira, A.D. (1965). Carbon-14 labelled photosynthates in wheat root exudates. *Nature* (Lond.), **207**, 1004.

Old, K.M. and Nicholson, T.H. (1975). Electron microscopical studies of the microflora of roots of sand dune grasses. *New Phytol.*, 74, 51.

Rovira, A.D. (1969). Diffusion of carbon compounds away from wheat roots. *Aust. J. Biol. Sci.*, 22, 1287.

Vancura, V. (1964). Root exudates of plants. I. Analysis of root exudates of barley and wheat in their initial phases of growth. *Plant Soil*, 21, 231.

Warembourg, F.R. (1975). Application de techniques radioisotopiques à l'étude de l'activité biologique dans la rhizosphère des plantes. *Rev. Ecol. Biol. Sol Coll. Ecologie du sol*, 12, 261.

Warembourg, F.R. and Paul, E.A. (1977). Seasonal transfers of assimilated ^{14}C in grassland: Plant production and turnover, soil and plant respiration. *Soil Biol. Biochem.*, 9, 295.

ECOPHYSIOLOGY OF VESICULAR-ARBUSCULAR MYCORRHIZA IN THE TROPICS

M. MOAWAD

Institut für Tropischen und Subtropischen Pflanzenbau, Göttingen, Germany

Summary

Mycorrhizal plants of *Eupatorium odoratum* supplied with the insoluble phosphate $Ca_5(PO_4)_3OH$ responded to soil temperature with a steep rise in growth from 20°C to 30°C and a fall from 30°C to 35°C, whereas non-mycorrhizal plants showed no response. Soil temperature seems to affect the physiological activity of the mycorrhiza more than its development, since plants kept at the optimum temperature (30°C) for 20 days before being moved to lower or higher temperatures showed the same inhibition of growth as plants kept permanently at the unfavourable temperatures. Diurnally changing soil temperatures with 35°C during the day only was not harmful to the functioning of mycorrhiza if followed by lower night temperatures. Contrary to *E. odoratum*, non-mycorrhizal *Guizotia abyssinica* and *Sorghum bicolor* did respond to temperature in the presence of insoluble phosphate although less so than mycorrhizal plants.

Shading had no effect on non-mycorrhizal plants, but decreased the efficacy of the mycorrhiza. In soils of different pH values, the influence of some types of V/A mycorrhiza on growth and P uptake of several plant species varied considerably.

V/A mycorrhiza improved the utilization of water by plants. The amount of water used to produce 1 g of dry matter in a soil fertilized with insoluble $Ca_5(PO_4)_3OH$ was much smaller in mycorrhizal than in non-mycorrhizal plants.

Introduction

Phosphorus in tropical soils - predominantly acidic latosols - is largely fixed as ferri- and/or aluminium phosphate and is not available to plants. Improved growth of crops in the presence of V/A mycorrhiza has been demonstrated many times and is greatest in soils of low fertility (Mosse, 1973) containing little available phosphate (Gerdemann, 1964; Daft and Nicolson, 1966; Murdoch, Jackobs and Gerdemann, 1967; Mosse and Hayman, 1971; Mosse, Hayman and Arnold, 1973; Sieverding, 1975; Mikhail, 1976); mycorrhizal plants usually contain more phosphorus than comparable non-mycorrhizal plants. Since V/A mycorrhizae are very widely distributed (Gerdemann, 1975), their study is most important to plant nutrition. The aim of our investigations was to find out which agronomic methods should be employed to exploit fully V/A mycorrhiza in phosphorus nutrition of tropical crops. We studied several factors controlling the activity of the fungi, especially soil temperature, soil pH, and water regime and the interactions of fertilization and light intensity with soil temperature. Only the effects on dry weight and P uptake are discussed in this paper.

Materials and Methods

Soil temperature

Experiments (Nyabyenda, 1977) were conducted in a glasshouse and in growth chambers. The plants *Eupatorium odoratum* L., a weed of warm regions, *Guizotia abyssinica* Cass., an oil plant, and *Sorghum bicolor* Moench were inoculated with surface sterilized roots of *Eupatorium* mother plants containing mycelium and vesicles of *Glomus macrocarpus* and transplanted into one-kg plastic pots filled with steamed (95°C for 2 h) acidic soil/sand mixture (pH 5.5 in 0.01 M

$CaCl_2$) with low phosphate availability. Besides the insoluble phosphates $Ca_5(PO_4)_3OH$, $FePO_4$, $AlPO_4$ and the soluble $Ca(H_2PO_4)_2 \cdot H_2O$, the soil was also fertilized with 66.6 mg N/kg for *E.odoratum* and 33.3 mg N/kg for *G.abyssinica* and *S.bicolor* and with 33 mg K/kg. Soil temperatures of 20°, 25°, 30°, 35°C were maintained in water baths regulated by thermostats.

After harvesting, representative root samples were stained with trypan blue in lactophenol (Phillips and Hayman, 1970) for the microscopic examination of mycorrhiza.

The dried plant material was digested in conc. H_2SO_4 with a few drops of H_2O_2. Phosphate was determined colorimetrically using molybdenum-blue (Fogg and Wilkinson, 1958).

Light intensity

In the glasshouse experiment (Nyabyenda, 1977), the above soil temperatures were combined with light intensities reduced by cloth to about 1/4 to 1/2 of the natural intensity.

Soil pH

Experiments were carried out (Graw, 1978) in the glasshouse with *Tagetes minuta* L. and *Guizotia abyssinica* Cass. A soil/quartz-sand mixture was treated with $CaCO_3$ to give pH values of 4.3, 5.6 and 6.6. The soils were then steamed, fertilized, and the plants inoculated as before.

Water regime

Experiments were carried out (Sieverding, 1975; and in preparation) with *Eupatorium odoratum* L., *Tagetes erecta* L., *Guizotia abyssinica* Cass. and *Sorghum bicolor* Moench in the glasshouse. The soil/sand mixture for investigating soil temperatures (field capacity 18%) and another acidic sandy soil treated with $CaCO_3$ (pH 5.4 in 0.01 M $CaCl_2$-field capacity 25%) were used. Fertilization and inoculation of the plants before transplanting into two-kg pots were carried out as for soil temperatures. The soils were watered to field capacity when the water content had dropped to 20% and 80% in the first experiment, and to 10%, 50% and 90% of available water in the second.

Results and Discussion

Effect of soil temperature

With insoluble phosphates, *E.odoratum* without mycorrhiza was unaffected by soil temperature between 20°C and 35°C (Figs. 1 and 2) and P uptake was uniformly low. Mycorrhizal plants, on the other hand, showed a very strong response with a steep rise from 20°C to 30°C and a fall from 30°C to 35°C. Growth of mycorrhizal plants was closely correlated with P uptake which was much higher than with non-mycorrhizal plants. When soluble phosphate was supplied, mycorrhiza showed a smaller effect on growth and P uptake. At the lowest temperature (20°) the non-mycorrhizal plants grew better and took up more phosphate than

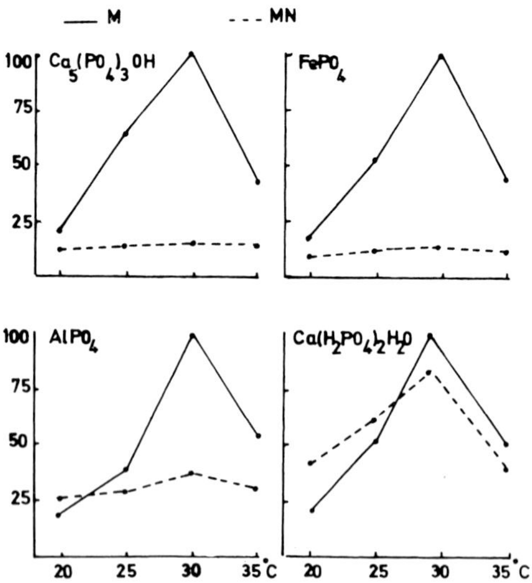

Fig. 1. Dry weight of *E.odoratum* with (M) and without (NM) mycorrhiza at different soil temperatures fertilized with several P forms. Relative values, dry weight of mycorrhizal plants at 30°C = 100.

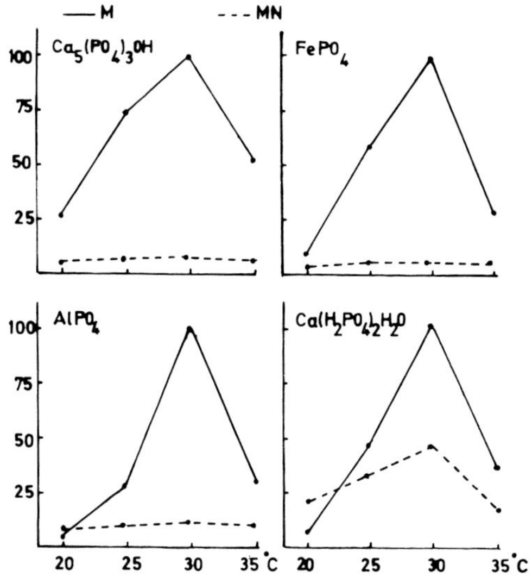

Fig. 2. P content of *E.odoratum* with (M) and without (NM) mycorrhiza at different soil temperatures fertilized with several P forms. Relative values, P content of mycorrhizal plants at 30°C = 100.

mycorrhizal, whereas, at the highest temperatures (30°C and 35°C), the mycorrhizal plants were always superior, although the effect was not as striking as with insoluble phosphate. Somewhat simalar results have been obtained by Hayman (1974) and Furlan and Fortin (1973) with onion and by Slankis (1974) and Schenk and Schroder (1974) with other plants.

In contrast to *E.odoratum*, Nyabyenda (1977) found that non-mycorrhizal plants of *G.abyssinica* and *S.bicolor* did respond to temperature in the presence of insoluble $Ca_5(PO_4)_3OH$, *G.abyssinica* less than *S.bicolor* (Figs. 3 and 4). The optimum soil temperature for both plants with and without mycorrhiza was 30°C. At temperatures of 25°C and higher, both species absorbed considerable amounts of phosphorus from insoluble compounds in the absence of mycorrhiza, sorghum more efficiently than *Guizotia* whilst mycorrhiza further increased P uptake and growth, particularly at 25°C and 30°C. There is an indication that two mechanisms with different response to temperature are in-

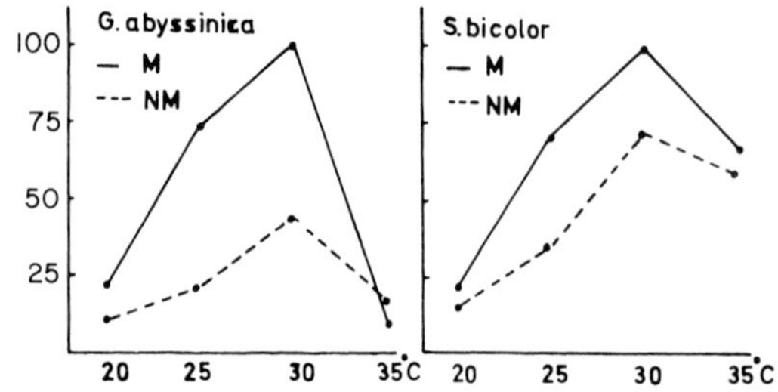

Fig. 3. Dry weight of *G.abyssinica* and *S.bicolor* with (M) and without (NM) mycorrhiza, fertilized with $Ca_5(PO_4)_3OH$ at different soil temperatures. Relative values, dry weight of mycorrhizal plants at 30°C = 100.

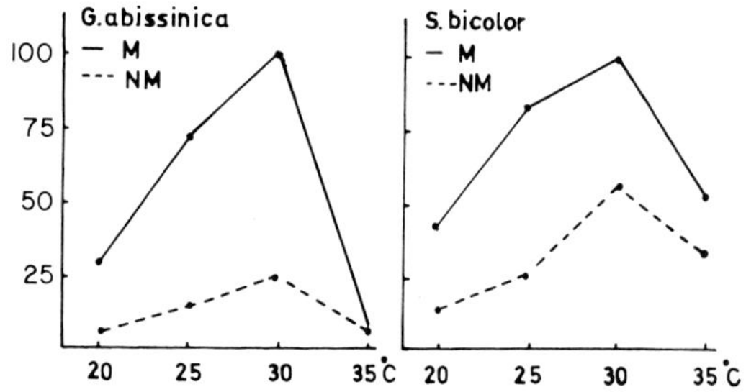

Fig. 4. P content of *G.abyssinica* and *S.bicolor* with (M) and without (NM) mycorrhiza fertilized with $Ca_5(PO_4)_3OH$ at different soil temperatures. Relative values, P content of mycorrhizal plants at 30°C = 100.

volved in P uptake since the increase in P uptake and growth from 20°C to 25°C was much steeper in mycorrhizal than in non-mycorrhizal plants. The much sharper drop in P uptake and growth of sorghum from 30°C to 35°C also points to differences between non-mycorrhizal and mycorrhizal plants. The adaptation of *G.abyssinica* to relatively low temperatures led to very poor growth at 35°C. Temperature seems to affect the physiological activity of the mycorrhiza more than its development, since plants kept at the optimum temperature (30°C) for 20 days before being moved to lower or higher temperatures showed the same influence of temperature on growth and P uptake as plants kept from the beginning at the different temperatures.

In an experiment with diurnally changing temperatures

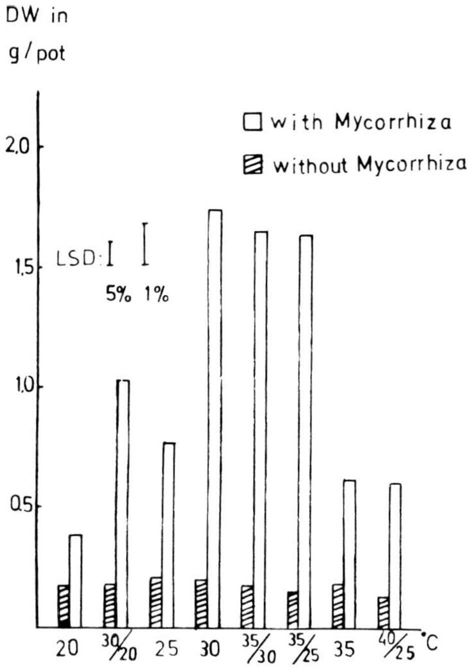

Fig. 5. Total dry weight of *E.odoratum* with and without mycorrhiza supplied with $Ca_5(PO_4)_3OH$ at constant and diurnally changed soil temperatures.

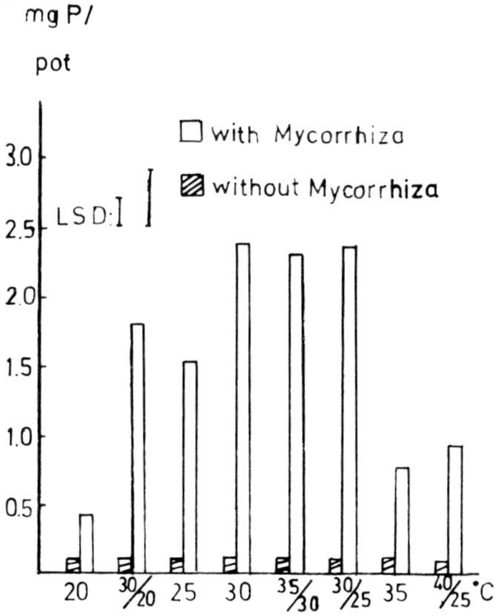

Fig. 6. Total P content of *E.odoratum* with and without mycorrhiza supplied with $Ca_5(PO_4)_3OH$ at constant and diurnally changed soil temperatures.

(30/20, 35/30, 35/25, 40/25) 35°C proved not to be harmful to the development and physiological activity of the mycorrhiza if this was given only during the day, followed by night temperatures of 25°C or 30°C (Figs. 5 and 6).

Effect of light intensity

Boullard (1963); Baylis (1964); Hayman (1974) and Redhead (1975) have reported contradictory results as to the effect of light on mycorrhiza activity. Nyabyenda (1977) found that shading had no effect on non-mycorrhizal, but decreased the efficiency of mycorrhizal *E. odoratum* plants. At 30°C, shading had less effect on mycorrhizal plants (growth and P uptake) than at 25°C and 35°C (Table 1).

TABLE 1

P content (mg/pot) of E.odoratum *without (NM) and with (M) mycorrhiza at different soil temperatures and light intensities NL = natural light intensity*

°C	light	total P content		
		NM	M	$\frac{M}{NM}$
	NL	0.09	1.73	19.2
25	1/2 NL	0.09	1.24	13.7
	1/4 NL	0.08	0.77	9.6
	NL	0.09	2.76	30.7
30	1/2 NL	0.09	2.15	23.9
	1/4 NL	0.09	2.16	24.0
	NL	0.09	1.09	12.1
35	1/2 NL	0.09	0.93	10.3
	1/4 NL	0.09	0.57	6.3
		LSD 5%	0.16	
		1%	0.22	

Effect of soil pH

Graw (1977) found in two species of the same family a different influence of soil pH on the effect of V/A mycorrhiza on growth and P uptake when P was supplied as $Ca_5(PO_4)_3OH$.

Mycorrhizal *Tagetes minuta* grew better and took up more phosphate at all pH's than non-mycorrhizal (Figs. 7 and 8), the difference being greatest at pH 4.3. The availability of $Ca_5(PO_4)_3OH$ decreases with rising soil pH (Fassbender and Ulrich, 1966; Fassbender *et al.*, 1966; Mattingly and Talibudeen, 1967); correspondingly, the growth and P uptake of mycorrhizal *T.minuta* decreased at higher pH-values.

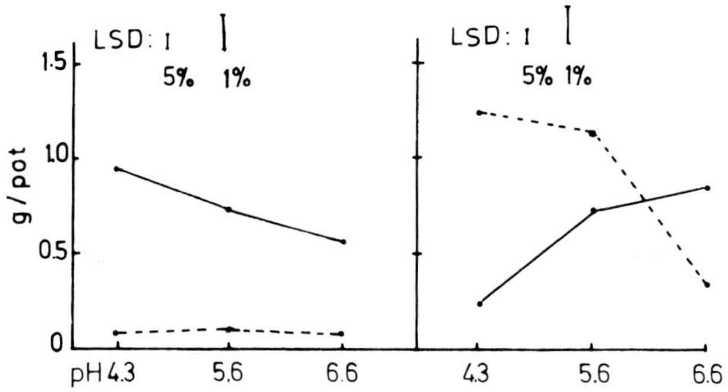

Fig. 7. Dry weight (g/pot) of non-mycorrhizal (NM) and mycorrhizal (M) *T.minuta* (left) and *G.abyssinica* (right) supplied with $Ca_5(PO_4)_3OH$ at different soil pH values. — M; --- NM.

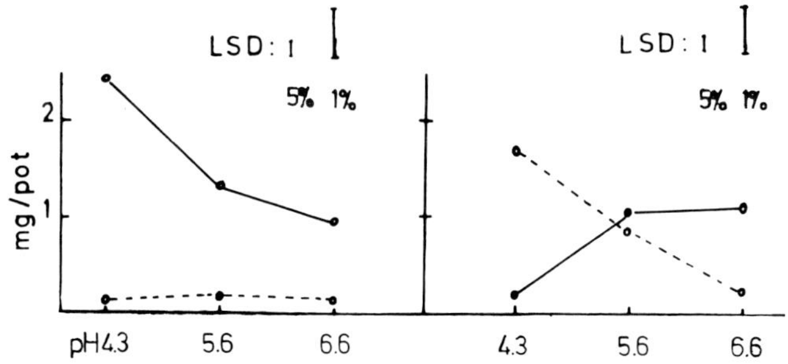

Fig. 8. P content (mg P/pot) of non-mycorrhizal (NM) and mycorrhizal (M) *T.minuta* (left) and *G.abyssinica* (right) supplied with $Ca_5(PO_4)_3OH$ at different soil pH values. — M; --- NM.

In contrast to *T.minuta*, mycorrhizal *G.abyssinica* grew better and took up more phosphate at pH 5.6 and 6.6 than at pH 4.3. Non-mycorrhizal *G.abyssinica* grew best at pH 4.3; at this pH, mycorrhizal plants were much inferior to non-mycorrhizal, but the reason for this is not clear.

Increasing the soil pH to 6.6 decreased the dry weight and P content of non-mycorrhizal plants, but the positive effect of the mycorrhiza was slightly increased.

The different reactions to soil pH in presence or absence of V/A mycorrhiza of the tropical species examined cannot be explained as yet. Apart from physiological processes in higher plant or fungus, changes in solubility or adsorption on soil particles of PO_4 play a role, as experiments with other P compounds demonstrated.

Effect of water regime

With *E.odoratum* and *T.erecta*, Sieverding (1975) found that the amount of water used to produce 1 g of dry matter in the soil/sand mixture fertilized with $Ca_5(PO_4)_3OH$ was much smaller in mycorrhizal than in non-mycorrhizal plants (Table 2). The plants generally used more water when $Ca_5(PO_4)_3OH$ was applied than with $Ca(H_2PO_4)_2 \cdot H_2O$. With $Ca_5(PO_4)_3OH$, the mycorrhizal effect was greatest in the drier soil. According to Safir, Boyer and Gerdemann (1971), V/A mycorrhiza decreased the resistance to water transport in intact soya bean plants, apparently in the roots (Safir, Boyer and Gerdemann, 1972). Daft and Okusanya (1973) found that mycorrhizal infection increased

TABLE 2

Water consumption in ml/g dry weight of E.odoratum *and* T.erecta *at two water regimes and with two P forms*

species	treatment	$Ca(H_2PO_4)_2 \cdot H_2O$ 80%	$Ca(H_2PO_4)_2 \cdot H_2O$ 20%	$Ca_5(PO_4)_3OH$ 80%	$Ca_5(PO_4)_3OH$ 20%
E.odoratum	NM	1208	1207	2860	4112
	M	1237	1177	1574	1436
T.erecta	NM	1073	1005	2563	3397
	M	923	1060	1180	1424

NM: not inoculated with V/A mycorrhiza; M: inoculated.
Soil water contents expressed as % available water before rewatering.

the amount of vascular tissue in tomato, petunia and maize stems; they believed that this was an indirect effect resulting from the greater uptake of phosphorus. The effect found by Sieverding need not be explained by improved uptake or transport of water; it may simply be due to the better utilization of water by plants not suffering from P deficiency.

In other experiments with *S.bicolor*, *G.abyssinica* and *E.odoratum* in the soil/sand mixture and in the acidic sandy soil, Sieverding (in preparation) found a similar trend to improved water utilization by mycorrhizal plants receiving insoluble $Ca_5(PO_4)_3OH$, particularly at 10% available water in the soil/sand mixture where the mycorrhiza was better developed than at the higher water regimes. This was not observed in the acidic sandy soil, indicating that the response of mycorrhiza to water supply is influenced by the soil type. Here, as with other factors, we are only at the beginning of understanding the complex system of soil/mycorrhiza/plant.

References

Baylis, G.T.S. (1967). Experiments on the ecological significance of phycomycetous mycorrhizas. *New Phytol.*, 66, 231.

Boullard, B. (1963). Licht und Mykorrhiza. *In* "Internationales Mykorrhiza-Symposium" (Ed. H. Lyr), 315pp. Weimar 1960. Fischer-verlag, Jena.

Daft, M.J. and Nicolson, T.H. (1966). Effect of endogone mycorrhiza on plant growth. *New Phytol.*, 65, 343.

Daft, M.J. and Okusanya, B.O. (1973). Effect of endogone mycorrhiza on plant growth. VI. Influence of infection on the anatomy and reproductive development in four hosts. *New Phytol.*, 72, 1333.

Fassbender, H.W. and Ulrich, B. (1966). Lösungsgeschwindigkeiten von Rohphosphaten. *Z. Pflanzenernähr. Düng. Bodenkunde*, 112, 212.

Fassbender, H.W., Lin, H.C. and Ulrich, B. (1966). Löslichkeit und Löslichkeitsprodukt von Hydroxylapatit und Rohphosphaten. *Z. Pflanzenernähr. Düng. Bodenkunde*, 112, 101.

Fogg, D.N. and Wilkinson, N.T. (1958). The colorimetric determination of phosphorus. *Analyst*, 83, 803.

Furlan, V. and Fortin, J.A. (1973). Formation of endomycorrhizae by *Endogone calospora* on *Allium cepa* under three temperature regimes. *Naturaliste Canad.*, 100, 467.

Gerdemann, J.W. (1964). The effect of mycorrhiza on growth of maize. *Mycologia*, 56, 342.

Gerdemann, J.W. (1975). Vesicular-arbuscular mycorrhizas. *In* "The Development and Function of Roots" (Ed. I.G. Torrey and D.T. Clarkson), p.575. Academic Press, London.

Graw, D. (1978). "Der Einfluss der vesikulär-arbuskulären Mykorrhiza auf das Wachstum tropischer und subtropischer Pflanzen unter Berücksichtigung von Wirtsspezifität und pH-Wert des Bodens". Dissertation, Göttingen.

Hayman, D.S. (1974). Plant growth responses to vesicular-arbuscular mycorrhiza. VI. Effect of light and temperature. *New Phytol.*, 73, 71.

Mattingly, G.E.G. and Talibudeen, O. (1967). Progress in the chemistry of fertilizer and soil phosphorus. *In* "Topics in Phosphorus Chemistry" (Ed. M. Grayson and E.J. Griffith), p.157. Interscience, New York.

Mikhail, N.B. (1976). "Der Einfluss symbiontischer Mikroorganismen auf Wachstum und Nährstoffaufnahme von *Eupatorium odoratum* L. bei Angebot verschiedener Phosphatformen". Dissertation, Göttingen.

Mosse, B. and Hayman, D.S. (1971). Plant growth responses to vesicular-arbuscular mycorrhiza. II. In unsterilized soils. *New Phytol.*, 70, 29.

Mosse, B., Hayman, D.S. and Arnold, D.J. (1973). Plant growth responses to vesicular-arbuscular mycorrhiza. V. Phosphate uptake by three plant species from P deficient soils labelled with ^{32}P. *New Phytol.*, 72, 809.

Mosse, B. (1973). Advances in the study of vesicular-arbuscular mycorrhiza. *Ann. Rev. Phytopathol.*, 11, 171.

Murdoch, C.L., Jackobs, J.A. and Gerdemann, J.W. (1967). Utilization of phosphorus sources of different availability by mycorrhizal and non-mycorrhizal maize. *Plant and Soil*, 27, 329.

Nyabyenda, P. (1977). "Einfluss der Bodentemperatur und organischer Stoffe im Boden auf die Wirkung der vesikulär-arbuskulären Mykorrhiza". Dissertation, Göttingen.

Phillips, J.M. and Hayman, D.S. (1970). Improved procedures for clearing roots and staining parasitic and vesicular-arbuscular mycorrhizal fungi for rapid assessment of infection. *Trans. Br. mycol. Soc.*, 55, 158.

Redhead, J.F. (1975). Endotrophic mycorrhizas in Nigeria: Some aspects of the ecology of endotrophic mycorrhizal association of *Khaya grandifoliola* C. DC. *In* "Endomycorrhizas" (Eds. F.E. Sanders, B. Mosse and P.B. Tinker), p.447. Academic Press, London.

Safir, G.R., Boyer, J.S. and Gerdemann, J.W. (1971). Mycorrhizal enhancement of water transport in soybean. *Science*, 172, 581.

Safir, G.R., Boyer, J.S. and Gerdemann, J.W. (1972). Nutrient status and mycorrhizal enhancement of water transport in soybean. *Plant Physiol.*, 49, 700.

Schenck, N.C. and Schroder, V.N. (1974). Temperature response of Endogone mycorrhiza on soybean roots. *Mycologia*, 66, 600.

Sieverding, E. (1975). "Einfluss von symbiontischen Pilzen auf Wachstum und Nährstoffaufnahme von *Eupatorium odoratum* L. und *Tagetes erecta* L. bei verschiedenen Wasserregimen und Phosphatformen im Boden". Diploma Thesis, Institut für Tropischen und Subtropischen Pflanzenbau, Göttingen.

Sieverding, E. "Influence of water regime and soil temperature on the efficiency of V/A mycorrhiza". Dissertation, Göttingen (in

preparation).
Slankis, V. (1974). Soil factors influencing formation of mycorrhizae. *Ann. Rev. Phytopathol.*, 12, 437.

EFFECTS OF SOIL MOISTURE ON V/A MYCORRHIZA FORMATION AND ROOT DEVELOPMENT IN *MEDICAGO*

C.P.P. REID* and G.D. BOWEN**

**Department of Forest and Wood Sciences, Colorado State University, Fort Collins, Colorado, USA*

Summary

Barrel medic (*Medicago truncatula*) was grown in Mallee Loam soil containing yellow-vacuolate spores of Endogonaceae at approximately constant soil moistures. Upon harvest at 8 days, primary root length, length of laterals, and number of laterals were all greatest at a water content of 0.22 g g^{-1} and decreased above and below this. The number of V/A infections per unit root length at 0.22 g g^{-1} was more than 4 times greater than at 0.15 g g^{-1} (0.43 -MPa) and almost twice as great as at 0.28 g g^{-1} (saturation). Although primary root diameter decreased with decreasing water content, root hair length increased. The volume of soil occupied by root hairs per unit length of root was calculated to be 3.3 times greater at a water content of 0.09 (1.4 -MPa) than at 0.22 (0.19 -MPa). Root hairs were absent at saturation.

Introduction

The effect of soil-moisture stress on the establishment and function of mycorrhizas has been little investigated.

The studies reported here examine the effect of specific

*** Division of Soils, CSIRO, Glen Osmond, S. Australia*

levels of soil moisture on V/A mycorrhiza formation and root growth when soil water contents of known soil Ψ were maintained relatively constant throughout the course of the experiments.

Methods

Soil and plant establishment

Air dry Mallee Loam soil containing yellow-vacuolate type (YV) (*Glomus mosseae* Gerd. and Trappe) spores was mixed to specific water contents (Θ) of 0.30, 0.20, 0.165, 0.12 and 0.09 g g^{-1} by spreading in thin layers, adding known quantities of water by atomizer, and thoroughly mixing. Water content of the soil was determined gravimetrically at time of placing soil in polythene tubes and again at time of plant harvest. Soil water potential (Ψ) was measured concurrently by using calibrated Spanner-type thermocouple psychrometer chambers in a constant temperature room (20°C). Polythene tubes (2.9 x 20 cm) were filled with each premixed soil to a depth of 12 cm and at a bulk density of 1.2. Seeds of barrel medic (Jemalong) (*Medicago truncatula* Gaertn.) were surface sterilized, germinated on nutrient agar, and transplanted into each soil tube when the radicles were about 5 mm in length (2 per tube). Planted tubes were maintained in a growth chamber under a 14-h photoperiod with a 20°C day, 15°C night.

Plant harvest

On the 8th day after planting, the plants from 3 soil tubes of each of the 5 moisture treatments were harvested (6 plants/treatment). Two, 2g soil samples, were collected for determinations of water content and soil Ψ. The developmental stage of the shoot was noted and various root parameters were measured (see Table 1). Vesicular-arbuscular (V/A) infection was determined by the method of Phillips and Hayman (1970). The number of infections in both primary and lateral roots of the entire root system of each plant was examined with a dissecting microscope.

Because of the lag in shoot development at the lower soil water contents, sets of 3 soil tubes with initial water contents of 0.12 g g^{-1} and of 0.09 g g^{-1} were harvested at 10 and 13 days, respectively when shoot development was similar to that at 8 days in the other water contents.

Results

Soil moisture

Moisture in the soil tubes varied little during the course of the experiments. A moisture-tension curve was constructed using all paired observations of water content (Θ) and soil Ψ and values of Ψ (Table 1) calculated from the linear regression: $\log \Psi = -2.24 - 2.29 \log \Theta$ (r^2 = .87).

Plant development

Shoot Opening of the spade leaf, and appearance of the first trifoliate leaf were similar for all water contents at the 8 day harvest except at 0.09 g g^{-1}. However, subjectively, shoot development appeared optimum at 0.22 g g^{-1}.

Roots Details of root growth and V/A mycorrhiza infection are shown in Table 1. For convenience, statistical analyses were performed on transformed data. At 8 days, lateral root production and total root length were greatest at 0.22 g g^{-1}, though not significantly different from those at 0.15 and 0.11 g g^{-1}. Total root length at these three moisture contents was some 60-95 per cent greater than at 0.28 and 0.09 g g^{-1}. Retardation of growth of primary and lateral roots at the two moisture extremes was marked, but not always significantly different from other treatments due to root variability. We consider the effects of soil moisture on lateral root production were related essen-

TABLE 1

Soil moisture effects on root development and V/A mycorrhiza formation. Statistically significant differences between soil moisture treatments for each measured characteristic are indicated by letters; values followed by the same letter are not significantly different ($P < 0.05$)

						Harvest at 10 days	Harvest at 13 days
Water Content (Θ, g g^{-1})*	0.28	0.22	0.15	0.11	0.09	0.13	0.10
Soil Water Potential (-MPa)	>0.10	0.19	0.43	0.91	1.4	0.63	1.1
Root Length (cm):							
Primary	6.0^{a+}	10.5^{ab}	8.3^{ab}	10.0^{ab}	7.2^{ab}	12.3^{b}	8.7^{ab}
Laterals	8.7^{ab}	18.6^{c}	15.5^{ac}	14.1^{ac}	5.4^{b}	23.4^{c}	13.5^{ac}
Total	15.5^{a}	30.2^{bc}	25.7^{bc}	24.5^{bc}	13.5^{a}	36.3^{b}	23.4^{c}
Number of Laterals	6.0^{a}	13.8^{b}	7.4^{ab}	13.2^{ab}	8.7^{ab}	16.6^{b}	13.2^{ab}
V/A Infection per cm root:							
Primary	1.7^{a}	4.1	1.3^{ab}	0.5^{bc}	0.1^{c}	1.3^{ab}	0.5^{bc}
Laterals	1.7^{ab}	2.8^{a}	0.9^{c}	0.1^{de}	0.1^{d}	1.1^{bc}	0.5^{ce}
Total	2.4^{a}	4.4	1.0^{b}	0.5^{bc}	0.1^{c}	1.5^{a}	0.7^{b}
Root Hair Length (μm)	-	80	150	220^{a}	240^{a}	220^{a}	210^{a}
Root Diameter (μm)	-	550^{a}	510^{ac}	430^{b}	450^{b}	470^{bc}	380
Root Hair Zone Distance from Apex (mm)	-	3.3^{ab}	11.2	3.8^{a}	3.6^{a}	3.4^{a}	2.4^{b}
Number of Nodules	1.7^{a}	6.8^{bc}	4.7^{bc}	3.6^{ab}	1.8^{a}	7.2^{c}	3.0^{a}

* Mean water content at time of harvest.

+ Statistical analysis was conducted on log transformations; values presented are back transformations (antilogs) of the means.

tially to primary root growth and fundamental changes in root initiation with moisture content did not occur. Since root length varied somewhat with soil moisture, V/A infections are expressed per unit root length to correct for this. It is readily apparent that a moisture content of 0.22 g g^{-1} was optimum for V/A mycorrhizal development and that both saturated and drier conditions markedly reduced infection. For the total root, the number of infections/cm was significantly greater at 0.22 g g^{-1} when compared to all other water contents, even for those plants harvested at 10 and 13 days. At 8 days, infections/cm at 0.22 g g^{-1} were more than four times as numerous as those at 0.15 g g^{-1} and almost twice as numerous as those at 0.28 g g^{-1}. A slightly greater number of infections/cm in the primary root as compared to the laterals probably reflects the time required for infection to take place.

There was a significant decrease in primary root diameter as moisture content decreased, whereas root hair length significantly increased. Root hairs were not present on roots in the saturated soil and root diameter was not measured. When correction is made for the overall size of the root system, the response of root diameter and root hair length to water content remains significant. The number of root nodules was also affected by water content with the maximum number occurring at 0.22 g g^{-1}.

Discussion

Although effects of soil moisture on V/A mycorrhizal infection have been suggested from field and greenhouse observations (Khan, 1974; Trinick, 1977), this is the first experimental study of this problem. The data show that quite moderate moisture changes have extremely large effects on infection (Table 1) and thus could affect mycorrhiza function considerably. Infection recorded here is the resultant of at least three processes, *viz.* spore germination, growth of the fungus to the root and the infection process. Currently we are developing methods to examine these three processes separately. Although little is known about the capacity of Endogonaceae to tolerate water stress, stress could reduce spore germination or hyphal growth in the soil and result in less infection of the root. However, it is known that some groups of fungi can tolerate quite low water potentials (c.f., Griffin, 1963; Mexal and Reid, 1973; Theodorou, 1978). The plant's water status might also affect fungal infection through alterations in the cortex which inhibit fungal penetration (*e.g.* suberization) or through changes in root-produced stimuli, *e.g.*, the quality and quantity of root exudates can differ with different substrate water potentials (Reid, 1974; Reid and Mexal, 1977). An increase in the secretion of mucilage and release of carbon by lysis from roots may also result from increases in soil water stress (Martin, 1977; Sprent, 1975).

If water uptake is an important function of hyphae of V/A mycorrhizas, the reduction of mycorrhizal establishment during periods of low soil Ψ could affect root water uptake. Regardless of whether or not V/A mycorrhizas have a significant impact on water uptake by the plant, the reduction in infection at lower soil Ψ could have a substantial effect on the uptake of mineral ions from the soil (especially immobile ions such as phosphate). Undoubtedly more information is required on the effect of water stress on spore germination, fungal growth in the soil, and the effect of soil moisture stress on the function of mycorrhizas previously established under favourable moisture conditions.

The effect of water content on root hair length and root diameter is of particular interest. When root and root hair cylinder volumes are calculated using root diameters and root hair lengths in Table 1, it is obvious that the effective volume occupied by the root plus root hairs has increased under reduced soil moisture even though root diameter has decreased (Fig. 1). The volume occupied by the root and root hairs at a water content of 0.09 g g^{-1} is 1.7 times greater than at 0.22 g g^{-1}, while the volume of the root hair cylinder alone is 3.3 times greater. Root hair density did not appear greatly different in the various treatments, but no measurements were made. Therefore, discussion assumes similar root hair numbers at each treatment. The reduction in root diameter with a decrease either in plant Ψ or soil Ψ has been observed previously (Cole and Alston, 1974; Huck, Klepper and Taylor, 1970). However, the expected response of root hairs to reductions in soil water content is less clear.

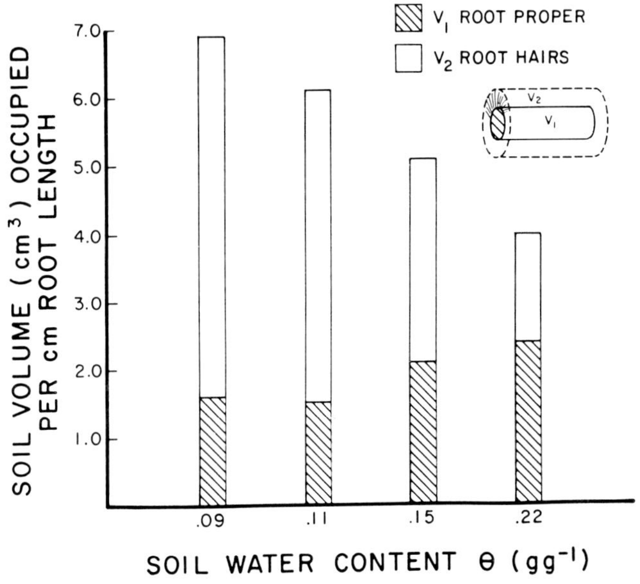

Fig. 1. Effect of water content on the volume of soil occupied by the root proper and root hair zone. Volumes calculated from data on root diameter and root hair length in Table 1.

One might expect greater root hair development in moist, well-aerated soils (Kramer, 1969; Reynolds, 1975), although Newman (1974) acknowledges that more direct evidence is needed. Increases in root hair length with reduced soil moisture might be an adaptation for enhancement of water uptake by the root under drier conditions, although the way in which root hairs contribute to water uptake from soil is unclear. There is no clear evidence that root hairs increase absorption of water directly (Kramer, 1969; Newman, 1974). However, it has been suggested that root hairs are important to water uptake, indirectly, by preventing the development of significant gaps between the root and soil, thereby maintaining liquid continuity across the soil-root interface (Tinker, 1976; Weatherly, 1975). Since roots may produce increased mucilage under reduced soil moisture, this may further aid in binding soil particles and reduce the probability for gap development around the root. Apart from the possible importance in water uptake, root hair length has certainly been shown to be important in ion uptake from soils (Bhat and Nye, 1973; Bhat, Nye and Baldwin, 1976). With poorly mobile ions, *e.g.* phosphate and zinc, increased root hair length with lower soil moistures demonstrated here may more than compensate for lower diffusion rates of ions with decreasing soil moisture.

Acknowledgements

This work was made possible, in part, by a grant to one of us (CPPR) from the National Science Foundation under the U.S.-Australia Cooperative Science Program. Statistical analyses by Dr. Ray Correll and the technical assistance of Rosemary Warren are gratefully acknowledged.

References

Bhat, K.K.S. and Nye, P.H. (1973). Diffusion of phosphate to plant roots in soil. I. Quantitative autoradiography of the depletion zone. *Pl. Soil*, 38, 161.

Bhat, K.K.S. and Nye, P.H. (1974). Diffusion of phosphate to plant roots in soil. II. Uptake along the roots at different times and the effect of different levels of phosphorus. *Pl. Soil*, 41, 365.

Bhat, K.K.S., Nye, P.H. and Baldwin, J.P. (1976). Diffusion of phosphate to plant roots in soil. IV. The concentration distance profile in the rhizosphere of roots with root hairs in a low-P soil. *Pl. Soil*, 44, 63.

Cole, P.J. and Alston, A.M. (1974). Effect of transient dehydration

on absorption by wheat roots. *Pl. Soil*, **40**, 243.
Griffin, D.M. (1963). Soil moisture and the ecology of soil fungi. *Biol. Rev.*, **38**, 141.
Huck, M.G., Klepper, B. and Taylor, H.M. (1970). Diurnal variations in root diameter. *Pl. Physiol.*, **45**, 529.
Khan, A.G. (1976). The occurrence of mycorrhizas in halophytes, hydrophytes and xerophytes and of Endogone spores in adjacent soils. *J. gen. Microbiol.*, **81**, 7.
Kramer, P.J. (1969). "Plant and Soil Water Relationships: A Modern Synthesis". McGraw-Hill, New York.
Martin, J.K. (1977). Effect of soil moisture on the release of organic carbon from wheat roots. *Soil Biol. Biochem.*, **9**, 303.
Mexal, J.G. and Reid, C.P.P. (1973). The growth of selected mycorrhizal fungi in response to induced water stress. *Can. J. Bot.*, **51**, 1579.
Newman, E.I. (1974). Root and soil water relations. *In* "The Plant Root and its Environment" (Ed. E.W. Carson), p.363. University Press of Virginia, Charlottesville.
Phillips, J.M. and Hayman, D.S. (1970). Improved procedures for clearing roots and staining parasitic and vesicular-arbuscular mycorrhizal fungi for rapid assessment of infection. *Trans. Br. mycol. Soc.*, **55**, 158.
Reid, C.P.P. (1974). Assimilation, distribution and root exudation of ^{14}C by ponderosa pine seedlings under induced water stress. *Pl. Physiol.*, **54**, 44.
Reid, C.P.P. and Mexal, J.G. (1977). Water stress effects on root exudation by lodgepole pine. *Soil Biol. Biochem.*, **9**, 417.
Reynolds, E.R.C. (1975). Tree rootlets and their distribution. *In* "The Development and Function of Roots" (Ed. J.G. Torrey and D.T. Clarkson), p.163. Academic Press, London.
Sprent, J.I. (1975). Adherence of sand particles to soybean roots under water stress. *New Phytol.*, **74**, 461.
Theodorou, C. (1978). Soil moisture and the mycorrhizal association of *Pinus radiata* D. Don. *Soil Biol. Biochem.* (in press)
Tinker, P.B. (1976). Roots and water. Transport of water to plant roots in soil. *Phil. Trans. R. Soc. Lond. B.*, **273**, 445.
Trinick, M.J. (1977). Vesicular-arbuscular infection and soil phosphorus utilization in *Lupinus* spp. *New Phytol.*, **78**, 297.
Weatherly, P.E. (1975). Water relations of the root system. *In* "The Development and Function of Roots" (Eds. J.G. Torrey and D.T. Clarkson), p.397. Academic Press, London.

THE INFLUENCE OF AMOEBAE ON THE UPTAKE OF NITROGEN BY PLANTS IN GNOTOBIOTIC SOIL

E.T. ELLIOTT, D.C. COLEMAN* and C.V. COLE**

Natural Resource Ecology Laboratory, Colorado State University, Fort Collins, Colorado, USA

Summary

Gnotobiotic associations of rhizosphere bacteria (*Pseudomonas*) with and without amoebae (*Acanthamoeba*) were grown with blue grama (*Bouteloua gracilis*) grass seedlings, in propylene oxide sterilized soil microcosms to test the hypothesis that when amoebae graze on bacteria plant growth and nitrogen content will increase.

Glucose (600 μg-C) and N amendments [as $(NH_4)_2SO_4$] were added with the microorganism inoculum to provide three initial levels of inorganic soil-nitrogen: low, 45; medium, 95; and high, 245 μg NH_4^+ - N · g^{-1} soil. Sixty days after seedling germination, soil and plant materials were sampled for levels of N and root and shoot biomass. The soils in microcosms containing amoebae with bacteria mineralized significantly (50-100%) more NH_4^+-N per gram of soil than those with bacteria alone, at the medium and high N levels. Shoot N concentration was significantly higher at all N levels when amoebae were present.

These results indicate a significant role of protozoa at

** *Phosphorus Laboratory, Science and Education Administration - Agricultural Research, U.S. Department of Agriculture, Fort Collins, Colorado, USA*

the soil-root interface by accelerating the mineralization of microbially immobilized nutrients.

Introduction

Microflora play an essential role in decomposing plant residues, but compete with primary producers for the uptake of mineralized nutrients. Therefore, the release of nutrients immobilized in microflora tissues by grazers such as amoebae could play an essential role maintaining nutrient levels for plant growth. Elliott and Coleman (1977) have shown that large numbers ($1\text{-}5 \times 10^4 \cdot g^{-1}$ of surface soil) of actively-feeding, bacteriophagic protozoa exist in grasslands, even in dry soils.

Recent experiments using sterilized soil reinoculated with known organisms have helped elucidate the roles of some rhizosphere microorganisms (Coleman *et al.*, 1977). Rates of CO_2 evolution and P_i and NH_4^+ mineralization were greater in microcosms where bacteria and amoebae or bacteria and nematodes were present together than when there were only bacteria (Coleman *et al.*, 1977, 1978; Cole *et al.*, 1978; Woods *et al.*, submitted). This tends to support Johannes' concept (1965, 1968) for aquatic ecosystems of accelerated release of nutrients by bacterial grazers.

There have been many studies of interactions between protozoa and higher plants (Geltzer, 1963; Nikoljuk and Geltzer, 1972), but few controlled studies of microfauna-bacteria-plant interactions have been made; thus, Rosswall, Lohm and Sohlenius (1977) grew pine seedlings in microcosms with and without microbial grazers, with inconclusive results, due to experimental variability.

The present study tests the hypothesis that plant growth and nutrient concentration would be greater when bacteria and grazing amoebae were present together in soil than when bacteria only were present. We also hypothesized that the difference in plant response between bacteria alone and bacteria plus amoebal treatments would be greater under the most N-limiting conditions.

Materials and Methods

Gnotobiotic soil microcosms were prepared by adding 100 g of soil to 250-ml Erlenmeyer flasks. The soil was moistened to field capacity, the flasks plugged with cotton, and after incubation for two days, were sterilized with propylene oxide for 3 days. The soil lots were dried and then stored.

Bacteria were inoculated into 18 of the sterile soil microcosms and amoebae added to one-half (9) of them. Seeds were added 5 days after bacterial inoculation.

The bacteria (*Pseudomonas cepacia*) and amoebae (*Acanthamoeba polyphaga*) used in this experiment were isolated from the rhizosphere of blue grama, as described by Anderson *et al.* (1978). Microcosms were inoculated with 1 x 10^7 bacteria per gram soil in 15 ml of solution and 1 x 10^4 amoebae per gram of soil in 5 ml of solution 1 day later. Seeds of *Bouteloua gracilis* (H.B.K.) Lag. ex. Steud. were surface sterilized in a 2% solution of sodium hypochlorite for 60 minutes followed by repeated rinsing with sterile distilled water. Ten to twelve seeds were added to each microcosm under aseptic conditions.

We used an Olney clay loam (17% at field capacity) collected from the Pawnee Site, of the Natural Resource Ecology Laboratory, Colorado State University, located on the USDA Science and Education Administration Central Plains Experimental Range in north-eastern Colorado. After sterilizing, the soil contained 45 μg · g^{-1} of NH_4^+-N. Glucose (500 μg C · g^{-1}) and 50 and 200 μg of NH_4^+-N · g^{-1} soil [as $(NH_4)_2SO_4$] were added with the bacterial inoculum. The amoebae were added in 5 ml of proteose-peptone, glucose, and mineral solution (PPGM) bringing the soil water to 20%. The PPGM addition, which was also pipetted into flasks with no amoebal additions, added 3 μg of inorganic

P, 70 μg of proteose-peptone-N, and 100 μg of glucose-C, all · g^{-1} soil. However, since the utilization of organic N in the PPGM was not determined, its amount was not included in the N-flow calculations.

The assembled flasks were maintained in an environmental chamber on a 12-h light (30°C) and 12-h dark cycle (22°C). Soil water was maintained with sterile distilled water. Several 1-cm diameter cores were removed from each flask, extracted with 2 M KCl, 10:1 (v/w), and assayed by the Conway microdiffusion technique (Stanford *et al.*, 1973). Plants were separated from the soil by gently washing roots in a fine mesh screen. Shoot weights were obtained after drying at 60°C for 48 h and root weights were determined as ash-free dry weights. Shoot nitrogen content was measured using a Kjeldahl digest in combination with the Conway microdiffusion technique.

The experiment was a randomized block design with each of three replicates handled as a block. Data were analyzed by an analysis of variance and all differences reported are significant at $P \leq 0.05$.

Results and Discussion

At the end of the growth period (60 days), we measured total plant yield (Table 1). Total yield (mg · $flask^{-1}$) was significantly affected by fertilizer levels but not biological interactions.

Glucose added initially with the inoculum-stimulated bacterial N uptake in agreement with previous microcosm studies without plants (Coleman *et al.*, 1977; Woods *et al.*, in preparation). Amoebal activity was evident, since at the end of the growth period there was significantly more soil NH_4^+-N in the presence of amoebae (AB treatment) at the two higher N levels than with bacteria alone (B treatments) (Fig. 1) (23.1 vs. 10.6 ppm and 145.1 vs. 111.6 ppm, respectively). The nitrogen content of shoots was increased when amoebae were present as compared to bacteria

TABLE 1

Mean plant yields per flask as mg dry matter (±SD) in the presence of bacteria (B), or bacteria and amoebae (AB)

	N level	Low	Medium	High
Shoot production	B	12.6 ± 2.5	21.7 ± 1.1	31.4 ± 3.5
	AB	11.8 ± 2.2	23.9 ± 4.6	25.3 ± 5.3
Root production	B	7.1 ± 1.0	11.0 ± 1.2	15.4 ± 3.4
	AB	5.9 ± 1.6	10.8 ± 3.3	11.3 ± 1.7
Total plant production	B	19.7 ± 3.1	32.7 ± 2.0	46.8 ± 6.9
	AB	17.7 ± 3.7	34.7 ± 7.6	36.6 ± 6.6

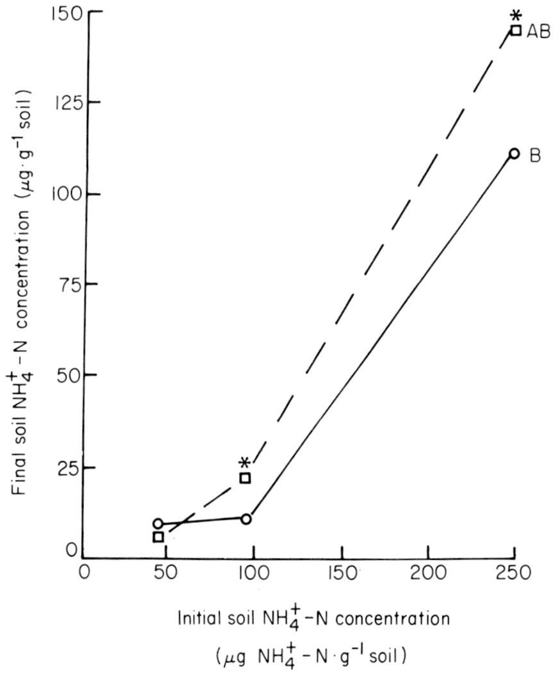

Fig. 1. The soil NH_4^+-N concentration in flasks at the end of the growth period with bacteria alone (B) or with amoebae (AB) at three initial soil NH_4^+-N concentrations. * Significant at $P < 0.05$.

alone at all three N levels (Fig. 2). The percentage N in shoots ranged from 0.8% for the low N, and bacteria alone treatment to 2.3% for the high N amoebae and bacteria treatment. In contrast, plant N uptake showed fewer dif-

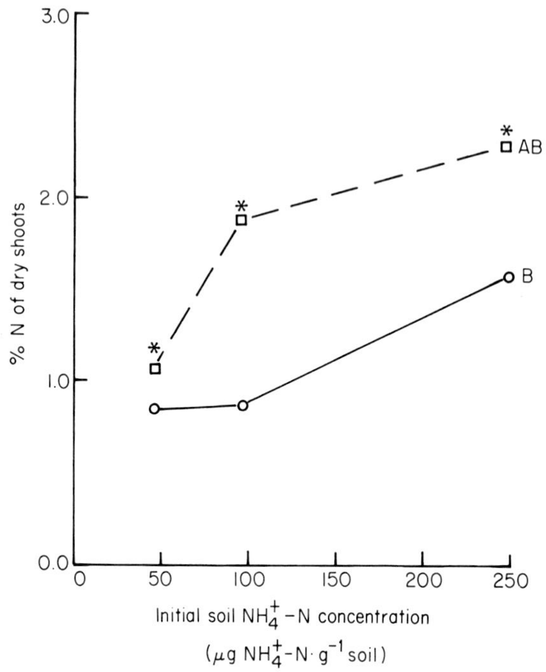

Fig. 2. The per cent N of dry shoots grown in soil in flasks with bacteria alone (B) or with amoebae (AB) at three initial soil NH_4^+-N concentrations. * Significant at $P < 0.05$.

ferences (Table 2) with only the medium N treatment showing significantly raised N levels with amoebae when contrasted to the bacteria-alone controls.

The percentage of net NH_4^+-N uptake by plants (Table 3) was calculated using the equation:

$$\% \text{ net uptake by plants} = \frac{PSNC \cdot (RW + SW) \cdot 100}{N_i - N_f}$$

where PSNC is the plant shoot nitrogen concentration ($\mu g \cdot g^{-1}$ dry wt), RW and SW are root and shoot weight ($\cdot\, g^{-1}$ soil), N_i and N_f are the initial and final soil NH_4^+-N level (μg NH_4^+-N $\cdot\, g^{-1}$ soil). There were no differences between the biological treatments at the low N level but at the medium and high levels plants had a greater per cent net uptake with amoebae.

TABLE 2

Total plant uptake of nitrogen (mg · g^{-1} soil ± SD) as affected by amoebae at different nitrogen levels. (B) bacteria, (AB) amoebae and bacteria present

N level	Low	Medium	High
B	1.65 ± 0.41	2.83 ± 0.24	7.27 ± 1.10
AB	1.93 ± 0.56	6.57 ± 1.40	8.37 ± 1.43

TABLE 3

The effect of amoebal presence on plant N uptake relative to total N uptake by plants and bacteria (± SD) (expressed in per cent). (B) bacteria, (AB) amoebae and bacteria present

N level	Low	Medium	High
B	4.63 ± 1.21	3.37 ± 0.31	5.40 ± 0.57
AB	4.87 ± 1.47	9.10 ± 1.55	8.47 ± 1.79

C/N/P ratios have been used to predict rates of mineralization by bacterial predators (Hunt *et al.*, 1977). The mechanisms for mineralization have been based on the assumptions that bacteria and amoebae have similar C/N ratios. The incorporation of C and N into amoebal tissue and the loss of amoebal respiratory CO_2 forces the excretion of N containing material with a narrower C/N ratio than the bacteria consumed by amoebae. The released NH_4^+-N may be reimmobilized by bacteria or plant roots. Since bacteria are better competitors for NH_4^+ than roots (Goring and Clark, 1948), bacteria will take up a larger portion of available N. When C sources are depleted in the soil, bacteria may be less efficient at NH_4^+ uptake and roots will remove a greater proportion of available N.

Our second hypothesis, which asserted plant N growth responses would be greatest at low N levels, was not supported. Our soil N data showed that NH_4^+-N was low in both biotic treatments (Fig. 2) so plant N uptake did not differ. At the high N level there was a large difference in final soil NH_4^+-N, but as the NH_4^+ in both amoebae-bacteria and bacteria treatments was beyond plant needs, perhaps the difference was inconsequential. However, the plant N production (Table 2) was sensitive to the differences in soil N at the intermediate level, as influenced by the two biotic treatments.

Final soil NH_4^+-N, per cent N of dry shoots, plant N production, and the proportion of the net NH_4^+-N uptake attributed to plants all benefited by amoebal presence. These results indicate a significant role for protozoa at the soil-root interface by accelerating the mineralization of microbially immobilized nutrients.

Acknowledgements

We thank M. Campion and A. Rubink for extensive statistical and technical assistance. This study was supported by NSF Grant DEB-7513468 A02.

References

Anderson, R.V., Elliott, E.T., McClellan, J.F., Coleman, D.C., Cole, C.V. and Hunt, H.W. (1978). Trophic interactions in soils as they affect energy and nutrient dynamics. III. Biotic interactions of bacteria, amoebae, and nematodes. *Microb. Ecol.*, 4 (in press).

Cole, C.V., Elliott, E.T., Hunt, H.W. and Coleman, D.C. (1978). Trophic interactions in soils as they affect energy and nutrient dynamics. V. Phosphorus transformations. *Microb. Ecol.*, 4 (in press).

Coleman, D.C., Cole, C.V., Anderson, R.V., Blaha, M., Campion, M.K., Clarholm, M., Elliott, E.T., Hunt, H.W., Schaefer, B. and Sinclair, J. (1977). An analysis of rhizosphere-saprophage interactions in terrestrial ecosystems. Soil Organisms as Components of Ecosystems. *Ecol. Bull. (Stockholm)*, 25, 299.

Coleman, D.C., Anderson, R.V., Cole, C.V., Elliott, E.T., Woods, L. and Campion, M.K. (1978). Trophic interactions in soils as they affect energy and nutrient dynamics. IV. Flows of metabolic and biomass carbon. *Microb. Ecol.*, 4 (in press).

Elliott, E.T. and Coleman, D.C. (1977). Soil protozoan dynamics in a shortgrass prairie. *Soil Biol. Biochem.*, 9, 113.

Geltzer, J.G. (1963). On the behaviour of soil amoebae in the rhizospheres of plants. *Pedobiologia*, 2, 249.

Goring, C.A.I. and Clark, F.E. (1948). Influence of crop growth on mineralization of nitrogen in the soil. *Soil Sci. Soc. Am. Proc.*, 13, 261.

Hunt, H.W., Cole, C.V., Klein, D.A. and Coleman, D.C. (1977). A simulation model for the effect of predation on bacteria in continuous culture. *Microb. Ecol.*, 3, 259.

Johannes, R.E. (1965). Influence of marine protozoa on nutrient regeneration. *Limnol. Oceanogr.*, 10, 434.

Johannes, R.E. (1968). Nutrient regeneration in lakes and oceans. *In* "Advances in the Microbiology of the Sea", Vol. I (Eds. M.R. Droop and E.J.F. Wood), p.203. Academic Press, London.

Nikoljuk, V.F. and Geltzer, J.G. (1972). "Soil Protozoa of the USSR". 'FAN' Press, Tashkent, Uzbek SSR. 320pp.

Rosswall, T., Lohm, U. and Sohlenius, B. (1977). Développement d'un microcosme pour l'étude de la minéralization et de l'absorption radiculaire de l'azote dans l'humus d'une forêt de coniferes (*Pinus sylvestris* L.). *Lejeunia*, 84, 1.

Stanford, G., Carter, J.N., Simpson, E.C., Jr. and Schwaninger, D.E. (1973). Nitrate determination by a modified Conway microdiffusion method. *J. Assoc. Off. Anal. Chem.*, 56, 1365.

Woods, L., Cole, C.V., Elliott, E.T., Anderson, R.V. and Coleman, D.C. (1978). Trophic interactions in soils as they affect energy and nutrient dynamics. VI. Nitrogen transformations. *Microb. Ecol.*, (submitted).

PHOSPHATASE ACTIVITY AND PHOSPHORUS AVAILABILITY IN THE RHIZOSPHERE OF CORN ROOTS

G. BOERO and S. THIEN

Kansas State University, Manhattan, Kansas, USA

Summary

The extent to which corn roots affect the availability of phosphorus by inducing changes in the rhizosphere was investigated. The entire root system of a corn plant was confined to a core, 2 cm in diameter, intended to simulate a macro-rhizosphere. Outside this core were four soil zones, each 5 mm thick and separated from each other by 43 μm mesh nylon fabric. After specified growing periods, the soil from the five areas was analyzed for phosphatase activity, pH, and available P content. Soil phosphatase was enriched up to 40% by corn roots in the central core and to a lesser extent 5 mm away. No decrease in organic P levels in these zones was observed suggesting that enzyme activity was not limiting the mineralization of phosphatase-hydrolyzable compounds. Available P tended to decrease in soil zones closest to the root core but these changes were not statistically significant. Root-induced pH changes of as much as 0.5 units were observed between the root core and the outer zones over the two-week growing period. Available P levels increased as the root-induced pH changes acidified the rhizosphere.

Introduction

As a plant root grows through the soil, it introduces

change to its surroundings. The affected zone, known as the rhizosphere, differs biologically, chemically, and physically from non-rhizosphere soil (Carson, 1974) and from it essential growth requirements are obtained by plant roots. Hence, through the soil-root interface a root exerts a degree of influence over its growth environment.

One such rhizosphere effect concerns changes in phosphate availability from organic and inorganic sources. The release of phosphate from organic sources is catalyzed by phosphatase, which has been shown to be located in the three rhizosphere components: soil (McLaren, 1975), microbes (Ridge and Rovira, 1971) and plant roots (Bieleski, 1973). The solubility, and hence availability, of inorganic phosphate is pH related (Lindsey and Moreno, 1960). Rhizosphere pH is regulated by: hydration of CO_2 to carbonic acid (Brady, 1974); release of cellular organic acids (Carson, 1974); and the imbalance of cation and anion uptake and the associated release of protons (H^+) or hydroxyl (OH^-) or bicarbonate (HCO_3^-) moieties (Kirby and Mengel, 1967).

In this study we investigated two effects, rhizosphere phosphatase activity and rhizosphere pH changes, and related them to the P nutrition of young corn (*Zea mays* L.) plants.

Methods

To circumvent the difficulties of sampling the rhizosphere of a single root, we devised a method to confine the entire root system of a plant in a small volume of soil so creating a macrorhizosphere where the reactions were assumed to approximate those in a rhizosphere of a single root.

Soils

Three soils were selected from uncultivated sites to provide a wide range of properties (Table 1), including:

TABLE 1

Analyses of soils

Characteristic	Ustoll	Udoll	Udalf
Sand-silt-clay, %	19-53-28	55-32-12	53-38-9
Texture	Silty clay loam	Sandy loam	Sandy loam
Moisture, 1/3 bar	26%	12%	10%
Cation exchange capacity (meq/100g)	18.9	9.9	5.7
Organic matter, %	2.3	1.6	0.6
pH, saturated paste	7.1	6.3	5.3
Available P, μg/g	14	36	38
NH_4-N, μg/g	7.0	4.2	2.1
NO_3-N, μg/g	4.2	8.2	0.7
Site vegetation	Brome grass	Native grass	Deciduous forest

a Pachic Argiustoll, fine montmorillonitic, mesic; a Fluventic Hapludoll, coarse-loamy, mixed, mesic; and a Typic Hapludalf, loamy, mixed, mesic. All soil samples were sieved and air-dried before being used.

Concentric-ring sampler

A stainless steel frame of concentric rings was constructed to have a central core of 2 cm diameter and three additional rings, each with a radius of 5 mm greater than the next (Fig. 1). Cylinders of 43 μm mesh nylon fabric were fastened to an upper and a lower set of concentric rings, creating five separable soil zones: the central core; zone 1, 0-5 mm; zone 2, 5-10mm; zone 3, 10-15 mm; and zone 4, greater than 15 mm from the central core. This sampler was centered in a plastic column 11 cm in diameter by 70 cm in height. Soil was placed in and around the sampler, compacted to a bulk density of 1.25 g cc^{-1}, and watered to 1/3 bar tension. Roots of one germinated corn seedling were directed into the central core. At the end of each experiment, the sampler was removed and soil

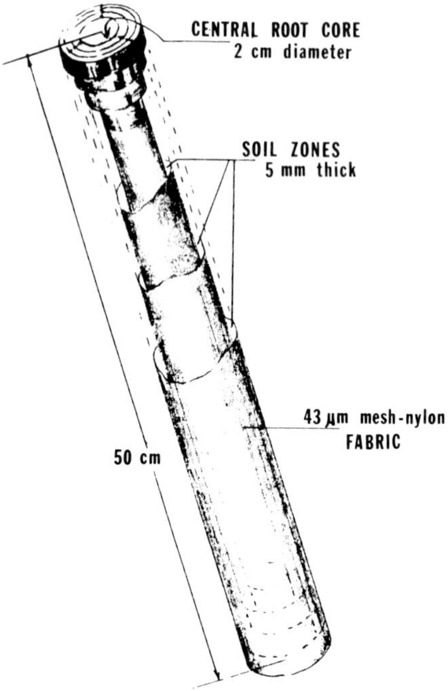

Fig. 1. A concentric ring sampling device used to create a macro-rhizosphere by confining the entire root system of one corn plant to the central core. Fabric cylinders enabled partitioning of soil samples in 5 mm increments away from the central root core.

samples were collected from the five zones. No roots were observed to penetrate the mesh of the nylon fabric.

Soil analysis

Moist soil removed from the sampler was seived and analyzed for phosphatase activity, pH, and moisture content. Portions of these soil samples were air-dried and stored at room temperature for further P determinations.

Phosphatase activity was performed using a modification of a method by Tabatabai and Bremner (1969). One gram of soil was mixed with 3 ml of modified universal buffer, MUB pH 6.5 (Skujins, Broal and McLaren, 1962) in a 50 ml volumetric flask and prewarmed to 37°C in a water bath. Two ml of 62.5 µmol ml^{-1} p-nitrophenyl phosphate (pNPP) buffered to pH 6.5 was added to start the reaction. After 55 minutes, 1 ml of 1N $CaCl_2$ was added and gently swirled. The reaction was stopped after 1 hour by adding 0.025N NaOH up to 50 ml volume. After shaking and filtering, the filtrate was measured for absorbance.

Organic P was measured by Olsen and Dean's (1965) ignition method and available P was measured using Bray's No. 1 extracting solution and the vanadomolybdate method according to Jackson (1970).

Fumigation

Fumigated treatments were prepared by injecting approximately 25 g of methylbromide containing 2% chlorpicrin (Dow MC2) into the prepared soil columns described above, but sealed on both ends. After 70 h, the fumigant was purged with filtered, compressed air for 3 h, covered, and incubated at 35°C for 4 days before planting with surface-sterilized seeds. At harvest, plate counts showed microbial numbers averaging 7×10^5 g^{-1} rhizosphere soil and 2×10^5 g^{-1} of bulk soil for the fumigated treatment and 2×10^7 g^{-1} and 1×10^7 g^{-1}, respectively, for the non-fumigated soil.

Growth conditions

A commercial hybrid, Pioneer 3195-A, was used in all experiments with a 16 h photoperiod and a 30/20°C diurnal cycle. Roots reached the bottom of the core in two weeks, but one three week treatment was included to determine if growing time increased the rhizosphere effect. Dry matter production of roots and tops for the three treatments is given in Table 2.

Results and Discussion

Phosphatase

Phosphatase activity was significantly higher in the central root core than in soil zones farther from the

TABLE 2

Growth chracteristics of plants in three soils

Growth conditions	Dry-matter production					
	Ustoll		Udoll		Udalf	
	Roots	Tops	Roots	Tops	Roots	Tops
	mg					
Two weeks	745	230	2,230	600	241	100
Two weeks (fumigated)	481	162	1,170	204	81	78
Three weeks	2,600	800	7,500	950	5,000	850

roots (Table 3). In most cases, phosphatase activity in the zone nearest (0-5 mm) the macrorhizosphere (i.e. central core) was also significantly enriched when compared to zones farther away, but less so than in the macrorhizosphere. While the general trend of phosphatase activity for the three soils studied was similar, the magnitude of this effect depended upon the soil and treatment.

The enrichment of phosphatase activity near corn roots is shown for the three soils studied in Figure 2. A 40%, 20% and 17% enrichment was observed in the central root core for the Udalf, Udoll and Ustoll, respectively. The degree of enrichment was inversely related to residual phosphatase activity and the levels of organic matter and clay, suggesting perhaps an enzyme-clay-humus complex of reduced activity as noted by McLaren (1970). Enrichment dropped to 15%, 3%, and 5% for the soils in the nearest root-free zone (0-5 mm) (Fig. 2). Little or no differences in phosphatase activity were found in soil zones farther away from the root core.

The K_m and V_{max} values found by applying Michaelis-Menten kinetics (White, Handler and Smith, 1964) to the phosphatase activity of the three soils studied are shown in Table 4. K_m values were inversely related to soil organic matter levels, suggesting that phosphatase and substrate binding is enhanced in the presence of organic colloids.

TABLE 3

Phosphatase activity, phosphorus level, and pH of three soils averaged for each growth condition studied

Growth conditions	Distance from root core	Phosphatase activity			Available P			Organic P			pH		
		Ustoll	Udoll	Udalf	Ustoll	Udoll	Udalf	Ustoll	Udoll	Udalf	Ustoll	Udoll	Udalf
	mm		μmol/g/h			μg/g			μg/g				
Two weeks	Root core	2.53	3.17	0.68	14.8	31.0	32.2	168	109	149	7.37	6.70	6.12
	0-5	2.42	2.73	0.55	15.2	33.2	31.9	197	168	147	7.47	6.40	6.03
	5-10	2.41	2.37	0.61	16.2	32.8	32.8	163	211	235	7.41	6.20	5.96
	10-15	2.24	2.62	0.57	14.0	34.4	34.4	159	162	62	7.42	6.18	5.88
	15-50	2.28	2.60	0.59	14.1	35.6	38.2	420	154	125	7.34	6.16	5.89
	LSD 0.05	0.17	0.21	0.06	NS	NS	NS	NS	NS	NS	NS	0.03	0.08
Three weeks	Root core	2.68	3.37	0.60	13.6	31.3	31.4	194	211	86	7.26	6.43	5.64
	0-5	2.52	2.95	0.62	13.4	34.0	30.9	159	182	93	7.24	6.23	5.90
	5-10	2.21	2.84	0.55	13.4	36.4	32.8	228	211	86	7.15	6.03	6.03
	10-15	2.14	2.83	0.56	14.5	39.6	35.2	220	178	69	7.17	6.00	5.92
	15-50	2.28	2.74	0.58	14.4	38.2	38.0	201	150	31	7.10	5.96	5.83
	LSD 0.05	0.15	0.23	NS	NS	NS	NS	NS	NS	NS	0.03	0.03	0.07
Two weeks (fumigated)	Root core	0.73	1.34	0.36	22.6	35.6	42.6	225	161	78	6.85	5.55	5.23
	0-5	0.89	1.02	0.02	21.8	39.5	34.8	166	180	166	6.91	5.75	5.21
	5-10	0.71	1.15	0.00	19.5	39.2	34.5	175	218	73	6.93	5.69	5.15
	10-15	0.68	0.99	0.06	19.4	38.0	38.5	201	185	128	6.87	5.70	5.16
	15-50	0.65	1.11	0.00	18.9	37.5	34.8	111	247	111	6.95	5.74	5.24
	LSD 0.05	0.14	0.15	NS	NS	NS	NS	NS	NS	NS	NS	0.05	0.06

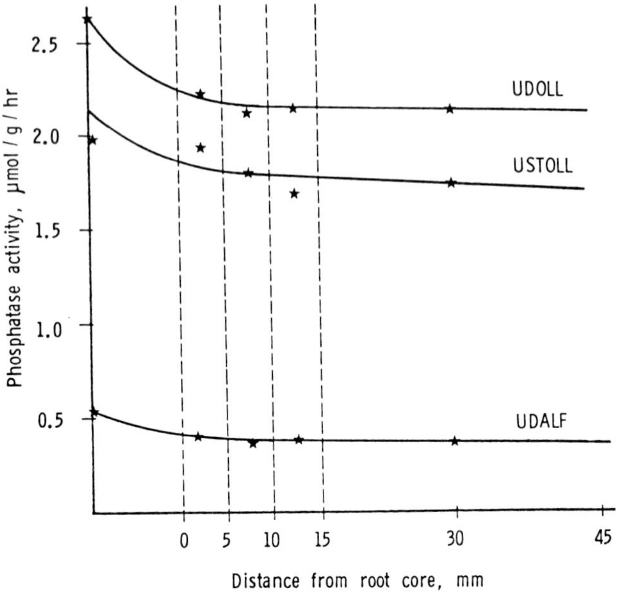

Fig. 2. Change in phosphatase activity in soil zones close to a central root core, averaged over three treatments.

TABLE 4

Phosphatase enzyme kinetic parameters of three soils

Parameter	Ustoll	Udoll	Udalf
K_m	2.36*	3.10*	6.81**
V_{max}	1.72*	2.44*	0.46**

*, ** Significant at the 0.05 and 0.01 levels, respectively.

After two weeks, the soils in the fumigation treatment showed phosphatase activities approximately one-half those in non-fumigated treatments. However, the enrichment pattern of phosphatase activity in soil zones near roots was still observed (Table 3). The drop in phosphatase activity could be due to decomposition of sulfhydrol groups by methyl bromide (D. Gould, personal communication). Because of the appreciable recolonization of the rhizosphere at the end of two weeks, determining the contribution of rhizosphere microflora to this enrichment was not possible.

Little phosphatase enrichment occurred during the third week of growth compared with that of the first two weeks. Average phosphatase activity in the central core rose to 119% of that observed in the bulk soil after three weeks compared with 117% after two weeks (Table 3).

pH

The ability of corn roots to influence the pH of soil near them is shown in Table 3. In the two, non-fumigated treatments, a significant rise in pH was observed in the rhizosphere, being most pronounced nearest the root. This change in pH of the rhizosphere of the individual soils was inversely correlated to their buffering (i.e. cation exchange) capacity (Table 1). This pH response suggested that nitrate was the predominant source of nitrogen utilized by the plant (Riley and Barber, 1971).

When the soil was fumigated, pH showed a slight reduction in soil zones nearest the root core. The reason for that pH change was not resolved, but might be explained in a number of ways such as a shift from nitrate to ammonium uptake by the plant, or that the roots preferentially absorbed an excess of cations. Both have been shown to cause reduced pH values near roots (Riley and Barber, 1971 and Kirby and Mengel, 1967). Alternately, the release of an S-H group as suggested above and its consequent oxidation to H_2SO_4 might also explain the pH drop observed with methyl bromide.

Soil phosphorus

For the two, non-fumigated treatments, available P in the soil of the central root core averaged across the three soils appeared to be 11% and 16% depleted after a two and three week growth period, respectively (Table 3), but these values were not significantly different. Available P levels generally decreased with increasing closeness to the central root core, but again, were not statistically significant. Fumigating the soil increased available P in the root core by 9% compared to that in the bulk soil.

Levels of organic P showed no uniform trend of enrichment or depletion for any of the treatments studied (Table 3).

Conclusions

Using a technique that created a macrorhizosphere, corn roots were observed to enrich the phosphatase activity of adjacent soil. Enrichment ranged between 40% and 17% in a 2 cm diameter core containing the entire root system of one corn plant. Fumigation to reduce rhizosphere microflora activity did not reduce that enrichment effect, but did lower the activity of the enzyme in all portions of the soil. But, because root growth was less (Table 2) in the fumigated treatment and because significant recolonization of the soil had occurred by harvest, the relative contribution of roots and microbes to the observed enrichment could not be estimated. Work by both Ridge and Rovira (1971) and Martin (1973) suggests that the presence of microflora in the rhizosphere does not increase phosphatase activity above that due to root enzymes.

The elevated enzymatic activity near roots suggests a possible mechanism for increasing utilization of soil organic phosphorus sources, but changes in organic-P levels did not correlate to phosphatase enrichment, a feature also reported by Thompson and Black (1970). Thus, even though plant roots raise the activity of phosphatase enzyme in the rhizosphere, use of organic-P does not appear to be stimulated. This can be explained under first-order kinetics; when the substrate available for enzymatic binding is significantly below the K_m value, increasing the enzyme activity does not increase the reaction velocity (Segel, 1975). Thus, it appears unlikely that enriching the rhizosphere with phosphatase will increase the availability of soil organic phosphorus because the organic-P moieties accessible for enzymatic hydrolysis are already saturated with indigenous soil enzyme. These data confirm the work of Martin (1973) and suggest that the dominant factor controlling the mineralization of phosphatase-hydrolyzable compounds is most probably their accessibility to enzymes. The concept is further supported by the work of Reid and Bieleski (1970) which showed that even when increased phosphatase activity was induced by a P deficiency, turnover of organic P was not increased. Apparently the role of plant-released phosphatase is merely to increase the reserve enzyme level in the soil.

Acknowledgements

This paper is Contribution No. 78-334-A, Department of Agronomy, Kansas Agricultural Experiment Station, Manhattan 66506.

References

Bieleski, R.L. (1973). Phosphate pools, phosphate transport and phosphate availability. *Ann. Rev. Plant Physiology*, 24, 225.
Brady, N.C. (1974). "The Nature and Properties of Soils". Mac-Millan, New York. p.469.
Carson, E.W. (Ed.), (1974). "The Plant Root and Its Environment". University Press of Virginia. Charlottesville.
Jackson, M.L. (1970). "Soil Chemical Analysis". Prentice Hall. Englewood Cliffs. New Jersey.
Kirby, E.A. and Mengel, K. (1967). Ionic balance in different tissues of the tomato plant in relation to nitrate, urea, and ammonium nutrition. *Plant Phys.*, 42, 6.
Lindsay, W.L. and Moreno, E.G. (1960). Phosphate phase equilibrium in soils. *Soil Sci. Soc. Amer. Proc.*, 24, 177.
Martin, J.K. (1973). The influence of rhizosphere microflora on the availability of ^{32}P *myo*inositol hexaphosphate, phosphorus to wheat. *Soil Biol. and Biochem.*, 5, 473.
McLaren, A.D. (1975). Soil as a system of humus and clay immobilized enzymes. *Chem. Scripta.*, 8, 97.
Olsen, S.R. and Dean, L.A. (1965). Phosphorus. *In* "Methods of Soil Analysis" (Ed. C.A. Black), p.1038. Agronomy IX. American Society of Agronomy, Madison, Wis.
Reid, M.S. and Bieleski, R.L. (1970). Changes in phosphatase activity in phosphorus deficient *Spirodela*. *Planta*, 94, 273.
Ridge, E.H. and Rovira, A.D. (1971). Phosphatase activity of intact young wheat roots under sterile and non-sterile conditions. *New*

Phytol., 70, 1017.
Riley, D. and Barber, S.A. (1971). Effect of ammonium and nitrate fertilization on phosphorus uptake as related to root-induced pH changes at the root-soil interface. *Soil Sci. Soc. Amer. Proc.*, 35, 301.
Segel, I.H. (1975). "Enzyme Kinetics". Wiley Publ., New York.
Skujins, J.J., Broal, L. and McLaren, A.D. (1962). Characterization of phosphatase in a terrestial soil sterilized with an electron beam. *Enzymologia*, 25, 125.
Tabatabai, M.A. and Bremner, J.M. (1969). The use of p-nitrophenyl phosphate for assay of soil phosphatase activity. *Soil Biol. and Biochem.*, 1, 307.
Thompson, E.J. and Black, C.A. (1970). Changes in extractable organic phosphorus in soil in the presence and absence of plants. III. Phosphatase effects. *Plant Soil*, 32, 335.
White, A., Handler, P. and Smith, E.L. (1964). "Principles of Biochemistry". McGraw-Hill Book Co., New York. p.219.

INFLUENCE OF TILLAGE AND RESIDUES ON WHEAT ROOT MICROFLORA AND ROOT COLONIZATION BY NITROGEN-FIXING BACTERIA

L.F. ELLIOTT, C.M. GILMOUR,* V.L. COCHRAN,**
C. COLEY[+] and D. BENNETT[+]

SEA-FR, Pullman, WA, USA

Summary

The effects of winter wheat straw placement on wheat rhizoplane microflora was studied under both greenhouse and field conditions. Results indicated that the greenhouse data cannot be readily compared with field observations. In related studies, tests on various wheat genotypes show that selected free-living, N_2-fixing cultures colonize wheat roots and also increase root nitrogenase activity.

Introduction

Studies on the influence of tillage practice, crop residue management and phytotoxicity on the rhizosphere-rhizoplane microflora deserve additional attention. It is recognized that the root microflora may exert either a beneficial or harmful effect on plant viability depending on the tillage or crop management system used.

It is not surprising that root exudates and related plant photosynthate levels influence bacterial colonization of root surfaces. Barber and Martin (1976) showed that 12-18% of wheat plant photosynthate was released as root

* *Department of Bacteriology and Biochemistry, U of I, Moscow, USA.*
** *SEA-FR, Pullman, WA, USA.*
\+ *University of Idaho, Moscow, USA.*

exudate into nonsterile soil. Barber and Lynch (1977) postulated that root microflora stimulation of soluble organic carbon loss from plant roots may have a significant influence on the activity of free-living, N_2-fixing bacteria in the rhizosphere. Undoubtedly, the mucilaginous sheath, as Rovira and Campbell (1974) reported for wheat roots, also contributes to root surface colonization. Other biotic and abiotic factors also influence root colonization, especially light, temperature and water stress which stimulate exudation from roots (Burstrom, 1965; and Rovira, 1969).

These studies were therefore initiated to determine the effect of residues and tillage on the wheat root microflora and whether the root systems were colonized by free-living N_2-fixing bacteria.

Materials and Methods

Greenhouse studies

Palouse silt loam (Pachic Ultic Haploxeroll) was fertilized to the equivalent of 224 kg N/ha, 20 kg P/ha, and 17 kg S/ha and 2900 g of the mix placed in metal cans. Winter wheat straw (*T. aestivum* L.) (2 to 2.5 cm) was added at 0 or 11.2 tonnes/ha equivalent on the soil surface or incorporated. Fourteen replications of the following treatments were prepared: no straw (NS), surface straw (SS), and straw mixed (SM). Each pot was seeded with 10 winter wheat seeds (*T. aestivum* L.), brought to approximately -0.3 bar water potential, incubated at 21°C until emergence, and vernalized. Then the plants were placed in the greenhouse (10 h day length, 4.5 to 10°C) and surface watered by misting. The pots were weighed for moisture control. The rhizoplane was sampled by gently washing the roots with Ringer's solution and then placing 5 g of blotted roots in a dilution bottle containing sterile glass beads and 45 ml of Ringer's solution. The mixture was shaken for 15 min. Dilutions of the soil and rhizoplane samples were placed in triplicate on starch, Frazier's yeast extract, soil extract agars (Allen, 1959), *Pseudomonas* agar (Simon, Rovira and Sands, 1973), and Martin's medium (Menzies, 1965). The plates were incubated at 21 °C and colonies were counted by conventional procedures.

Field studies

In the fall of 1976 and 1977, winter wheat (Nugaines) was planted into field plots (2.5 m x 9 m) having heavy wheat straw residue with conventionally tilled seedbeds (T), with no-tillage directly into the surface residues (NT), with no-tillage where the residues were burned (B), and with no tillage where the residue was removed from the seed row (SA). The plots received 135 kg N/ha, 20 kg P/ha, and 15 kg S/ha. Each treatment was replicated four times. In mid-February and mid-April 1977, and the first week of November 1977, winter wheat roots and non-rhizosphere soil were sampled to a depth of 30 cm from each treatment. The rhizoplane and soil were plated as described previously. In January and February 1978, the replicates were randomly sampled to observe bacteria on the wheat roots with scanning electron microscopy (SEM). The critical drying procedure consisted of the use of phosphate and cacodylate buffers, fixation with OsO_4, dehydration in alcohol and Freon[113], mounting and gold plating (A. Cohen, Washington State University, personal communication).

Nitrogenase activity and root colonization

Isolation of bacterial cultures was carried out using a modified mineral salts medium (Line and Loutit, 1971). Seeds were surface sterilized by soaking in water for 1 h, in hypochlorite (1000 ppm) for 4 h, and then transferred to glucose nutrient agar plates to test seed sterility.

In micro-simulation studies sterile seeds were placed on the surface of a (0.4% agar) N-free mineral medium (Line and Loutit, 1971) contained between two glass plates.

The medium, except for the controls, was inoculated with a 1 ml suspension of individual bacterial isolates. The double-plate growth chamber was placed in a Sherer "Controlled Temperature Lab" with a 12-h-day length at 22°C. The root samples were subjected to critical-point drying and preparation for SEM (see Campbell and Rovira (1973), procedure 2. Conventional microscopy was accomplished with a conventional microscope-camera assembly.

In macro-simulation experiments pots containing 500 g of a non-sterile sand-soil mixture adjusted to -0.3 bar water potential were planted with inoculated or uninoculated seeds and incubated as described previously. The inoculum was mixed with finely ground peat to which 10% $CaCO_3$ had been added to control pH (4.7 x 10^8 bacteria/g). Seed bacterization was accomplished using the aforementioned peat inoculum plus 20% PEL GEL (Dr. J.C. Burton, Nitragin Co., Milwaukee, Wisconsin). The control peat for uninoculated seeds was prepared as described except that sterile H_2O was added in place of the culture.

Periodically, plants were removed and the roots were washed in Ringer's solution and excised for acetylene reduction assay (Hardy *et al.*, 1968). For the acetylene reduction assay, root pieces or pure cultures were transferred to 32-ml aluminium-capped bottles containing nitrogen-free medium (semi-solid) (Lind and Loutit, 1971). After 2 to 3 days incubation at 20°C, the aluminium cap was replaced with a sterile serum bottle cap to facilitate injection of acetylene and measurement of ethylene production (Hardy *et al.*, 1968).

Results and Discussion

Greenhouse and field studies

The data shown in Table 1 were representative of those obtained from all samplings from the greenhouse. They show a marked rhizoplane effect except with Martin's medium. In all cases, the trend was that the rhizoplane microflora

TABLE 1

Observed influence of wheat straw placement on winter wheat rhizoplane microflora counts

Plating Medium	Rhizoplane and soil counts/g x 10^5*											
	3 weeks				6 weeks				9 weeks			
	SS[+]	NS	SM	Soil	SS	NS	SM	Soil	SS	NS	SM	Soil
Frazier's	0.6	0.6	0.4	0.1	5.7	2.9	4.0	0.8	10.6	8.0	13.4	0.3
Pseudomonas	7.0	0.7	2.5	0.5	3.4	3.1	12.0a[#]	0.6	18.4	22.4	33.3	0.7
Starch	9.0	8.2	2.7a	3.1	128.0	95.0	151.0a	3.8	22.6	49.5	110.0	2.1
Yeast extract	30.0	21.8	23.4	17.9	26.4	20.6	22.0	16.0	113.8	86.0	164.0	9.7
Soil extract	30.0a	42.0a	111.5a	14.3	42.4	29.5	306.0	19.3	63.5	93.5	125.0	12.8
Martin's	0.1	0.07	1.0	1.6	0.5	0.3	0.8	0.9	0.2	0.1	0.2	2.4

* Microbial numbers are expressed as per g dry wt of root or soil.

+ SS = surface straw; SM = straw mixed; NS = no straw; these treatments were subject to rhizoplane analysis.

Microbial numbers designated by an a were significantly different, at the 5% level, from others counted on the same medium at the same sampling date.

from the SM treatment was greater than the other treatments when plated on soil extract agar, but was significantly greater only at the 3-week sampling. On starch agar, rhizoplane micro-organisms were significantly less at 3 weeks with the SM treatment and greater at 6 weeks on both starch and *Pseudomonas* agar. Plant yields (dry wt) were significantly lower from the SM than from the other two treatments (data not presented) indicating plant stress occurred with this treatment. When rhizoplane:soil ratios were calculated, the trend with soil extract agar remained. (Numbers of rhizoplane micro-organisms counted on a specific medium per g dry wt of root material: numbers of micro-organisms counted on the same medium per g dry wt of soil.)

In the field, bacterial numbers on the rhizoplane and in the soil were 10 to 100 times greater than those found in the greenhouse studies (Table 2) and distinct trends in rhizoplane microbial counts were observed which were affected by tillage and/or residue management. In February (Table 2), the NT treatment consistently showed the highest rhizoplane bacterial count and the T treatment the lowest. Microbial numbers, as measured by *Pseudomonas*, starch, yeast extract, and soil extract agars, were significantly greater for the NT treatment. With the exception of Martin's medium (which selects fungi), an obvious rhizosphere effect was obtained with all other media used. As Rovira (1965) pointed out, fungi and actinomycetes do not respond to plant-root exudates as dramatically as bacteria. In April, these rhizoplane trends disappeared except that the SA treatment showed consistently lower rhizoplane bacterial numbers; and rhizoplane micro-organisms, as counted on yeast extract agar, were significantly higher for the NT treatment. In this case, there was a rhizoplane effect with Martin's medium. If rhizoplane:soil ratios were calculated from these field data, the same trends were obtained. The data for the November 1977 sampling

TABLE 2

Winter wheat rhizoplane microbial counts as affected by tillage and residue management in the field

Plating Medium	Wheat root rhizoplane and soil microorganisms/g x 10^6*									
	February 1977					April 1977				
	NT+	T	SA	B	Soil	NT	T	SA	B	Soil
Frazier's	0.9	0.2a#	1.7	1.6	0.3	-	-	-	-	-
Pseudomonas	13.5a	2.2	3.1	3.3	0.5	27.1	27.0	10.5	16.4	0.4
Starch	18.0a	2.2	3.6	5.8	1.1	18.8	37.0	22.0	31.0	3.1
Yeast extract	52.0a	10.4	29.6	20.8	18.8	161.0a	91.0	85.0	94.0	12.1
Soil extract	43.5a	10.4	8.6	13.5	21.1	99.3	78.6	67.7	91.3	11.0
Martin's	0.1	0.08	0.06	0.07	0.2	0.8	1.0	5.6	3.7	0.3

* Microbial counts are expressed as per g dry wt of root or soil.

+ NT = no tilled; T = tilled; SA = no tilled, straw away from seedling row; B = no tilled, residue burned; these treatments were subject to rhizoplane analysis.

Microbial numbers designated by an a were significantly different, at the 5% level, from others counted on the same medium at the same sampling date.

are not presented because reliable soil counts were not obtained. Root rhizoplane bacterial trends with the T treatment showed the highest numbers while the NT and SA showed the lowest. All the aforementioned bacterial counts represent the mean of 6 replicate determinations.

Scanning electron microscope studies on wheat root sections taken from the NT and T field treatments during the winter of 1977-78 verify the data shown for the February plate counts (Table 2). Wheat roots from the conventionally tilled plots were sparsely populated and showed little adhering soil debris (Fig. 1). Figure 1 also shows maximum colonization observed with the T treatment. Wheat roots from the NT plots were more densely populated by bacteria and there were large quantities of soil particles and bacteria clinging to the roots, indicating greater root exudation than in the T treatment (Fig. 2). Figure 2 also shows typical colonization patterns observed with the NT treatment. The electron micrographs presented in Figures 1 and 2 were typical of root colonization patterns observed for the T and NT treatments during the late winter sampling period.

The literature indicates that plant stress increases root exudation which would itself increase the associated bacterial flora (Burstrom, 1965; Katznelson *et al.*, 1954, 1955; Rovira, 1965; Rovira, 1969; Rovira and Campbell, 1975). In late fall, the soil in the T plots would be expected to be cooler than the soil in the NT plots and so show increased exudation. In mid-April, all treatments were growing vigorously and perhaps root exudation would be similar in all treatments. Phytotoxicity as mentioned by Cochran *et al.* (1977) could play a role. For instance, reports indicate that freshly incorporated residues cause apparent phytotoxicities but usually of limited duration, so that freshly incorporated residues could have an effect on conventionally tilled wheat plantings in the fall (Elliott *et al.*, 1978), whereas surface residues could be

Fig. 1. Colonization of wheat roots with conventional tillage planting. x 1600.

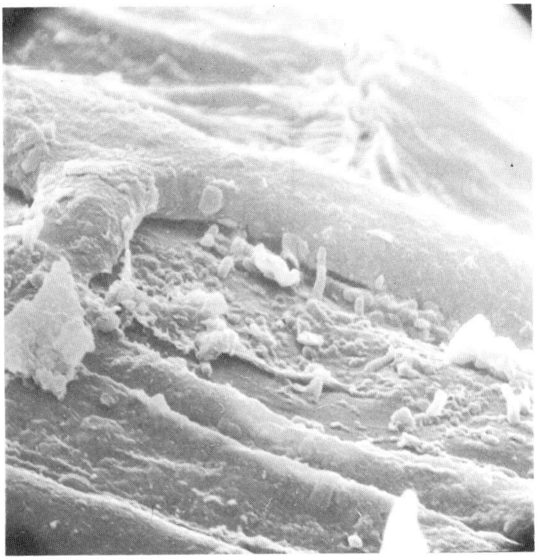

Fig. 2. Wheat root microbial colonization with no-tillage planting. x 3150.

phytotoxic in February under Pacific Northwest climatic conditions, when surface wheat residue water extracts can also be very toxic (Cochran *et al.*, 1977). However, these stress relationships require additional study.

Nitrogenase activity and root colonization potential

With all wheat cultivars tested (Table 3), inoculation with the five bacterial species significantly increased root nitrogenase activity with treatment A (seed coated with inoculated peat) as compared to treatment B (seed coated with non-inoculated peat) and treatment C (seed alone). In the case of *T. spelta*, *T. aestivum* and *T. carthlicum*, both total ethylene production and rate of ethylene production were markedly increased (3 to 12 times) by inoculation. Presumably, inoculation enriched the indigenous nitrogen-fixing root flora by the addition of N-fixing isolates. The low values shown for *T. aegilopoides* may

TABLE 3

Influence of inoculation of wheat with selected free-living N_2-fixing bacteria on potential nitrogenase activity

Wheat species	Chromosome number	Treat-ment*	Growth period	Total n mol C_2H_4/vial	n mol C_2H_4/h
Triticum spelta CI1774	6N	A+	4 weeks	1986.1	152.8
		B		302.2	23.2
		C		86.9	6.7
Triticum aestivum var. Nugaines	6N	A+	6 weeks	2857.1	168.1
		B		972.9	57.2
		C		92.0	5.4
Triticum carthlicum CI7385	4N	A+	7 weeks	2480.1	155.0
		B		210.5	13.2
		C		393.5	24.6
Triticum aegilopoides PI330527	2N	A+	11 weeks	242.4	15.2
		B		1.8	0.1
		C		33.7	2.1

NOTE: All data are based on the mean of 4 replicates (roots in medium II semi-solid).

* A = seed coated with inoculated peat; B = seed coated with non-inoculated peat; C = seed alone.

\+ Significant differences between A and B and A and C, P = 0.05.

reflect the inability of the inoculum to colonize the root system of that wheat variety.

Examination of excised roots by conventional microscopy (microsimulation study) shows the high density of bacterial cells which may be obtained by inoculation of wheat roots with free-living, N_2-fixing bacteria (Fig. 3). The zoning effect adjacent to the main root stem was observed with the mixed-culture inoculum on roots of *T. aestivum* var. Nugaines.

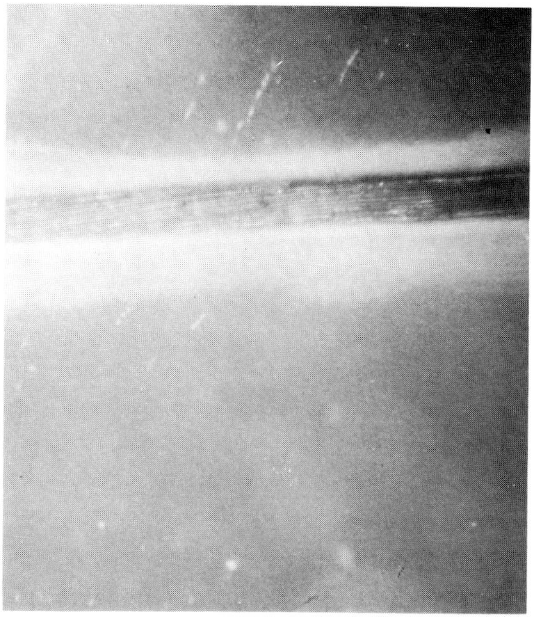

Fig. 3. Root stem and root hairs showing zone of bacterial growth at or near the zone of wheat root exudate production. x 36.

Scanning electron microscopy studies of root pieces from the microsimulation experiments showed positive and negative colonization by the test isolates on *T. aestivum* roots. Reference to Figure 4 shows that the sterile control seedling roots exhibited no bacterial colonization. On the other hand, Figure 5 shows the cell distribution pattern and micro-colonies of a bacterial community on inoculated roots. Undoubtedly, the colony developed as a consequence of gum or mucilaginous material production

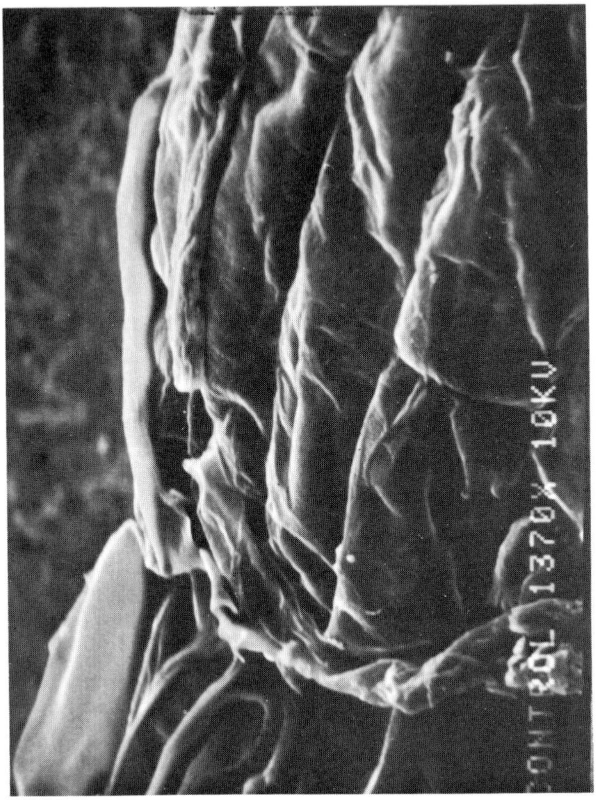

Fig. 4. Photomicrograph of uninoculated wheat seedling root section. S.E.M. x 1050.

(Fig. 6) as described by Campbell and Rovira (1973). The individual cells shown in Figures 5 and 6 were 2.5μ - 3.0μ in length x 1.25μ - 2.0μ in diameter and were tentatively placed in the family, Azotobacteraceae.

The unidentified gram negative bacillus shown in Figure 7 measured 2.0μ x 0.75μ. Again the tendency of the individual cells to clump at the root-hair surface and initiate colony formation was evident.

As with all SEM studies, it is difficult to obtain a completely representative illustration of cell-colony distribution. However, the photomicrographs presented were selected as typical of the general trends observed.

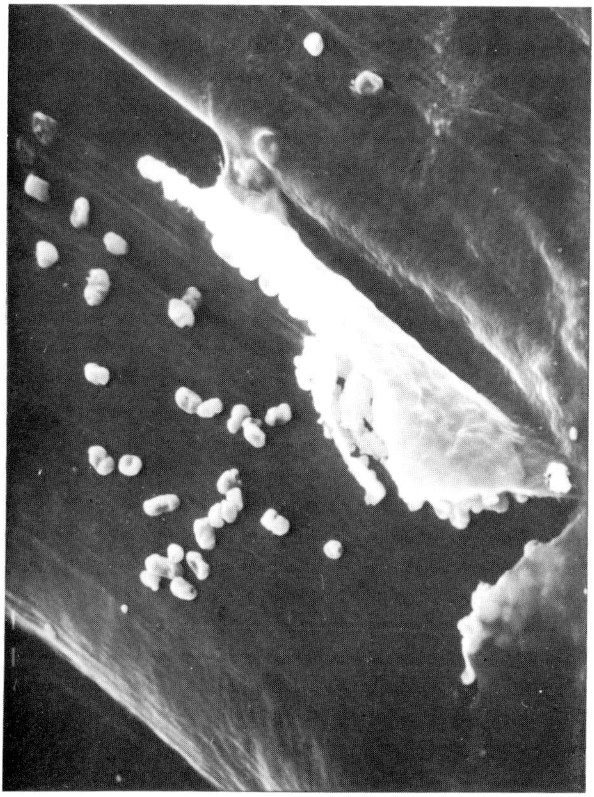

Fig. 5. Photomicrograph of wheat seedling root section inoculated with nitrogenase-positive isolate. Unfolding of the mucilaginous sheath discloses layered cells comprising a micro-colony. Micro-colony and individual ovoid cells attest to bacterial community development. S.E.M. x 2100.

Conclusions

These studies show greenhouse experiments and will be of questionable use for determining the effect of crop residue and tillage management on wheat-root microbial colonization. However, the greenhouse studies may have shown a correlation between the SM treatment and plant stress. In the field, tillage and/or residue treatments affect bacteria on the rhizoplane of winter wheat but the relationship is dynamic. Further investigation is required to determine the consequences of these effects, pathogen and

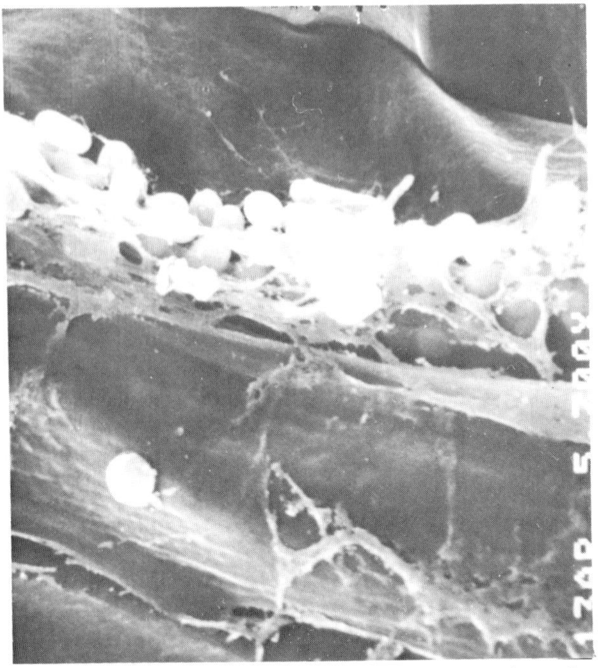

Fig. 6. Photomicrograph of wheat seedling root section showing mucilaginous complex and adhering cells. S.E.M. x 4600.

nutrient relationships, and plant-efficiency effects.

The data given in Table 3 and the related illustrative material presented in Figures 3 to 7 show that the nitrogenase-positive micro-organisms used have the capacity to colonize the root system of wheat plants. Admittedly, the micro- and macro-simulation approach used does not simulate field conditions. Yet the procedure has proved valuable in testing the colonization potential of the nitrogenase-positive isolates prior to greenhouse and field application. Field studies are now in progress, with winter wheat, to test the potential of the inoculum.

Acknowledgements

This paper is a contribution from Agricultural Research Science and Education Administration, U.S. Department of Agriculture, in cooperation with the College of Agriculture Research Center, Washington State University, Pullman; and the University of Idaho, Moscow, USA. WSU Scientific Paper No. SP 5073.

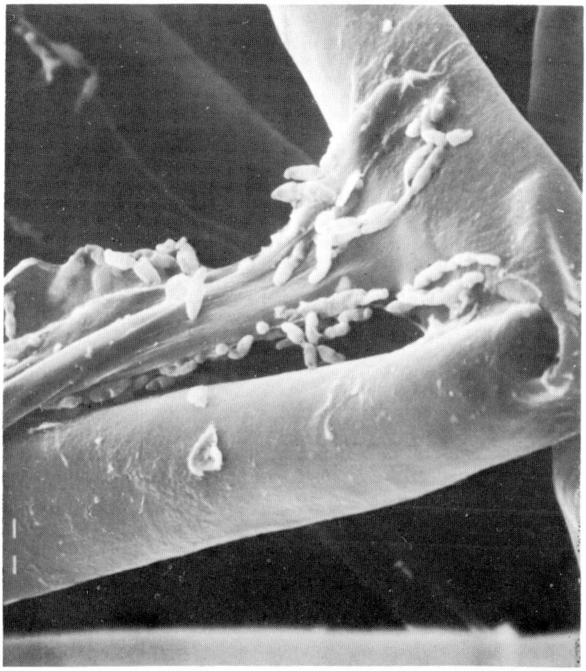

Fig. 7. Photomicrograph of wheat seedlings showing adhering bacillus-shaped cells and initiation of colony formation. S.E.M. x 2100.

References

Allen, O.N. (1959). "Experiments in Soil Bacteriology". University of WI., Madison, WI. 117pp.

Barber, D.A. and Lynch, J.M. (1977). Microbial growth in the rhizosphere. *Soil Biol. and Biochem.*, 9, 305.

Barber, D.A. and Martin, J.K. (1976). The release of organic substances by cereal roots into soil. *New Phytol.*, 76, 69.

Burstrom, H.G. (1965). *In* "Ecology of Soilborne Plant Pathogens" (Eds. K.F. Baker and W.C. Snyder), p.154. University of Calif. Press, Berkeley, CA.

Campbell, R. and Rovira, A.D. (1973). The study of rhizosphere by scanning electron microscopy. *Soil Biol. and Biochem.*, 5, 747.

Cochran, V.L., Elliott, L.F. and Papendick, R.I. (1977). The production of phytotoxins from surface crop residues. *Soil Sci. Soc. Am. J.*, 41, 903.

Elliott, L.F., McCalla, T.M. and Waiss, D., Jr. (1978). Phytotoxicity associated with residue management. *In* "Crop Residue Management Systems" (Ed. W.O. Oschwald). *Soil Sci. Soc. Am. J.*, (in press).

Hardy, R.W., Holsten, R.D., Jackson, E.K. and Burns, R.C. (1968). The acetylene reduction assay for N_2-fixation: laboratory and field evaluation. *Plant Physiol.*, 43, 1185.

Katznelson, H., Rouatt, J.W. and Payne, J.N.B. (1954). Liberation of amino-acids by plant roots in relation to desiccation. *Nature* (London), 174, 1110.

Katznelson, H. Rouatt, J.W. and Payne, J.N.B. (1955). The liberation of amino-acids and reducing compounds by plant roots. *Plant and Soil*, 7, 35.

Line, M.A. and Loutit, M.W. (1971). Non-symbiotic N_2 fixing organisms from some New Zealand Tussock - grassland soils. *J. Gen. Microbiol.*, 66, 309.

Menzies, J.D. (1965). Fungi. *In* "Methods of Soil Analysis", Part 2 (Ed. C.A. Black). Am. Soc. Agron., Madison, WI. *Agronomy*, 9, 1504.

Rovira, A.D. (1965). Interaction between plant roots and soil microorganisms. *In* "Annual Review of Microbiology", Vol. 19 (Ed. C.E. Clifton), pp.241-266. Annual Reviews, Inc., Palo Alto, CA.

Rovira, A.D. (1969). Plant root exudates. *Bot. Rev.*, 35, 35.

Rovira, A.D. and Campbell, R. (1974). Scanning electron microscopy of microorganisms on the roots of wheat. *Microbial Ecology*, 1, 15.

Rovira, A.D. and Campbell, R. (1975). A scanning electron microscope study of interactions between micro-organisms and *Gaeumannomyces graminis* (Syn. *Ophiobolus graminis*) on wheat roots. *Microbial Ecology*, 2, 177.

Simon, A., Rovira, A.D. and Sands, D.C. (1973). An improved selective method for isolating fluorescent pseudomonads. *J. Appl. Bact.*, 36, 141.

Mention of a trademark does not constitute a guarantee or warranty of the product by the USDA nor its approval to the exclusion of other products which may be suitable.

THE ROOT-SOIL INTERFACE AND ITS SIGNIFICANCE FOR FRUIT TREE ROOTS OF DIFFERENT AGES

D. ATKINSON and SANDRA A. WILSON

East Malling Research Station, Maidstone, Kent, U.K.

Summary

The fruit tree root system includes roots which differ in age, diameter and degree of suberization. The ability of cherry roots of different types to absorb water and phosphorus was measured. Both white (unsuberized) and woody roots absorbed both water and ^{32}P. Uptake by white roots and by woody roots, which were of a greater mean diameter, were similar on a basis of surface area so that uptake per unit volume by the white roots was higher. This result was unexpected and needs further investigation to determine whether it is true over a wider range of root types and environmental conditions.

Time lapse cinematography has shown that the activities of the soil fauna, particularly around unsuberized roots, can have a major effect on root-soil contact. The influence of root-soil contact on the relative importance of different root types is discussed.

Introduction

The young roots of fruit trees as seen behind the observation panels of an underground root laboratory are initially white. After about 3 weeks during the summer months, or longer in winter, browning of the cortex spreads from the older regions towards the tip. Cortical browning is followed by decay and disintegration which is due largely to the soil fauna feeding on them (Head, 1968a). After the loss of the cortex, secondary thickening becomes obvious in some roots which then become part of the permanent root system. Other roots either remain unthickened or disappear completely (Head, 1968b). The tree root system, therefore, comprises a range of roots of different types, ages, diameters and life histories.

It has often been assumed that the absorption of water and minerals occurs entirely through the younger regions of roots such as the root tip and the root hair zone. However, in a comparison of the uptake of ^{85}Sr and ^{32}P by apical and basal segments of seminal roots of barley, Clarkson, Sanderson and Russell (1968) found that both segments were able to absorb both tracers. Subsequent studies (i.e. Harrison-Murray and Clarkson, 1973; Graham, Clarkson and Sanderson, 1974) showed that, although the absorption of most minerals and water was greatest in the apical region, most of the root was able to function as an absorbing surface. The ability of excised suberized tree roots to absorb water has also been shown by Kramer and Bullock (1966).

There have been few direct comparisons of the abilities of attached unsuberized and woody roots of trees to absorb water and minerals so preliminary data on this are presented here. These were obtained in an underground root laboratory using time lapse cinematography. It will be shown that the extent of contact between root and soil is dynamic and can vary with time.

Materials and Methods

Plant material

Cherry rootstocks were used as test plants because of their ease of propagation and the rapid increase in diameter of roots caused by secondary thickening. F12/1 (*Prunus avium*) cherry rootstocks were produced by a micropropagation technique (Jones, Hopgood and O'Farrell, 1977; Jones and Hopgood, 1978), to give a continuous supply of uniform material throughout the year. The micropropagules were grown in a sand-loam mix and used experimentally at 3 months old.

Measurement of the absorption of water and nutrients by parts of intact root systems

Root systems were washed free from soil, rinsed in dilute sodium metabisulphite and then transferred to solution culture in a growth room which provided a temperature of 20-25°C, a relative humidity of 40-50% RH and constant illumination of 24 Wm^{-2} for water uptake experiments and similar temperature and humidity but illumination of 40 Wm^{-2} for ion uptake experiments. Plants were grown in a sterilized nutrient solution which provided the following nutrients in millimolar concentrations: KNO_3, 2; $Ca(NO_3)_2$, 2; $MgSO_4.7H_2O$, 0.75; $KH_2PO_4 7H_2O$, 1.33; $Fe(Na_2)EDTA$, 0.05; $MnSO_4 4H_2O$, 0.01; $CuSO_4 5H_2O$, 0.001; $ZnSO_4 7H_2O$, 0.001; H_3BO_3, 0.03; NaCl, 0.1; $Na_2MOO_4 2H_2O$, 0.0005.

Phosphate absorption

Phosphate uptake was measured using a modified version of the equipment described by Russell and Sanderson (1967). Short (3.5 mm) lengths of attached root were sealed across the diameter of a vinyl tube. The tube was connected to a peristaltic pump which circulated a solution containing ^{32}P (as $KH_2\ ^{32}PO_4$ at 1μCi ml^{-1}). The rest of the root system was kept in unlabelled solution.

After uptake the treated segment was identified by circulating aqueous toluidine blue dye and its dimensions, structure and the extent of lignification recorded. The remainder of the plants were divided into roots, stems and leaves, dried at 80°C, weighed and ground. Samples were prepared for the determination of ^{32}P by means of Cerenkov radiation as described by Atkinson (1974).

Water absorption

The uptake of water was measured using apparatus similar to that used in studying the uptake of phosphate except that one or more roots were sealed into a polythene tube connected at one end to a 2 ml glass syringe which acted as a reservoir of solution and at the other end to a 10 μl glass microsyringe which acted as a manometer (Clarkson and Sanderson, 1971). At the end of the experiment the treated segments were identified as described for phosphate.

Observations in the root laboratory

For experiments to compare root growth and water use trees of Worcester/MM.104 (*Malus pumila*) were planted in 1960, c.f. Head (1967), adjacent to an underground root laboratory (Rogers, 1969). Soil water depletion was measured as described by Atkinson (1977) and the length of white root as described by Head (1966).

Data on root behaviour in the soil and on root-soil contact were obtained by the analysis of time lapse films (Atkinson, Lewis and Jones, 1977) using trees of Cox and Golden Delicious/M.9 described by Atkinson, Naylor and Coldrick (1976). Measurements of root diameter were made on 11th August 1976 by the method of Head (1968b) using trees of Merton Glory/F12/1 planted in March 1975.

Results

Nutrient and water absorption

Uptake of ^{32}P was measured on segments within 6 mm of the apices of white roots 0.8 to 1 mm in diameter. Similar segments were examined on woody roots 1 to 3 mm in diameter. Rates of uptake are shown in Table 1. Both types of root absorbed phosphate although the relative rates depended upon the basis of expression. In both root types uptake per unit area was similar while uptake per unit volume was greater in the white roots. The proportion of absorbed ^{32}P translocated from the treated segment of woody roots (range 49-54%) was at least as high as that from white roots (range 27-64%).

A similar pattern of absorption was shown for water (Table 2). In both sets of experiments variation in absorption between replicates was higher for woody roots than white roots on a basis of volume (SE % mean 33 compared 27) but lower when expressed on a basis of surface area (7 compared 22).

TABLE 1

The absorption of phosphate by unsuberized and woody cherry roots

Root type	Uptake per unit		
	Length	Surface Area	Volume
	ng P $mm^{-1}h^{-1}$	ng P $mm^{-2}h^{-1}$	ng P $mm^{-3}h^{-1}$
White	15.6 ± 3.3	6.0 ± 1.4	29.2 ± 7.7
Woody	31.7 ± 12.0	4.9 ± 0.2	12.8 ± 4.2

All results are means for 3 replicate experiments.

TABLE 2

The absorption of water by unsuberized and woody cherry roots

Root type	Uptake per unit		
	Length	Surface Area	Volume
	μl $mm^{-1}h^{-1}$	μl $mm^{-2}h^{-1}$	μl $mm^{-3}h^{-1}$
White	0.42 ± 0.12	0.18 ± 0.04	1.21 ± 0.33
Woody	0.51 ± 0.26	0.14 ± 0.01	0.72 ± 0.24

Results for white roots are means for 6 replicate experiments and for 3 replicate experiments for woody roots.

Water absorption under field conditions

Conditions for absorption under field conditions can be very different to those in water culture. Although it is difficult to measure the uptake of water by specific roots in these circumstances some evidence for the role of different types of root is given by a comparison of the seasonal periodicity of root growth and soil moisture depletion. The periodicity of white root length and the depletion of soil moisture for two mature apple trees in two consecutive years is shown in Figure 1.

In both years rapid depletion of soil moisture occurred when there was little or no white root present. Similar results to those in Figure 1 have been obtained for most years between 1972 and 1976.

Estimates of soil root contact and root density

Contact between roots and the soil will be reduced if soil is removed from the region adjacent to the root surface or if root diameter is decreased. Data on these possibilities has been obtained using the root observation laboratories. Measurements made on a time lapse film showing an apple root growing through soil, indicated that the proportion of the root surface clearly visible, i.e. not in contact with soil, could increase with time (Fig. 2). While part of the apparent decrease in contact could have been due to growth in root diameter, most was due to the movement over the root surface of non-parasitic nematodes whose activities separated root from soil. Measurements in the early morning and at midday, during the dry conditions of August 1976, of the diameter of a number of cherry roots visible in the root laboratory showed (Fig. 3) that, although a few roots showed large (20-25%) changes in diameter, most showed little or no change.

The importance of root soil contact to water absorption will be influenced by root density. Values for the density of apple roots, expressed as root length per unit soil surface (LA), obtained from a number of experiments at East Malling are given in Table 3. Estimates were obtained either by using a variety of field methods (1) soil monoliths, 2) soil trenches or 3) complete excavations), or from measurements made in the root laboratory. For root laboratory data LA was calculated using the assumption of Taylor and Klepper (1971) that all roots within 2 mm of

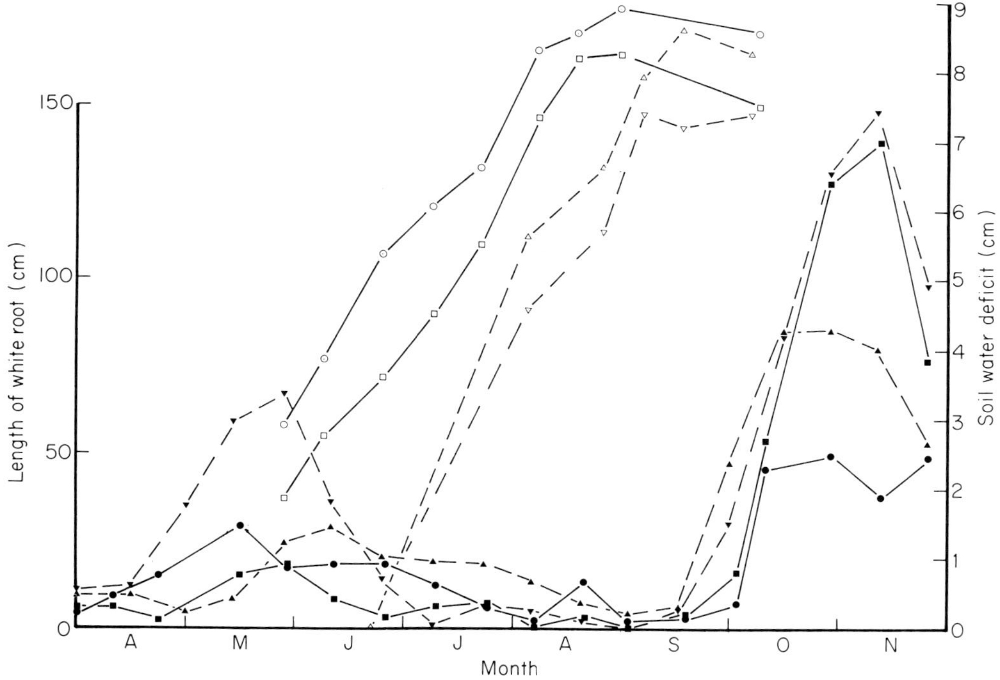

Fig. 1. The seasonal periodicity of white root length and of the depletion of soil moisture in 1970. ●Tree 1 root growth, ○Tree 1 water depletion; ■ Tree 2 root growth, □ water depletion; and 1971 ▲Tree 1 root growth, △ water depletion; ▼ Tree 2 root growth, ▽ water depletion.

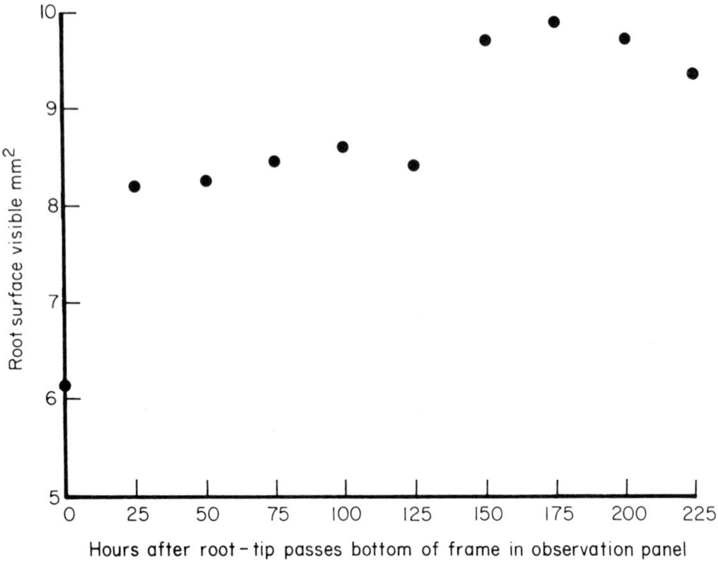

Fig. 2. Variation in the area of root surface of an individual root visible, i.e. not in contact with soil, as viewed through the observation panel of a root laboratory.

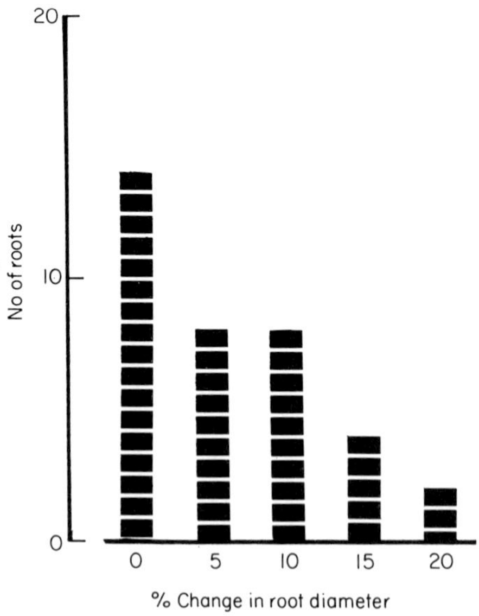

Fig. 3. The number of roots showing different % changes in root diameter between early morning and mid-day, as observed through the observation panel of a root laboratory.

TABLE 3

Values of LA (root length under unit area of ground surface) for field grown apple trees under herbicide management

Tree Age	Depth of Sampling (cm)	LA	Source of data
Field measurements			
26 year	75	2.81	Atkinson (1973)
12 year	30	1.85-2.16	White (1977)
5 year	120	0.79-4.33	Atkinson *et al.* (1976)
Root Laboratory measurements - maximum values			
1-3 year	120	3.6-23.8	Atkinson (1978)
4 year	120	5.23	Atkinson (1977)

the glass would be recorded and assuming that root density at the glass was similar to that at a distance. Even allowing possible losses during processes such as excavation apple trees often do have values of LA below 10 cm cm^{-2}. In addition, LA will vary during the year and maximum values as seen in the root laboratory will be transitory and may occur at times inappropriate for water supply (see Fig. 1).

Discussion

Both unsuberized and woody roots of cherry (and presumably of other fruit tree species) can absorb both water and phosphate. The relative contribution of the two types of root will depend upon the relative amounts present (which will vary with tree age and during the season), the substance being absorbed (which will influence the relative importance of root length, surface area and volume) and any factors which differentially modify the inherent ability of the two types to absorb.

The data presented here supports the conclusion of Kramer and Bullock (1966) who found that excised suberized roots of Loblolly pine were permeable to water. They also concluded that this type of root probably provided the major absorbing surface of the tree. For roots of the diameters used here they found permeabilities of 0.04 - 0.35 $\mu l\ mm^{-2}h^{-1}$ which are comparable with values in Table 1. They also found an increase in absorption per unit length with increasing root diameter. There was a similar relationship in the present experiment.

Addoms (1945) suggested that the main paths of water entry into suberized roots were via lenticels, breaks around root branches, and wounds. The low variability observed in absorption of water per unit area of root surface (Table 2) suggests that if water enters roots this way the uptake sites must be widely distributed on their surface.

The ability of woody roots to absorb ^{32}P is consistent with the results of Crinder (1933) and Nightingale (1935) who found that woody roots could absorb minerals and with those of Hartley *et al.* (1956) who demonstrated the movement of elements such as zinc through tree bark.

The relative importance of length, area and volume in the absorption of nutrients has been discussed by Russell and Sanderson (1967) who found that ion uptake by young cereal roots was more consistently related to their volume than to surface area. This latter relationship however becomes important when diffusion to the root surface limits the rate to which slowly diffusible ions, such as phosphate, are absorbed (Bhat and Nye, 1974). However, the similarity in uptake per unit area found in this study could imply that for the tree root system as a whole uptake would be proportional to surface area. This clearly needs further investigation to determine whether it is true over a wider range of root sizes and environmental conditions.

The degree of contact between tree roots and soil varies with root type and age. Initially, particularly if the root deformed the soil during its growth rather than using an existing channel, contact with the soil can be good. Sanders (1971) found that 40% of apple roots visible in an observation laboratory were not in contact with soil while others were in incomplete contact and as shown in Figure 2, contact can be reduced by the activities of the soil fauna. Analysis of time lapse films (Head, 1968a; Atkinson, Lewis and Jones, 1977) indicates that this type of activity is greatest around white and newly suberized roots. Root soil contact could also be reduced by diurnal changes in root diameter due to water stress (Huck, Klepper and Taylor, 1970). However, measurements on white cherry roots (Fig. 3) suggested that even in dry years large diurnal changes in diameter were rare and, therefore, unlikely to have as large an effect on contact as some other factors.

After browning and the decay of the cortex contact between root and soil is generally poor; the loss of the cortex can cause a 50% reduction in root diameter (Rogers, 1968) so that the unthickened stele is often suspended in the root cavity with minimal contact (Head, 1968a, Plate 3; Head, 1968b, Plate 2). If the root becomes secondarily thickened root soil contact improves. The maximum rate of root thickening found by Head (1968b) was of 3.7 mm per season. Therefore, within a season an average apple root with a diameter of 1.5 mm which decreased to 0.75 mm on the loss of the primary cortex could re-establish contact with the soil which probably would remain good (Head, 1968b, Plate 1). Root soil contact will always be dynamic and will vary between different root types.

The importance of root soil contact in water absorption has been discussed by Tinker (1976) who suggested that when uptake per unit root length (Iw) was low ($< 0.015\ cm^3cm^{-1}d^{-1}$) the degree of root soil contact would be unimportant, but that when Iw was large ($0.3\ cm^3cm^{-1}d^{-1}$) the water potential at the root surface would be critically dependent upon contact. In such a case an appreciable gradient in potential between the root surface and bulk soil would occur.

Among other factors, Iw will be influenced by root length. On the assumption that LA (root length per unit soil surface) was rarely as low as 10 cm cm^{-2} Newman (1969) concluded that appreciable gradients in water potential at the root surface would also be rare. Table 3 indicates that apple trees often have values of LA below 10, and, therefore, appreciable gradients at the root surface and high values of Iw may occur.

If transpiration occurs at 0.5 cm d^{-1} (Tinker, 1976) then water flow per unit length would vary from 0.12 - 0.31 $cm^3cm^{-1}d^{-1}$ through the roots of a high density planting of apple (Atkinson *et al.*, 1976) which had an LAI of over 3. In these conditions variations in root-soil contact would be expected to have an important influence on absorption.

Acknowledgements

This work was carried out while Sandra A. Wilson was in receipt of an Agricultural Research Council Research studentship.

References

Addams, R.M. (1946). Entrance of water into suberized roots of trees. *Plant Physiol.*, 21, 109.

Atkinson, D. (1973). Field studies on root systems and root activity. Rep. E. Malling Res. Stn for 1972, 56.

Atkinson, D. (1974). Some observations on the distribution of root activity in apple trees. *Plant and Soil*, 40, 333.

Atkinson, D. (1977). Some observations on the root growth of young apple trees and their uptake of nutrients when grown in herbicided strips in grassed orchards. *Plant and Soil*, 49, 459.

Atkinson, D. (1978). The use of soil resources in high density planting systems. *Acta Horticulturae*, (in press).

Atkinson, D., Lewis, J.K. and Jones, E.Y. (1977). Time-lapse cinematographic studies of root growth using an underground observation laboratory. *Zeszyty Problemowe Poslepow Nauk Rolniczych*, 188, 293.

Atkinson, D., Naylor, D. and Coldrick, G.A. (1976). The effect of tree spacing on the apple root system. *Hort. Res.*, 16, 89.

Bhat, K.K.S. and Nye, P.H. (1974). Diffusion of phosphate to plant roots in soil. III. Depletion around onion roots without root hairs. *Plant and Soil*, 41, 383.

Clarkson, D.T., Sanderson, J. and Russell, R.S. (1968). Ion uptake

and root age. *Nature*, 220, 805.
Clarkson, D.T. and Sanderson, J. (1971). Relationship between the anatomy of cereal roots and the absorption of nutrients and water. A.R.C. Letcombe Lab. A. Rept. 1970, 16.
Crinder, F.J. (1933). Selective absorption of ions not confined to young roots. *Science*, 78, 169.
Graham, J., Clarkson, D.T. and Sanderson, J. (1974). Water uptake by the roots of marrow and barley plants. A.R.C. Letcombe Lab. A. Rept. 1973, 9.
Harrison-Murray, R.S. and Clarkson, D.T. (1973). Relationships between structural development and the absorption of ions by the root system of *Cucurbita pepo*. *Planta*, 114, 1.
Hartley, C.P., Regeimbal, L.Q. and Moon, H.H. (1956). Absorption of nutrient salts by bark and woody tissues of apple and subsequent translocation. *Amer. Soc. Hort. Sci. Proc.*, 67, 417.
Head, G.C. (1966). Estimating seasonal changes in the quantity of white unsuberized roots on fruit trees. *J. Hort. Sci.*, 41, 197.
Head, G.C. (1967). Effects of seasonal changes in shoot growth on the amount of unsuberized root on apple and plum trees. *J. Hort. Sci.*, 42, 169.
Head, G.C. (1968a). Studies of growing roots by time lapse cinematography. *9th Int. Cong. Soil Sci. Trans.*, 1, 751.
Head, G.C. (1968b). Seasonal changes in the diameter of secondarily thickened roots of fruit trees in relation to growth of other parts of the tree. *J. Hort. Sci.*, 43, 275.
Huck, M.G., Klepper, B. and Taylor, H.M. (1970). Diurnal variations in root diameter. *Plant Physiol.*, 45, 529.
Jones, O.P., Hopgood, M.E. and O'Farrell, D. (1977). Propagation *in vitro* of M.26 apple rootstocks. *J. Hort. Sci.*, 52, 235.
Jones, O.P. and Hopgood, M.E. (1978). Propagation *in vitro* of fruit trees: Rootstocks of cherry and plum. *J. Hort. Sci.*, 53, (in press)
Kramer, P.J. and Bullock, H.C. (1966). Seasonal variations in the proportions of suberized and unsuberized roots of trees in relation to the absorption of water. *Amer. J. Bot.*, 53, 200.
Newman, E.I. (1969). Resistance to water flow in soil and plant. I. Soil resistance in relation to amounts of root: theoretical estimates. *J. appl. Ecol.*, 6, 1.
Nightingale, G.T. (1935). Effects of temperature on growth anatomy and metabolism of apple and peach roots. *Bot. Gaz.*, 96, 581.
Rogers, W.S. (1968). Amount of cortical and epidermal tissue shed from roots of apple. *J. Hort. Sci.*, 43, 527.
Rogers, W.S. (1969). The East Malling root observation laboratories. *In* "Root Growth" (Ed. W.J. Whittington), pp.361-376. Butterworth, London.
Russell, R.S. and Sanderson, J. (1967). Nutrient uptake by different parts of the intact roots of plants. *J. Exp. Bot.*, 18, 491.
Sanders, F.E.T. (1971). "The Effect of Root and Soil Properties on the Uptake of Nutrients by Competing Roots". D.Phil. Thesis, Oxford.
Taylor, H.M. and Klepper, B. (1971). Water uptake by cotton roots during an irrigation cycle. *Aust. J. Biol. Sci.*, 24, 853.
Tinker, P.B. (1976). Roots and Water: Transport of water to plant roots in soil. *Phil. Trans. R. Soc. Lond. B.*, 273, 445.
White, G.C. (1977). Herbicide strip width/nitrogen trial. Rep. East Malling Res. Stn for 1976, 106.

A PHOTOGRAPHIC VIEW OF MICROSCOPIC PROCESSES AT THE ROOT-SOIL INTERFACE*

MORRIS G. HUCK

A.R.S., U.S. Dept. of Agr., Dept. of Agronomy and Soils, Auburn University, Auburn, Alabama, U.S.A.

Roots of cotton and other crops growing in soil behind a vertical glass window were photographed at intervals of 1 to 5 minutes with a motion-picture camera. When germinating seeds were used as a carbon-source, elongation of the emerging radicle was maintained at a constant rate over intervals of several days in a moisture-tight box at constant temperature. Roots were observed as they grew along naturally-occurring channels or as they pushed soil particles aside to create new root-channels in the soil mass. Morphological development, from initial passage of the root cap until emergence of root hairs is visible. Removal of molecular oxygen from soil pore-spaces by flushing the soil container with N_2 gas resulted in an immediate interruption of root elongation processes.

Root ageing, branch-root formation, and growth throughout a larger soil volume were photographed in the Auburn rhizotron using photosynthetic leaves as a carbon-source. Roots supplying water to transpiring leaves were observed to shrink away from the walls of their root-channels during the day and recover their turgor at night. The amount of root shrinkage in response to plant water stress was much

* Dr Huck was prevented from submitting a fuller account of his contribution because of ill-health.

less with older roots. Under rhizotron conditions, water droplets appeared to condense on root hairs and soil particles within 10-30 cm of the soil surface each night as the soil cooled below the temperature of saturated vapour in equilibrium with deeper soil layers. With the onset of transpiration next morning, the condensed droplets disappeared before any root shrinkage was observed.

THE HYDRAULIC RESISTANCE OF THE SOIL-ROOT INTERFACE - A CAUSE OF WATER STRESS IN PLANTS

P.E. WEATHERLEY

Department of Botany, University of Aberdeen, UK

Summary

Experiments with sunflower plants indicate that at rapid rates of transpiration a perirhizal drop in water potential arises of about 8 bars in sand or soil, twice as large as that through the plant. The perirhizal zone was estimated to extend 0.1 cm from the root surface, giving a perirhizal gradient of 80 bars cm^{-1}. Such a steep gradient would require the hydraulic conductivity of the sand or soil to be several orders of magnitude less than estimates obtained experimentally. It is concluded that the resistance to water flow is at the soil-root interface.

Introduction

The origin of water stress in transpiring plants can be understood with references to a simple physical model (Weatherley, 1970, 1976). Two porous pots filled with water are joined by a vertical tube. The upper one in the air represents the leaf, the lower one which is semipermeable is buried in the soil and represents the roots. A manometer attached to the vertical tube represents the vacuoles of the cells adjacent to the pathway e.g. the parenchyma of the xylem of the stem. At a steady rate of transpiration the manometer will register a reduction of pressure arising in the pathway, which is analogous to the water stress in the tissues of the plant. If the hydraulic resistance of the tissues is constant as is the porous pot and vertical tube in the model, the stress in the plant will be a linear function of the rate of transpiration.

The validity of this prediction can be tested by measuring steady state values of depression of leaf water potential ($\Delta\psi_\ell$) at various steady rates of transpiration (F) contrived by maintaining different levels of humidity in the atmosphere. Plotting ($\Delta\psi_\ell$) against F a straight line should be obtained. Experiments of this kind using water culture grown plants to avoid any complications introduced by the movement of water through the soil, have yielded widely different results for different species and for the same species at the hands of different experimenters (Macklon and Weatherley, 1965; Tinklin and Weatherley, 1966; Stoker and Weatherley, 1971; Barrs, 1973; Newman, Thurtell and Stevenson, 1974; Black, 1974). In this department we have always found with *Ricinus*, *Gossypium* and *Helianthus* that $\Delta\psi_\ell$ is in fact constant over a wide range of transpiration rates. Such constancy means that the hydraulic conductance of the transpiration pathway in these plants is, over this range, a linear function of the rate of flow through the plant.

Whatever the explanation of such an adjustment in conductance, and one hypothesis has recently been put forward (Powell, 1978), the plateau value of $\Delta\psi_\ell$ over a wide range of transpiration rates makes it relatively easy to detect any departure from this pattern when plants are rooted in unsaturated sand or soil. In studies by Macklon, Tinklin, Stoker and Black referred to above, it has been shown that while at low rates of transpiration the values of $\Delta\psi_\ell$ with unsaturated soil were identical with those of water culture plants or with saturated soil (say 7 bars), at higher rates of transpiration $\Delta\psi_\ell$ rises considerably (say to 15 bars). On reducing the rate of transpiration to a low value again, $\Delta\psi_\ell$ falls to its former value. This cycle can be repeated several times with similar results.

The rise in $\Delta\psi_\ell$ above the plateau value is interpreted as a rise in the depression of water potential in the soil ($\Delta\psi_s$) at the root surface. In the example cited above $\Delta\psi_s$ would be 8 bars. This cannot be the result of a general drying out of the soil since there is complete 'recovery' on reducing the rate of transpiration. Also it seems that this rise in $\Delta\psi_s$ occurs near to the surface of the individual roots since such cycles can be obtained in pots in which the root system is dense and ramifies throughout the soil mass (Faiz and Weatherley, 1977, 1978).

Tinklin and Weatherley (1968) put forward the hypothesis that the hydraulic conductance of the soil was sufficiently low to allow steep gradients of soil water potential to arise dynamically round the individual roots (perirhizal gradients). This is consistent with the models put forward by Phillip (1957), Gardner (1960) and Cowan (1965) which predict the existence of such gradients.

It is evident that the drier the soil and the greater the rate of transpiration the more likely is it that such gradients will arise. However, Newman (1969a, b) and Andrews and Newman (1969) argue forcibly that under conditions normally found in the field, perirhizal gradients will be negligible until the water content of the soil approaches the wilting point. How then are the results of Macklon and Tinklin to be explained? Newman suggested that the greatest drop in water potential within the soil is not round each individual root but between the rooting zone as a whole and the region of soil untapped by roots which acts as a reservoir from which water is drawn. The resistance encountered by the water in this movement from one region to another Newman calls the *pararhizal* resistance. In our experiments the pararhizal resistance between the water table and the rooting zone was, it is suggested, much greater than the perirhizal resistance i.e. the resistance *within* the rooting zone. On this hypothesis 'recovery' from water stress on reducing the rate of transpiration results from a flow of water into the rooting zone from these supplying regions, whilst on the hypothesis of perirhizal gradients, recovery represents a readjustment of water *within* the rooting zone.

Experiments and Discussion

To decide between these two possibilities two kinds of experiments were carried out (Faiz and Weatherley, 1977). In the first, plants were grown in large containers in which the roots ramified throughout the soil volume so that there could not have been a pararhizal flow. It was found that a rise in $\Delta\psi_\ell$ of around 8 bars followed a rise in rate of transpiration and this was dynamic in origin since 'recovery' occurred when the rate of transpiration was reduced. In the second kind of experiment, pots were used which were divided into an upper part to which the roots were confined by a fine nylon gauze which separated it from a lower root-free part containing moist soil. The roots could not penetrate the gauze but there was capillary continuity through it so that water could move freely into the upper part from the lower part. A water table was maintained at a fixed level in the lower part. As in the previous experiments when the rate of transpiration was raised $\Delta\psi_\ell$ increased steadily. If now the rate of transpiration was lowered by raising the atmosphere humidity 'recovery' occurred even when the lower water supplying pot was simultaneously removed, thus the changes in soil moisture involved were within the rooting zone i.e. perirhizal.

Having established that gradients of water potential appear to arise round the roots of transpiring plants, Faiz and Weatherley (1978) carried out further experiments in which additional data on the dimensions of the root system and distribution of water potentials within the rooting zone were obtained. Sunflower plants were grown in pots each of which consisted of an upper and lower part separated by nylon gauze which restricted the roots to the upper part and a constant water table was maintained in the lower pot. The experiments were carried out in a wind tunnel in which different levels of humidity favouring different rates of transpiration, could be maintained. Rates of transpiration were measured continuously on a recording balance. Leaf water potentials were measured psychrometrically and with the pressure bomb and continuous recordings of leaf water status were obtained using a beta gauge. Soil water potentials ($\Delta\psi_s$) in samples and *in situ* were measured psychrometrically. Estimates of the total length of the root system and root diameters were based on direct measurements.

The results of an experiment using sand grown plants are shown in Figure 1. During the first period the relative humidity of the air stream was maintained at 95% giving a rate of transpiration of about 0.75 $gm^{-1}\ dm^{-2}$. $\Delta\psi_\ell$ remained steady at about 6 bars and $\Delta\psi_s$ was 0.5-1.0 bar at various locations in the root zone and at or near zero below the gauze. When the transpiration rate was increased by decreasing the humidity to 40% (period 2) $\Delta\psi_\ell$ increased to a steady value of about 15 bars accompanied by an increase in $\Delta\psi_s$ to 2-3 bars in the main root zone. Just above the nylon gauze it was slightly greater than 1 bar and below the gauze it remained at or near to zero. The increase in rate of transpiration at the beginning of period 2 was not maintained but fell away to a lower but steady value by the end of the period. This decline was presumably in response to a rise in stomatal resistance resulting from the rise in $\Delta\psi_\ell$. In period 3 the sand was saturated with water and the rate of transpiration kept as nearly as possible at the same level as the steady rate at the end of period 2, by regulating the humidity of the air. In response to this, $\Delta\psi_\ell$ fell to around 4 bars. It may be assumed that during period 3 $\Delta\psi_s$ was virtually zero and so

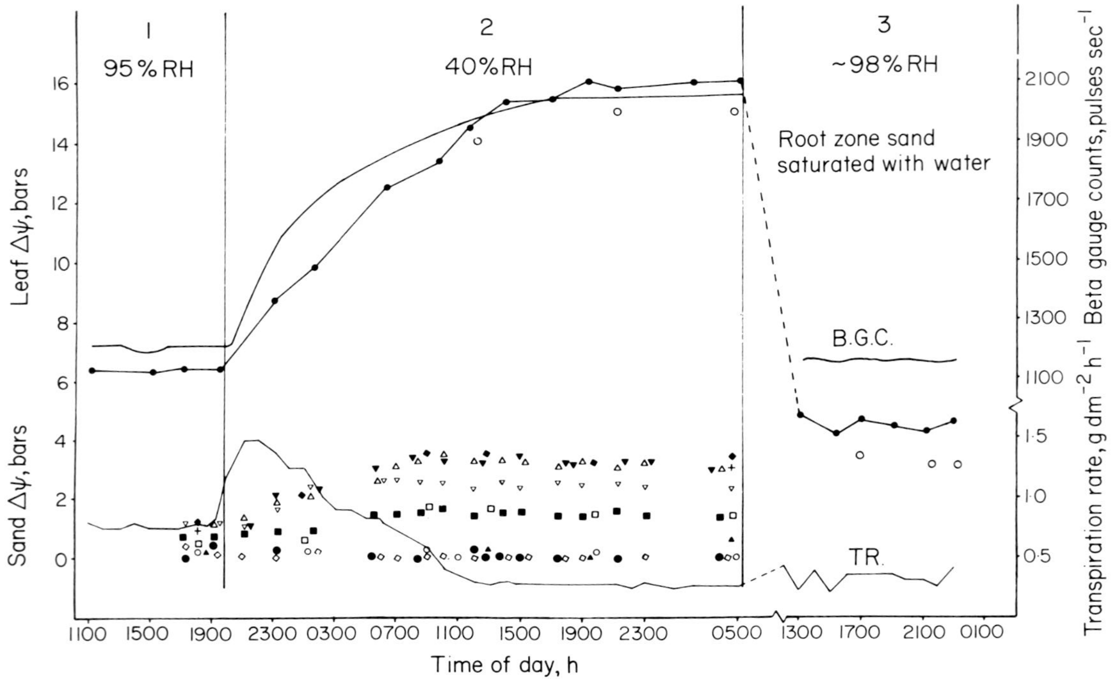

1
95% RH
2
40% RH
3
~98% RH
Root zone sand
saturated with water
B.G.C.
TR.
Leaf Δψ, bars
Sand Δψ, bars
Time of day, h
Transpiration rate, g dm⁻² h⁻¹
Beta gauge counts, pulses sec⁻¹

$\Delta\psi_\ell$ represents the drop in water potential through the plant alone. This value will also apply to the final steady state in period 2 where the rate of flow of water through the system is the same. Thus, subtracting $\Delta\psi_\ell$ in period 3 from that in period 2 gives a value for $\Delta\psi_s$ at the root surface. Furthermore, subtracting the $\Delta\psi_s$ of the bulk sand in the rooting zone from this value of $\Delta\psi_s$ at the root surface gives the magnitude of the perirhizal drop in water potential. Data from the above experiment together with those from an experiment with soil, are given in Table 1. Separate figures obtained with the psychrometer and pressure bomb are given. It will be seen that the perirhizal drop in water potential with both sand and soil is around twice the drop through the plant. Thus in these experiments the major resistance to water flow in the soil-plant system lay in soil round the roots.

How far are these perirhizal differences in water potential consistent with the magnitude of the hydraulic conduc-

---

Fig. 1. (opposite)
Plants were grown in sand in pots with upper and lower compartments separated by fine nylon gauze which confined the roots to the upper compartments but allowed upward capillary rise of water from a water-table maintained at 27 cm below the gauze. The curves show the effect of changing the humidity of the airstream on the rate of transpiration (TR) and the depression of leaf water potential $\Delta\psi_\ell$ (measured with the psychrometer , and with the pressure bomb . β gauge readings are also given for comparison, BGC.). The depressions of water potential of the sand were measured at the following locations above and below the gauze:

From samples
◆ 3 - 7 cm below surface, 6 cm from centre
+ 10 cm below surface, 2 - 4 cm from centre
□ just above the gauze
▲ just below the gauze
○ 10 cm below the gauze

Measured in situ
▼ 4 cm below surface, 3 cm from centre
△ 10 cm below surface, 6 cm from centre
▽ 13 cm below surface, 1 cm from centre
■ just above the gauze
● just below the gauze
◇ 10 cm below the gauze

(after Faiz and Weatherley, 1978: by permission of Blackwell Sci.Pub.)

TABLE 1

Data for calculating the perirhizal drop in water potential. All figures in bars.

(From Faiz and Weatherley, 1978: by permission of Blackwell Sci.Pub.)

	Sand			Soil		
	Psychrometer	P.bomb	mean	Psychrometer	P.bomb	mean
a) Drop in Ψ through the plant ($\Delta\Psi_\ell$ period 3)	4.3	3.2	3.8	5.5	3.8	4.7
b) $\Delta\Psi_\ell$ (period 2)	15.9	15.0		15.0	14.4	
c) $\Delta\Psi_s$ at root surface in period 2 (b-a)	11.6	11.8		9.5	10.6	
d) $\Delta\Psi_s$ in root zone (period 2)	3.2	3.2		2.3	2.3	
e) Perirhizal drop in Ψ_s (c-d)	8.4	8.6	8.5	7.2	8.3	7.8

tivity of the sand or soil? This question can be examined in the following way. Under the steady state conditions at the end of period 2 in Figure 1 the volume rate of flow from the water table through the gauze (pararhizal flow) and the flow through the total surface of the roots (perirhizal flow) were the same. Now the cross sectional area of the gauze was 177 cm^2 whilst that of the whole root system was 10,715 cm^2. This means that the root surface was 60 times the area of the gauze. Thus under steady state conditions the *velocity* of flow through the root surface was only 1/60 that through the pararhizal pathway. From the soil psychrometer readings as shown in Figure 1 it can be seen that in the region of the gauze the gradient in water potential was about 1 bar cm^{-1}. The perirhizal gradient would then be expected to be 1/60 of this i.e. it would be negligibly small. However, as shown in

Table 1, the perirhizal drop in water potential was about 8 bars. The distance b, through which this drop occurred may be taken as half the distance between adjacent roots and was obtained from the identity $b = 1/(L_v)^{\frac{1}{2}}$ where L_v is the root density (Andrews and Newman, 1969). It was found to be 0.1 cm in both our sand and soil experiments. Thus the perirhizal gradient was about 80 bars cm^{-1}. Such a steep gradient would require the hydraulic conductivity of the soil in the perirhizal zone to be 6 orders of magnitude less than that in the pararhizal pathway. In the experiments with sand the discrepancy was 4 orders of magnitude. Admittedly the soil within the rooting zone was somewhat drier ($\Delta\psi \simeq$ 2 bars) than that in the pararhizal zone, but it is not thought that this would account for such a large discrepancy. At all events $\Delta\psi_s$ in the root zone was less than 1 bar at the beginning of period 2, at which time the rate of transpiration was rising, but it is difficult to see how a perirhizal drop in water potential greater than the pararhizal drop can have developed when the perirhizal velocity of flow was so small relative to the pararhizal velocity.

It seems that these data are not after all consistent with the development of marked perirhizal gradients and indeed lend support to Newman's contention that such gradients are unlikely when $\Delta\psi_s$ is well below the wilting point. Yet our experiments provide clear evidence (Table 1) of a perirhizal drop in water potential of around 8 bars. If this does not reside in the soil round the roots we suggest that it is the *interface* between the soil and the surface of the root which presents a high resistance to water flow. There may be no obvious reason why the soil-root interface should have a higher resistance to water flow than that within the soil itself. But the interface does present a sharp change in the architecture of the hydraulic pathway, about which we know very little and the characteristics of which remain to be worked out.

The preliminary experiments described above suggest two features which may be important. First, the interfacial drop in water potential was rather similar with sand and soil (8.5 bars and 7.8 bars respectively). The presumably wide difference in texture between the sand and soil might have been expected to have given rise to a more marked difference in root-interface properties. Second, the magnitude of the interfacial resistance seems to depend on the rate of flow across the interface and the water content of the soil or sand at the interface. The interaction between these factors is complex however. This is evident from a comparison between period 1 and the final steady state in period 2 in Figure 1. In period 1 the interfacial drop in water potential was zero (i.e. $\Delta\psi_\ell$ at a plateau value). At the end of period 2 it was 7-8 bars although now the rate of transpiration (rate of flow across the interface) was only 1/3 of the rate in period 1. There is a suggestion here of a positive feed-back from the water potential within the plant. Such a feed-back system might be the contraction of the roots with increasing water stress in the root tissues which could lead to a gap opening up between the soil and the root i.e. an attentuation of the interface, with a corresponding increase in its hydraulic resistance. The occurrence of such gaps has been envisaged by Philip (1957, 1958), Bonner (1959) and de Roo (1969) and Huck, Klepper and Taylor (1970) found that during a dry sunny day the roots of cotton plants shrank by as much as 40%. In the course of the experiments described here, Faiz (1973) showed that the relative water content of the roots fell from 94% to 70% between the end of period 1 and the end of period 2. This considerable decrease in tissue volume entailed a contraction of the roots of about 20% which undoubtedly led to the formation of a gap, or at least a loosening of the soil particles round the roots.

Whilst root contraction may form the basis of a feed-back mechanism, it cannot be the cause of the increased resistance to flow at the onset of a rise in rate of transpiration. The roots will have shrunk only after they have been initially to some degree stressed and this initial stress would appear to result from the original interfacial resistance.

To sum up the present position: There is very strong evidence that a large drop in water potential arises round the roots of rapidly transpiring plants growing in soil or sand. This is dynamic in origin in that 'recovery' occurs on reducing the rate of transpiration. There is some evidence that this drop in water potential is not in the form of a perirhizal gradient in the soil - the hydraulic conductivity of the soil would appear to be too great for such a gradient to arise. It is concluded therefore that the resistance is probably at the soil-root interface. This resistance causes a fall in water potential in the plant at the onset of an increase in rate of transpiration which results in a contraction of the roots and a rise in interfacial resistance giving a positive feed-back.

References

Andrews, R.E. and Newman, E.I. (1969). Resistance to water flow in soil and plant. III. Evidence from experiments with wheat. *New Phytol.*, 68, 1051.

Barrs, H.D. (1973). Controlled environment studies of the effects of water stress on photosynthesis, transpiration and water status. *In* "Plant Response to Climatic Factors" (Ed. R.O. Slatyer). pp. 249-258. Paris: U.N.E.S.C.O.

Black, C.R. (1974). "The Resistance to Transpirational Water Flow in Sunflower Plants." Ph.D. Thesis, University of Aberdeen.

Bonner, J. (1959). Water transport. *Science*, 129, 447.

Cowan, I.R. (1965). Transport of water in the soil-plant-atmosphere system. *J. appl. Ecol.*, 2, 221.

De Roo, M.C. (1969). Water stress gradients in plants and soil-root systems. *Agron. J.*, 61, 511.

Faiz, S.M.A. (1973). "Soil-root Water Relations". Ph.D. Thesis, University of Aberdeen.

Faiz, S.M.A. and Weatherley, P.E. (1977). The location of the resistance to water movement in the soil supplying the roots of transpiring plants. *New Phytol.*, 78, 337.

Faiz, S.M.A. and Weatherley, P.E. (1978). Further investigations into the location and magnitude of the hydraulic resistance in the soil-plant system. *New Phytol.*, 77 (in press).

Gardner, W.R. (1960). Dynamic aspects of water availability to plants. *Soil Science*, **89**, 63.

Huck, M.G., Klepper, B. and Taylor, H.M. (1970). Diurnal variations in root diameter. *Plant Physiol.*, 45, 529.

Macklon, A.E.S. and Weatherley, P.E. (1965). Controlled environment studies of the nature and origin of water deficits in plants. *New Phytol.*, 64, 414.

Neumann, H.H., Thurtell, G.W. and Stevens, K.R. (1974). *In situ* measurements of leaf water potential and resistance to water flow. *Can. J. Pl. Sci.*, 54, 175.

Newman, E.I. (1969a). Resistance to water flow in soil and plant. I. Soil resistance in relation to amounts of root: theoretical estimates. *J. appl. Ecol.*, 3, 139.

Newman, E.I. (1969b). Resistance to water flow in soil and plant. II. A review of experimental evidence on the rhizosphere resistance. *J. appl. Ecol.*, 6, 261.

Philip, J.R. (1957). The physical principles of soil water movement during the irrigation cycle. *Proc. Int. Congr. Irrig. Drain.*, 8, 125.

Philip, J.R. (1958). The osmotic cell, solute diffusibility and the plant water economy. *Plant Physiol.*, 33, 264.

Powell, D.B.B. (1978). Regulation of plant water potential by membranes of the endodermis in young roots. *Plant, Cell and Environment*, 1, 69.

Stoker, R. and Weatherley, P.E. (1971). The influence of the root system on the relationship between the rate of transpiration and depression of leaf water potential. *New Phytol.*, 70, 547.

Tinklin, R. and Weatherley, P.E. (1968). The effect of transpiration rate on the leaf water potential of sand and soil rooted plants. *New Phytol.*, 67, 605.

Weatherley, P.E. (1970). *In* "Advances in Botanical Research" (Ed. R.D. Preston), 3, 171. Academic Press, New York.

Weatherley, P.E. (1976). A discussion on water relations of plants. *In* "Phil. Trans. R. Soc. Lond. B." (Ed. J.L. Monteith and P.E. Weatherley), 273, 435.

CORN AND COTTON ROOT GROWTH IN RESPONSE TO OSMOTIC POTENTIAL AND OXYGEN AND NITRATE CONCENTRATIONS IN NUTRIENT SOLUTIONS

B. BAR-YOSEF and J.R. LAMBERT

Institute of Soils and Water, Agricultural Research Organization, The Volcani Center, Bet Dagan, Israel

Summary

Corn and cotton plants were grown in nutrient solution in acrylic containers constructed of two co-axial cylinders, the annular volume of which was partitioned into six compartments. The root system of a single plant located in the internal cylinder was divided into the compartments, where the solutions differed in osmotic potential, oxygen concentration or nitrate concentration. Root length (L) in each compartment was measured as a function of growth time (t) to study the validity of equation (1) for the various parts of the root system and to determine the relationship between GRC (the specific root growth rate) and the concentration of each of the three variables in the solution.

$$dL/dt = GRC \cdot L \qquad (1)$$

Evidence is presented for the validity of equation (1) for the growth of various parts of the root system during 20 to 25 days.

While equation (1) held, GRC was affected by the different environmental factors, varying between 0.1 and 0.6 day^{-1}. Nitrate concentration in the range of 0 to 30 meq/liter had little effect on GRC at 24°C, but an increase in GRC with increasing nitrate concentration was observed at a temperature of 35°C.

Increasing the dissolved O_2 concentration from 5 to 25

ppm increased GRC, but the response diminished above 10 ppm O_2. A difference between GRC of a split and a whole root system was observed only in different osmotic potentials induced by adding carbowax. While in a split root system the minimum GRC was obtained for roots found at a potential of 1.6 bars, whole corn roots showed an almost continuous, though moderate decrease in GRC when the osmotic potential increased from that in the control solution to about 5 bars.

Introduction

We have suggested (Bar-Yosef and Lambert, 1978) that as long as carbohydrate supply from the shoots to the roots is non-limiting, the rate of root growth in various volumes of a root system can be described by

$$dL/dt = GRC \cdot L \qquad (1)$$

where L is the root length (cm) in a defined soil volume, t is time (days) and GRC is a specific rate coefficient (day^{-1}). When the various regions of a root system were under different water potentials and soil impedance, root elongation rate could still be predicted by equation (1), but GRC in each region varied with the water potential and soil impedance. The validity of equation (1) and the fact that various physical and chemical factors affecting root growth can be accounted for through their effect on GRC allow simulation of root growth under variable conditions. To apply this concept, relationships between GRC and environmental factors must be investigated and GRC found to be constant under fixed conditions throughout the plant's main growth period. A prerequisite to a study of GRC and its relationships is to separate the various interactive factors, such as water potential, soil impedance to root penetration, and nutrient and oxygen concentrations in the soil solution. Taylor *et al.* (1972) reported that the independent effect of water potential on root growth in the soil was negligibly small in a range of -0.18 to about -10.

bars. Hemsath *et al.* (1974) found, however, a response (varying in direction and magnitude with soil bulk density) to matric potential between -0.15 and -5 bar. In both reports, the possible interactions and effects of oxygen and nutrient concentrations in the soil solution on the results were not analyzed. Grable *et al.* (1968) and Huck (1970) studied the effect of O_2 concentration on root elongation. While Grable *et al.* (1968) reported almost linear response in corn when the oxygen concentration in the soil gas phase was increased from 0 to 21%, Huck (1970) reported no permanent change in the growth rate of cotton roots when O_2 concentration was varied between 21 and 2-3%. Part of the discrepancy can be attributed, as before, to interaction of factors in the two experiments. The effect of the concentration of nutrients on root elongation rate is evident from several sources. Pearson (1971) reported retarding effects due to low Ca^{++} and increasing Al^{3+} and NH_3 concentrations in soil and nutrient solutions. Brouwer (1962) and May *et al.* (1965) found an enhanced root growth when nitrogen concentration decreased, while Drew *et al.* (1973) showed that an increase in nitrate concentration promoted lateral initiation which subsequently increased the total extension rate of a barley root system.

The objectives of the present work were: a) to confirm the validity of equation (1) in nutrient solutions and to compare results of GRC obtained in split and whole root systems, and b) to determine GRC as a function of various levels of osmotic potential and nitrate and oxygen concentrations in nutrient solutions. A nutrient solution technique was chosen because it seemed to be the only way these relationships could be determined independently.

Experimental Procedure

The growth container shown schematically in Figure 1 was constructed of two coaxial acrylic cylinders, the annular volume of which was partitioned into six compart-

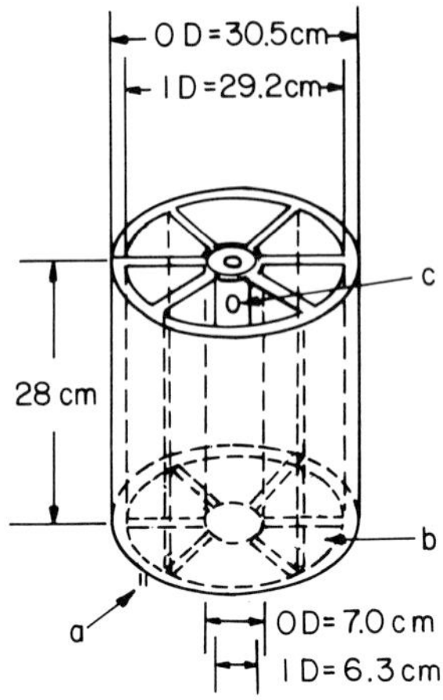

Fig. 1. Schematic description of the growth container. C = window (D = 1.2 cm) for roots. Partitions making up to 6 compartments are 0.6 cm thick.

ments. The volume of each compartment from the bottom to window was 2550 cm^3. Seedlings about 10 days old with at least six roots of about 5-cm length were supported in a rubber stopper, and inserted in the internal cylinder. The roots were separated and pulled through the windows to the outside compartments, which were filled with appropriate solutions. The solution level in the compartments was regulated so that the roots passing through the windows were covered with a capillary water film to prevent drying. Preliminary studies showed that contamination of the solution in the internal cylinder was negligibly small when all the solutions were replaced every second day. Plants were grown in chambers in which the temperature, humidity and photoperiod were automatically controlled. They were illuminated at 2500 foot candles for 16 hours daily. The "day" and "night" temperatures were 24 ± 2 and 17 ± 1, or 35 ± 1 and 26 ± 2°C, depending on the experiment. The "day" and "night" relative humidities were 70 ± 2 or 90 ± 2% respectively. The solutions were aerated and mixed by compressed air passed through a water trap. The roots length in each compartment was determined at two to four-day intervals. In each measurement the total length of all the components of all the roots were determined nondestructively. This was accomplished by comparing the length of individual roots with the length of a calibrated, flexible wire which was immersed into the solution. The comparison was made through the transparent walls of the growth container. Maximum root length per compartment was about 300 cm. To compare effects on split and whole root systems, whole plants shared the same compartment with split roots.

Three experiments (A,B,C) were conducted. In each, the composition of the basic nutrient solution, later termed solution H, was (in mM) 6.0 KNO_3, 4.0 Ca $(NO_3)_2$, 0.8 KH_2PO_4, 0.4 $MgSO_4$, 0.0245 H_3BO_3, 0.020 $FeSO_4$, 0.002 $MnSO_4$ and $ZnSO_4$ and 0.0005 $CuSO_4$ and MoO_3. The solutions were prepared with tap water of pH 8.40 and the above concentrations do not include the water background. The final pH of all the nutrient solutions was adjusted with 0.1 N H_2SO_4 to be 5.2 ± 0.2.

Experiment A

The independent variable was NO_3^- concentration. KNO_3 and $Ca(NO_3)_2$ concentrations were varied in the various compartments to give twice the NO_3 concentration in the H solution (termed TH), half (HH) and a quarter of its value (QH). Two other treatments were 0.01 mM or less NO_3 concentration (AH) and the H solution itself. Except for KNO_3 and $Ca(NO_3)_2$, the nutrient concentrations in all the compartments were identical but of course the total ionic strength and Ca and K concentrations varied. The two cations are assumed to be nonlimiting due to their concentration in tap water and presence as KH_2PO_4. The sixth compartment in each container was used as a replicate that varied for each repeat. Experiment A was repeated five times for both corn and cotton and took about 20 to 25 days to complete.

Experiment B

The independent variable was the osmotic potential of H solutions. The potentials were adjusted by adding to H solutions varying amounts of dialyzed polyethylene glycol (carbowax) 20,000 M (Fisher). The dialysis procedure, prevention of excessive foaming and the calibration curve relating percentage of carbowax in the equilibrium solution to osmotic potential were taken from Lagerwerff *et al.* (1961). The osmotic potential of the H solution was calculated from Van't Hoff's equation to be about 0.65 bar. Experiment B had two replicates with both corn and cotton and lasted five days.

Experiment C

The independent variable was dissolved O_2 concentration. Oxygen gas flowed from a pressure tank *via* a water trap to a dividing manifold, regulated and bubbled into the various compartments at desired flow rates. To obtain O_2 concentrations below 8 ppm, which is approximately the concentration in fresh, unstirred water, N_2 gas was bubbled instead of O_2. By this means, 3 ppm was the minimum O_2 concentration that could be obtained in our open system. Dissolved O_2 concentration was determined by polarography (portable YSI 5UA analyzer). Experiment C had two replicates with both corn and cotton and lasted 20 days.

Results

In order to test the validity of equation (1) during the growth period, data for root length versus time were analyzed to yield elongation rates (dL/dt) versus root length (L). In doing so for the various compartments, the time increments for the differentiation chosen were one to two days. The root elongation rate coefficient, GRC, was estimated from the slope of a linear regression (passing through the origin) relating the observed $\Delta L/\Delta t$ and L. The significance of the correlation coefficient of this regression can be used as an estimate of the validity of equation (1).

Results of experiments A and C showed that in more than 80% of the experimental compartments, the correlation coefficient of the linear regression was significant at the P = 0.05 level, or better. This indicates that equation (1) successfully described growth rates of both split and whole root systems grown in nutrient solutions for periods of time of 20 to 25 days. In Experiment B, which lasted 5 days only and consequently included fewer observations per compartment, equation (1) could not be validated.

The relationship between GRC and nitrate concentration in the nutrient solution is summarized in Figure 2. At 24°C and concentration range lower than 10 meq/liter, a

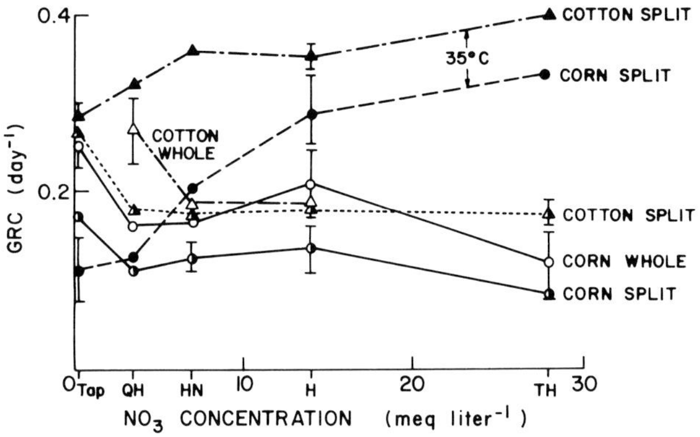

Fig. 2. Root growth rate coefficient (GRC day^{-1}) in whole and split root systems as a function of nitrate concentration in the nutrient solution. If not otherwise stated, the temperature was 24°C. The vertical lines show the average deviations of GRC from the mean of the replicates. Where not given, the deviation is near that of the next point to the right.

trend of increasing GRC values with decreasing nitrate concentration is observed in both whole and split root systems of corn and cotton. GRC for corn is shown to be generally lower than for cotton at any nitrate concentration. GRC values for whole root systems appear to be consistently, though not significantly, higher than those of the parts of the split root systems under identical conditions. Increasing the temperature to 35°C during the light hours resulted in an inverse relationship between GRC and concentration, namely, a decrease in GRC of both corn and cotton with decreased nitrate concentrations. That is, GRC values in the high concentration range were much higher at 35 than at 24°C, but not at the lower concentrations.

The relationship between osmotic potential at 24°C and GRC is given in Figure 3. Due to the short experimental period, GRC here was simply calculated according to equation (1). It can be seen that GRC in split root systems is lowest in the partial root system which is in a solution of -1.6 bar, and is highest in roots in solutions of -4.6 bars. As in Figure 2, the GRC value at any potential was higher in cotton than in corn, the difference becoming maximal at -4.6 bars. When whole corn plants were grown in the same solutions, GRC values decreased with increased osmotic potential but no minimum point was observed.

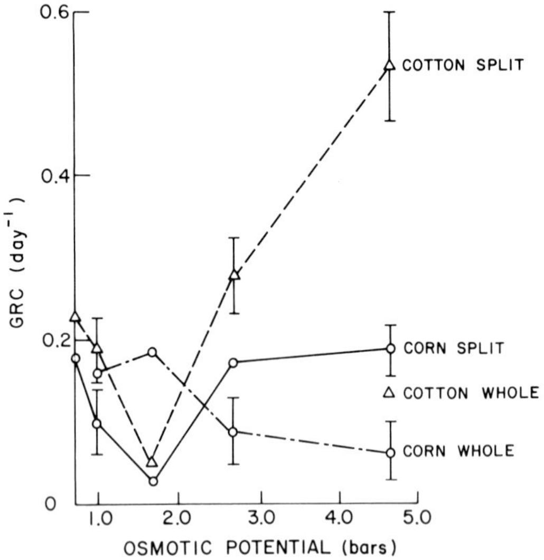

Fig. 3. Root growth rate coefficient (GRC day^{-1}) in whole and split root systems as a function of the solution osmotic potential. The temperature was 24°C. The vertical lines indicate confidence bands. Where not given, the band equals that of the next point to the right. All compartments had solution H.

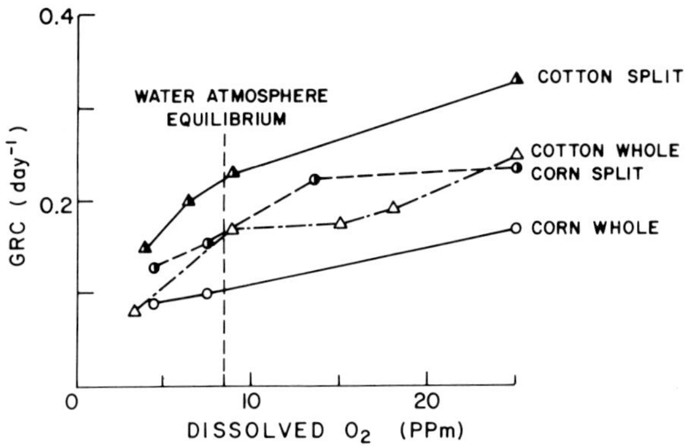

Fig. 4. Root growth rate coefficient (GRC day^{-1}) in whole and split root systems as a function of dissolved O_2 concentration. The temperature was 24°C. The maximum deviation in GRC between the two replicates was 30% of the mean, the average deviation being 15%.

The results of Experiment C, designed to relate GRC to the oxygen concentrations in nutrient solutions at 24°C, are shown in Figure 4. An observed increase in GRC with increasing oxygen concentration is common to split and whole root systems of both cotton and corn plants. The response of GRC to oxygen diminishes with increasing O_2 concentrations, and saturation seems to be attained at/or about 25 ppm O_2. At low oxygen concentration the various curves tend to converge and extrapolate towards zero GRC.

Discussion

In a previous work (Bar-Yosef and Lambert, 1978) evidence was presented for the validity of equation (1) under specific soil conditions. The results shown here indicate that the same is true in nutrient solutions. In both cases different environmental conditions in the root volume affected the growth rate coefficient, GRC, but not the relation given by equation (1). Among the environmental factors, the effect of nitrate concentration (C) on GRC at 24°C (Fig. 2) is similar to that reported by May *et al.* (1965) for whole barley root systems, and by Brouwer (1962) for single barley roots. The data in Figure 2 indicate that the responses of GRC to variations in concentration were similar in whole and parts of a root system under identical conditions though the rate of elongation of a whole root system was higher at each concentration. The inverse response of GRC to nitrate concentration at 35°C, namely, the decrease in GRC with decreasing concentrations, might be explained by the different behaviour of various components of the root system under the combined effect of temperature and nitrate concentration. Drew *et al.* (1973) reported that increasing concentration from 0.01 to 1. meq/liter resulted in a promotion of the initiation and subsequent rate of extension of lateral roots. If extrapolation to higher nitrate concentration is legitimate, and accepting that at 24°C GRC did not increase with increasing concentration this could imply that at 35°C the main mechanism of root elongation is enhancement of lateral initiation, while at 24°C elongation is due mainly to extension of main roots. Another possible explanation of the increase in GRC with increasing nitrate concentration at 35°C is the higher demand for N due to the increased rate of metabolic processes. Brouwer (1962) reported that under conditions of relative nitrogen deficiency in the plant, as might have been the case at 35°C and low N concentration, enrichment of N concentration in part of the root system enhanced root growth rate in that root volume.

The relation between osmotic potential and GRC (Fig. 3) enables the study of the independent effect of water potential on root growth. According to Taylor (1974), a water potential between -0.18 and -7 bars, has a negligibly small effect on root growth. The present data indicate a gradual decrease in GRC with decreasing osmotic potentials for whole root systems of corn plants and a variable relationship for corn and cotton split root systems. A decrease in root growth-rate with decreasing osmotic potentials is qualitatively reported by Brouwer (1962). The decrease in GRC in parts of the root system exposed to osmotic potentials decreasing from -0.65 to -1.6 bars (Fig. 2) is consistent with the decrease in GRC with water potential in the soil, observed in a previous work (Bar-Yosef and Lambert, 1978). This is apparent if an osmotic potential of about -0.6 bar, which existed in the soil solution, is added to the reported matric potential. Lower osmotic potentials are outside the range of the data in this previous work. Observations during the experiment showed that lower osmotic potentials resulted in the initiation of a greater number of thinner roots. This effect was more pronounced for cotton than for corn. It may be assumed that, in the case of split root systems, when only about 2/3 of the root volume is under limiting environmental conditions, carbohydrate supply from the shoots to the roots is not limiting root growth. Consequently, the decreased osmotic potential from -1.6 to -2.6 and to -4.6 bars might have resulted in proliferation of secondary and higher order roots which caused the increase in GRC with decreasing osmotic potentials. A similar phenomenon was reported by Voorhees *et al.* (1975) who found with pea seedlings that increased soil impedance promoted production of first order laterals. The whole plant may not be able to adjust root growth when exposed to different environmental conditions, as the partial root systems do, since carbohydrate supply may become limiting. This might explain the gradual decrease in GRC with decreased osmotic potential for a whole corn plant (Fig. 3).

Oxygen concentration is another environmental factor whose effect on GRC can be easily masked when studied in a soil system. Unlike results of Huck (1970), our data (Fig. 3) show an increase in GRC with increasing concentrations of dissolved oxygen in the nutrient solution. Assuming the roots to be uniform cylinders of a known radius (say 0.035 cm), the system to be at steady state, the roots surrounded by an immobile solution film of known thickness and O_2 concentration at the root surface to be zero, the oxygen uptake flux can be calculated for various O_2 concentrations in the bulk, stirred solution (Letey and Stolzy, 1967). Accepting 0.2 μg O_2 cm^{-2} min^{-1} as a growth limiting flux (Letey and Stolzy, 1967), oxygen concentration in the bulk solution must be at least 5 ppm to provide that flux when the immobile film is 1 mm thick. This concentration is attainable when O_2 concentration in the gas phase in equilibrium with the solution at 24°C is about 12%. If the water film is 2 mm thick, the required dissolved O_2 solution concentration is 8.5 ppm, corresponding to 21% O_2 in the gas phase. Assuming the film thickness to be between these two values, root elongation rate would respond to O_2 concentration in the range 0 to 6-7 ppm with a diminishing response above that concentration. The experimental results for corn and cotton (Fig. 3) follow the anticipated trend. Grable *et al.* (1968) found a linear relationship between root elongation and O_2 concentration in the soil atmosphere over the range of 0 to 21%. However, they also report a decrease in growth rate at higher O_2 concentrations, which we did not find.

Throughout this experiment the specific root growth rate at any level of the factors studied, is greater for cotton than for corn. The opposite was found when relating GRC to water potential in the soil (Bar-Yosef and Lambert, 1978), where GRC for corn was higher by up to 50% than for cotton. While values for corn in the soil and in the solution were close, that for cotton in the soil is appreciably smaller. It should be noted, however, that the range of total water potential in the soil (matric plus osmotic) was limited to the water potential range to the left of the low point in Figure 2, where the difference in GRC between corn and cotton was relatively small. The different ratio of GRC for corn to GRC for cotton in soil and solution implies that root growth of cotton is far more sensitive to impedance of root penetration by the soil than is that of corn and that the threshold impedance for cotton roots is much lower than that for corn.

References

Bar-Yosef, B. and Lambert, J.R. (1978). Corn and cotton root growth in response to environment: Soil water potential and impedance. Submitted to *Soil Sci. Soc. Amer. J.*

Brouwer, R. (1962). Nutritive influences on the distribution of dry matter in the plant. Neth. *J. Agric. Sci.*, 10, 399.

Drew, M.C., Saker, L.R. and Ashley, T.W. (1973). Nutrient supply and the growth of seminal root system in barley. *J. Exp. Bot.*, 24, 1189.

Grable, A.R. and Seimer, E.G. (1968). Effects of bulk density, aggregate size and soil water suction on oxygen diffusion, redox potential and elongation of corn roots. *Soil Sci. Soc. Amer. Proc.*, 32, 180.

Hemsath, D.L. and Mazurak, A.P. (1974). Seedling growth of sorghum in clay sand mixtures at various compactions and water contents. *Soil Sci. Soc. Amer. Proc.*, 38, 387.

Huck, M.G. (1970). Variation in tap root elongation rate as influenced by composition of soil air. *Agron. J.*, 62, 815.

Lagerwerff, J.V., Ogata, G. and Eagle, H.E. (1961). Control of osmotic pressure of culture solutions with polyethylene glycol. *Science*, **131**, 1486.

Letey, J. and Stolzy, L.H. (1967). Limiting distance between root and gas phase for adequate oxygen supply. *Soil Sci.*, 103, 404.

May, L.H., Chapman, F.H. and Aspinall, D. (1965). Quantitative studies of root development. I. The influence of nutrient concentration. *Aust. J. Biol. Sci.*, 18, 25.

Pearson, R.W. (1971). Introduction to symposium - the soil solution. *Soil Sci. Soc. Amer. Proc.*, 35, 417.

Taylor, H.M., Huck, M.G. and Klepper, B. (1972). Root development in relation to soil physical conditions. *In* "Optimizing Soil Conditions for Maximum Yields" (Ed. D. Hille), p.57. Wiley, New York.

Taylor, H.M. (1974). Root behavior as affected by soil structure and strength. *In* "The Plant Root and its Environment" (Ed. E.W. Carson), p.271. University Press of Virginia, Charlottesville, Virginia.

Voorhees, W.B., Farrell, D.A. and Larson, W.E. (1975). Soil strength and aeration effects on root elongation. *Soil Sci. Soc. Amer. Proc.*, 39, 948.

THE SOIL-ROOT INTERFACE IN RELATION TO DISEASE

S.D. GARRETT

Botany School, University of Cambridge, Cambridge, U.K.

Summary

The most direct effect of environmental factors at the soil-root interface on development of soil-borne diseases is that which determines the success or failure of primary infections of the root system or collar region by the pathogen. Examples illustrating this statement have been drawn from the pathogenic root-infecting fungi, which can be divided into three groups: unspecialized pathogens; specialized pathogens able to spread systemically through the vascular tracts of their host plants; specialized pathogens with an ectotrophic infection-habit. The critical importance of primary root infections has been assessed according to the relative number required for establishing a progressive ontogeny of disease in the individual host plant; this increases up the series: vascular-wilt pathogens → ectotrophic pathogens → unspecialized pathogens. So also does the dependence of full disease development on favourable conditions for primary infections at the soil-root interface.

Second for discussion has been the infectivity of fungal inocula in the soil, as expressed by their inoculum potential; this is determined by fungal energy of growth, which is related both to volume and nutrient status of the inoculum and to environmental factors at the soil-root interface. Examples drawn from work with *Gaeumannomyces graminis* var. *tritici*, causing the take-all disease of wheat and barley, show that soil inocula of low potential take longer and are less likely to establish primary root infections, and that the pre-penetration stage may take a different course. Such inoculum factors are important in the interpretation of take-all decline, which develops with continued monoculture of wheat or barley and appears to result from the encouragement of soil bacteria suppressive towards *G.graminis*, as a direct result of high populations of the pathogen maintained by cereal monoculture. The effect of such decline soils in reducing the severity of take-all may be due chiefly to suppression of primary root infections from inocula only marginally adequate for infection in non-decline soils.

Introduction

Soil-borne pathogens of crop plants, as usually defined, include fungi, bacteria, actinomycetes, some viruses and nematodes, but not insect pests, which are the province of entomologists. I shall restrict myself to examples drawn from my own speciality of pathogenic root-infecting fungi. Even so, my subject remains as wide as the establishment of their parasitic phase by all these pathogens and I shall discuss here only two interlinked questions of general interest. The first question is how the influence of factors, both abiotic and biotic, at the soil-root interface is modified and limited by the infection-habit of the fungal pathogen; the second is how this influence is affected by the degree of infectivity, or inoculum potential, of fungal inocula lying in the soil.

Infection-habits of Pathogenic Root-infecting Fungi

There are three main types of infection-habit among these fungi, which can be separated according to degree and type of parasitic specialization.

Unspecialized root-infecting fungi

In this group, infection is limited by host resistance, at least in the early and mid-stages of disease development, to juvenile tissues that have not yet acquired mature-tissue resistance, and to tissues predisposed to infection by some environmental factor adverse to general health of the host crop. Such pathogens usually cause disease in seedlings and young plants, but pathogenicity is not restricted to these, because a plant goes on producing juvenile root tissues, in the form of nutrient-absorbing rootlets, throughout its life. If the population of pathogen inocula in the soil is high, and especially if adverse environmental factors also impair host resistance, then mature trees can be killed by a progressive destruction of absorbing rootlets. Death may be dramatic, as exemplified by the 'sudden death' of cloves in Zanzibar, a problem that was finally settled by Nutman and Roberts (1954), who determined the pathogen as *Valsa eugeniae* Nutman and Roberts. The suddenness of the death syndrome is due to the remarkable drought-resistance of the clove tree. The gummosis disease of citrus, caused by *Phytophthora citrophthora* (R. and E. Smith) Leonian similarly develops through a progressive killing of the feeding rootlets until the resistance of the tree to further infection becomes exhausted; then collar canker and final death of the tree follow (Fraser, 1942). Through a similar sequence of events, *Phytophthora cinnamomi* Rands causes widespread destruction of avocado orchards (Zentmyer, Paulus and Burns, 1962) and has devastated large tracts of eucalypt forest in Australia (Podger, 1975).

Because progress of infection by this group of fungal pathogens is restricted by host-defence mechanisms, at least until the penultimate stage of disease development, a severe attack of disease can result only from the continued establishment of many primary infections. Primary infection is the most critical stage in the ontogeny of the pathogen-host relationship, because it is the one most subject to interference by adverse factors, both abiotic and biotic, at the soil-root interface. So these unspecialized pathogens appear to be the group most subject to control by the soil environment, as well as by host resistance. The species of *Phytophthora* cited above require a high soil moisture content for primary infection; reduction of soil moisture can greatly reduce incidence and severity of such diseases. But their lack of parasitic specialization seems to have permitted some at least of these pathogens to retain a high degree of competitiveness against other soil micro-organisms (Garrett, 1970).

Vascular-wilt fungi

These form a natural group of highly specialized root-infecting fungi, mainly in the genera *Fusarium* and *Verticillium*. Although highly specialized in their infection-habit, they share one limitation with the unspecialized pathogens, inasmuch as entry is restricted to juvenile tissues of young roots and rootlets by host-defence mechanisms. The infection route is usually limited to the immature cortex of the apical region of rootlets and the pathogen is able to pass thence through the endodermis, not yet fully differentiated, and so into the xylem tracts. Its further passage up the vascular tracts may be retarded, or even halted, by host-defence mechanisms and particularly by tylosis of the vessels. But this fact of host resistance to the extension of infection does not affect the main conclusion here, which is that, from relatively few primary infections of the root system, the pathogen is capable of systemic infection of the entire vascular system, assisted by rapid transport of its conidia up the xylem vessels in the transpiration stream. So in this group of diseases, primary infections, which are susceptible to interference by factors at the soil-root interface, are numerically of less importance than in either of the other groups. Once a progressive systemic infection has been established, the pathogen is then independent of any further direct effect of the soil environment, though still susceptible to indirect effects as mediated by host physiology and defence mechanisms.

One of the most spectacular and devastating amongst vascular wilts is (or was) the Panama disease of bananas, caused by *Fusarium oxysporum* Fr. f. sp. *cubense* (E.F. Smith) Snyder and Hansen. Some soils in Jamaica and Central America, which are now called 'suppressive' or 'long life' soils, continue for many years to support more or less healthy plantations of bananas, despite the presence of the pathogen in them. From his soil survey in Jamaica and subsequent experiments, Rishbeth (1957, 1960) concluded that these suppressive soils were characterized by good physical texture and by a high content of available potash, thus promoting vigorous growth of bananas. Nevertheless, some factor(s) present in these soils may also help to suppress primary root infections. Thus, in a survey of 67 banana soils in Central America, Stotzky and Martin (1963) found that all long-life soils except one were characterized by a sufficient proportion of montmorillonite clay, and that all short-life soils except one were characterized by its absence. A sufficient proportion of this clay renders a soil favourable for bacterial activity which tends to suppress fungal activity (Stotzky and Rem, 1966, 1967). So although the establishment of primary root infections by vascular-wilt pathogens may be numerically of smaller importance for the vascular-wilt diseases than for those caused by the two other groups of root pathogens, yet such infections are still an essential link in the chain of disease ontogeny.

Ectotrophic root-infecting fungi

In this specialized group of pathogens, advance of the

fungus is at first epiphytic over the surface of the host root; from the epiphytic mycelium, infection hyphae, either single or in aggregates, invade the underlying host tissues. On herbaceous hosts, the epiphytic mycelium consists of a network of individual hyphae, as in the take-all disease of wheat and barley caused by *Gaeumannomyces graminis* (Sacc.) Arx and Olivier var. *tritici* Walker. On woody perennial hosts, the epiphytic mycelium is organized into mycelial sheets, as in the root and butt rot of conifers caused by *Fomes annosus* (Fr.) Karst. or into rhizomorphs, as in the tree-root disease caused by *Armillaria mellea* (Fr.) Kummer. Defence mechanisms of arborescent hosts, both passive and active, are more substantial than those of herbaceous hosts; this greater resistance to fungal invasion is matched by an appropriate increase in the infectivity of the fungus, through mycelial aggregation in the form of sheets, strands or rhizomorphs. Such aggregation increases the inoculum potential of the pathogen, i.e. its total energy of growth at the host surface, though this is also affected by soil conditions, favourable or not to fungal growth, at the soil-root interface (Garrett, 1970). The ectotrophic infection habit appears to function as a mechanism whereby host resistance to invasion can be continuously overcome as the initially epiphytic mycelium travels along a root from a focus of primary infection. From this travelling epiphytic mycelium, infection hyphae or mycelial aggregates penetrate the host at short intervals; synergism between these infection attempts at short intervals of space and time eventually overwhelms the active defences of the host and so a progressive infection of the root results (Garrett, 1970). The ectotrophic infection-habit is not always obligate for such a pathogen; it is enforced upon *Fomes annosus* by the high resistance to infection of *Pinus* spp. but on some other conifers root-infection can progress internally (Rishbeth, 1951). Similarly, after the defence mechanisms of wheat roots have been sufficiently impaired by the effects of a progressive infection, the runner hyphae of *Gaeumannomyces graminis* abandon their epiphytic growth and travel several cells deep inside the root cortex (Garrett, 1934); with a still further reduction in tissue resistance, the hyphae are able to run longitudinally through the stele as a rapidly progressing internal infection (Holden, 1976).

So on some hosts, and under some conditions of reduced host-resistance, normally ectotrophic pathogens are able to spread internally within roots, as if they were vascular-wilt fungi, though less quickly. In so doing, they similarly avoid conditions often adverse to spread at the soil-root interface. In truly ectotrophic spread, on the other hand, the pathogen is fully exposed to adverse factors of the soil environment. But the effect of adverse conditions in this situation can be overemphasized because an infection, once established and progressive, generates a momentum that carries it along; spread may be retarded but is more rarely halted, except by extremely adverse soil conditions. The progress of an established infection thus contrasts strongly with attempts at primary root infection, where inoculum potential of the fungus is often lower and so establishment of an infection is all the more susceptible to adverse conditions at the soil-root interface.

So we can conclude that effects of adverse physico-chemical factors and of competing and antagonistic micro-organisms at the soil-root interface are likely to become more critical with an increase in the numerical importance of primary root infections for disease ontogeny. This importance increases up the series: vascular-wilt pathogens → ectotrophic pathogens → unspecialized pathogens. So it is interesting to see that increasing specialization in the pathogenic root-infecting fungi has brought with it a decreasing degree of dependence on soil conditions, often adverse to the establishment of primary root infections, at the soil-root interface.

Inoculum Potential of Pathogen Inocula

I have chosen to say something here about inoculum potential, which expresses the degree of infectivity of fungal inocula, because it is a factor critical for the establishment of primary root infections. Inoculum potential expresses the fungal energy of growth from an inoculum; this is related to both the volume and the nutrient status of an inoculum, and also to the environmental factors affecting fungal growth rate at the soil-root interface. Examples will be taken from work on the take-all disease of grasses and cereals, caused by *Gaeumannomyces graminis* var. *tritici* (hereafter referred to as *G.graminis*), which has become one of the most studied soil-borne diseases in the world today. In a field with well dispersed but not too abundant inocula of this fungus surviving in the soil, *G.graminis* slowly progresses during the growing season of the wheat or barley crop from primary infections, often on outlying parts of the root system, towards the crown with its attached tiller-bases; if these become infected early enough, their ears turn into 'whiteheads', empty of harvestable grain.

After harvest of such a crop with take-all, the infected along with the healthy stubble is, in conventional farming practice (though not in a sequence of direct-drilled crops), ploughed under the soil. An infected crown with its tiller-bases constitutes the most massive part of the take-all inoculum that then lies in wait for a following host crop. A single infected tiller-base constitutes an inoculum with ample potential for infection of any cereal root that passes close enough to it (Garrett, 1940). Similarly, an 8 mm disk cut from a colony of *G.graminis* growing on potato-dextrose agar (2% glucose) provides 'instant' inoculum potential for infection of wheat seminal roots within 1 week at 20°C of planting a wheat seed thereupon (Garrett, 1936). But some inocula of *G.graminis*, e.g. broken segments of infected roots and fragments of infected leaf-sheaths detached from tiller-bases, are smaller and less well nourished; establishment of primary root infections from such inocula is more uncertain, and takes a slower and somewhat different course. This situation has been studied by Hornby (1969), who separated organic debris from a field soil infested by *G.graminis* into two fractions by wet sieving. Coarse debris (retained by a screen with 0.42 mm apertures) was the more infective and usually caused lesions on test wheat seedlings within 3 weeks; finer debris (passing through 0.42 mm apertures) caused fewer and less extensive lesions, which often took longer than 3 weeks to become apparent. The development of infections from such minimal inocula has been further followed by Brown and Hornby (1971). Washed mycelial mats of *G.graminis* grown on a weak nutrient solution (0.4% malt extract, *w/v*) were teased into portions *ca* 1 mm^2 in area and these were employed as individual inocula; hyphae grew out from them towards adjacent seminal roots of wheat seedlings, forming clumps of mycelial cells in the root-hair region and infecting root hairs before penetrating the root cortex. When organic debris infested by *G.graminis* was screened out of a field soil, as done earlier by Hornby (1969), the frequency of mycelial clump formation and root-hair infection as a precursor of cortical infection was found to be *inversely* correlated with the volume of the inoculum fragments. Brown and Hornby therefore concluded that minimal-volume inocula require a preliminary saprophytic feeding stage on seminal-root exudates, accompanied by root-hair infection, before inoculum potential reaches a level adequate for invasion of root cortices. This interpretation is likely to be correct, because it is now established that the macroconidia and chlamydospores of some *Fusarium* species are dependent on exogenous nutrients supplied by root exudates for the establishment of primary infections (Garrett, 1970).

These observations on inoculum potential can help in

the interpretation of a remarkable phenomenon known as 'take-all decline', which was first named by Slope and Cox (1964) at the Rothamsted Experimental Station. It seemed remarkable because previous experience of various soil-borne diseases had shown that, under crop monoculture, incidence and severity of the disease usually increased to a peak level and remained there, though with some fluctuation, for as long as monoculture of the crop continued. But in winter wheat at Rothamsted, Slope and Cox found that peak severity of take-all was reached with the third or fourth consecutive crop, and thereafter declined to a lower level, which was maintained. In continuous, spring-sown wheat and barley, Shipton (1972) found peak severity of take-all to occur within 3-7 years, and this was followed by a progressive but limited decline in disease level. Although monoculture of wheat or barley is necessary for the development of take-all decline, it was evidence first produced by Gerlagh (1968) in the Netherlands that showed that the immediate cause of decline was the build-up of the soil population of *G.graminis* which follows cereal monoculture, rather than the continued presence of cereal roots in the absence of *G.graminis*. The agent responsible for decline is transmissible to non-decline soils (Pope and Jackson, 1973). Cook and Rovira (1976) have concluded, from a wide variety of evidence, that fluorescent pseudomonad bacteria are the most likely agents responsible for decline. It seems probable that the reduced severity of take-all at harvest in decline soils is largely due to a reduced level of primary root infections in such soils, within a month or more after the crop has been sown. Thus Lester and Shipton (1967) determined percentage incidence of seminal-root infection in wheat seedlings grown for 3 weeks in decline and non-decline soils inoculated with a finely dispersed culture of *G.graminis* on sand + 3% wheatmeal. In eight pairs of soils collected from widely scattered localities in England, the proportion of seminal roots lesioned by *G.graminis* in the decline soils varied from one-fifth to one-half of the parallel figures obtained for non-decline soils. Pope and Jackson (1973) have ascribed this reduced infection in decline soils to a reduction in infectivity of small inocula. Employing 5 mm segments of infected seminal roots as inocula, they demonstrated that emerging hyphae of *G.graminis* showed a tropic response to exudates from healthy seminal roots growing sufficiently close to them; in decline soils, hyphae grew out more slowly from the inocula than in non-decline soils, and the maximum distance over which a tropic response to nearby roots could be observed was shorter.

The evidence summarized above suggests that the agent responsible for take-all decline operates chiefly by inhibiting infection from minimal inocula, i.e. inocula with potential only marginally adequate to establish primary infections in non-decline soils soon after the crop is sown. Such minimal inocula are provided by survival of *G.graminis* in infected parts of root systems left in the soil by the preceding cereal crop. It seems unlikely that infection by the more massive inocula provided by infected tiller-bases in the stubble could be inhibited by the decline agent, unless it is a more powerful fungistat than a degree of soil acidity corresponding to pH5 or less. In such acid soils, infection from an adequate inoculum regularly occurs, though subsequent ectotrophic spread of *G. graminis* along roots is much retarded (Garrett, 1936; Ferraz, 1973; Smiley and Cook, 1973). Such massive inocula are likely to remain fully infective until sowing time of a spring cereal; this can explain why cereal monoculture results in a decline of take-all rather than in a temporary elimination of the kind observed after a 1-year break under a non-host crop when the cereal may be take-all-free.

For the solution of problems as difficult as that of take-all decline, collaboration between plant pathologists and soil microbiologists is now essential. The particular contribution that the plant pathologist can make is through his understanding of the pathogen-host relationship, a balance determined by the infection potential of the pathogen countered by the defence mechanisms of the host.

References

Brown, Margaret E. and Hornby, D. (1971). Behaviour of *Ophiobolus graminis* on slides buried in soil in the presence or absence of wheat seedlings. *Trans. Br. mycol. Soc.*, 56, 95.

Cook, R.J. and Rovira, A.D. (1976). The role of bacteria in the biological control of *Gaeumannomyces graminis* by suppressive soils. *Soil Biol. Biochem.*, 8, 269.

Ferraz, J.F.P. (1973). Influence of the soil atmosphere on spread of *Ophiobolus graminis* along wheat roots. *Trans. Br. mycol. Soc.*, 61, 237.

Fraser, Lilian (1942). Phytophthora root rot of citrus. *J. Aust. Inst. agric. Sci.*, 8, 101.

Garrett, S.D. (1934). Factors affecting the severity of take-all. I. The importance of micro-organisms. *J. Agric. S. Aust.*, 37, 664.

Garrett, S.D. (1936). Soil conditions and the take-all disease of wheat. *Ann. appl. Biol.*, 23, 667.

Garrett, S.D. (1940). Soil conditions and the take-all disease of wheat. V. Further experiments on the survival of *Ophiobolus graminis* in infected wheat stubble buried in the soil. *Ann. appl. Biol.*, 27, 199.

Garrett, S.D. (1970). "Pathogenic Root-infecting Fungi". Cambridge University Press, Cambridge.

Gerlagh, M. (1968). Introduction of *Ophiobolus graminis* into new polders and its decline. *Meded. Lab. Phytopath. Wageningen* No. 241.

Holden, J. (1976). Infection of wheat seminal roots by varieties of *Phialophora radicicola* and *Gaeumannomyces graminis*. *Soil Biol. Biochem.*, 8, 109.

Hornby, D. (1969). Methods of investigating populations of the take-all fungus (*Ophiobolus graminis*) in soils. *Ann. appl. Biol.*, 64, 503.

Lester, E. and Shipton, P.J. (1967). A technique for studying inhibition of the parasitic activity of *Ophiobolus graminis* (Sacc.) Sacc. in field soils. *Pl. Path.*, 16, 121.

Nutman, F.J. and Roberts, F.M. (1954). *Valsa eugeniae* in relation to the sudden-death disease of the clove tree (*Eugenia aromatica*). *Ann. appl. Biol.*, 41, 23.

Podger, F.D. (1975). The role of *Phytophthora cinnamomi* in dieback diseases of Australian eucalypt forests. *In* "Biology and Control of Soil-borne Plant Pathogens" (Ed. G.W. Bruehl), p.27. American Phytopathological Society, St. Paul, Minnesota.

Pope, A.M.S. and Jackson, R.M. (1973). Effects of wheatfield soil on inocula of *Gaeumannomyces graminis* (Sacc.) Arx and Olivier var. *tritici* J. Walker in relation to take-all decline. *Soil Biol.*

Biochem., 5, 881.
Rishbeth, J. (1951). Butt rot by *Fomes annosus* Fr. in East Anglian conifer plantations and its relation to tree killing. *Forestry*, 24, 114.
Rishbeth, J. (1957). Fusarium wilt of bananas in Jamaica. II. Some aspects of host-parasite relationships. *Ann. Bot.*, 21, 215.
Rishbeth, J. (1960). Factors affecting the incidence of banana wilt ('Panama disease'). *Emp. J. exp. Agric.*, 28, 109.
Shipton, P.J. (1972). Take-all in spring-sown cereals under continuous cultivation: disease progress and decline in relation to crop succession and nitrogen. *Ann. appl. Biol.*, 71, 33.
Slope, D.B. and Cox, Judith (1964). Continuous wheat growing and the decline of take-all. *Rep. Rothamsted Exp. Sta. for 1963*, p.108.
Smiley, R.W. and Cook, R.J. (1973). Relationship between take-all of wheat and rhizosphere pH in soils fertilized with ammonium vs. nitrate-nitrogen. *Phytopathology*, 63, 882.
Stotzky, G. and Martin, R.T. (1963). Soil mineralogy in relation to the spread of Fusarium wilt of banana in Central America. *Pl. Soil*, 18, 317.
Stotzky, G. and Rem, L.T. (1966). Influence of clay minerals on micro-organisms. I. Montmorillonite and kaolinite on bacteria. *Can. J. Microbiol.*, 12, 547.
Stotzky, G. and Rem, L.T. (1967). Influence of clay minerals on micro-organisms. IV. Montmorillonite and kaolinite on fungi. *Can. J. Microbiol.*, 13, 1535.
Zentmyer, G.A., Paulus, A.O. and Burns, R.M. (1962). Avocado root rot. *Circ. Calif. Agr. Exp. Sta.* No. 511.

EFFECT OF PHYSICAL FACTORS, HOST RESISTANCE AND FUNGICIDES ON ROOT INFECTION AT THE SOIL-ROOT INTERFACE

GEORGE A. ZENTMYER

University of California, Riverside, U.S.A.

Summary

Root rots incited by species of *Phytophthora* provide unique systems for examining factors of the soil involved in pathogenesis. Sporangium production is influenced by the microbial population, by the osmotic and matric potentials, by soil temperature and other soil factors. Chemotaxis of zoospores, the major mechanism for attraction to roots, is followed by invasion by the large numbers of germinating zoospores, and then by root infection. Infection, but not necessarily chemotaxis, can be influenced by host resistance, and by organic fungicides applied either to soil or as systemics to the foliage. In species of *Persea* resistant to *P. cinnamomi*, chemotaxis may be active but root invasion is limited. A preformed toxicant, borbonol, is involved. Two new organic fungicides, the acyl alanines (CIBA-Geigy), and organic phosphites (aluminum ethyl phosphite, Rhone-Poulenc) greatly reduce root invasion and infection when applied to the soil, or when applied to the foliage, the latter indicating downward translocation. The acyl alanines drastically reduce growth and sporangium production of the pathogen *in vitro*, and probably similarly in soil. The phosphites have little influence on *Phytophthora in vitro* but are equally effective to the acyl alanines *in vivo*.

Resistance apparently affects invasion more than the interactions at the soil-root interface. The systemic fungicides may affect both chemotaxis and invasion. Moisture potential (matric and osmotic) affects the pathogen by favouring or reducing sporangium production by *P.cinnamomi*. This obviously affects zoospore production, chemotaxis, and invasion; and variously influences other components of the microbial population. Thus there is a complex system of interactions in relation to root diseases that is variously influenced by biological, physical, and chemical factors. Identification of the precise relationships will provide valuable information on disease development and control.

Introduction

Root rots induced by plant pathogenic fungi in the genus *Phytophthora* provide unique systems for examining many of the biological and physical factors of the soil involved in pathogenesis. The situation is complicated by the intricate life cycle of *Phytophthora*, including four spore stages: sporangia, motile zoospores, chlamydospores, and oospores. Any of these spores as well as vegetative mycelium can infect roots.

This paper considers the effect of various physical and biological factors of the soil on the pathogen and on the infection process at the soil-root interface, and will emphasize two major protective mechanisms that can influence infection and disease development: host resistance, and the use of new non-phytotoxic organic fungicides. The principal pathogen that will be considered is *P.cinnamomi*, a very destructive and widely distributed fungus, attacking over 900 plants, primarily trees and woody perennials (Zentmyer, 1978).

Physical and biological factors influencing Phytophthora

Sporangia are the most important spores of *Phytophthora* in relation to pathogenesis. There has been considerable research on various factors influencing their development and germination. Sporangium production is influenced by the osmotic and matric potentials of the soil, by soil temperature, pH, and aeration, as well as by the microbial population of the soil. Interaction with other soil micro-organisms is especially important in the case of *Phytophthora cinnamomi*.

Soil moisture is probably the most important physical factor of the environment for *P.cinnamomi*, in relation to growth and sporulation as well as the resultant infection process. This fungus forms sporangia only in liquid culture, not on solid media in contrast to many other species of *Phytophthora*. The fungus makes its best growth in the range of tolerance of many economically important plants, from -2 to -8 bars osmotic potential (Sterne, Zentmyer and Bingham, 1976). A growth reduction of 60% occurs at potentials higher than -10 bars for most osmotica; the pathogen does not grow at osmotic potentials of -20 to -30 bars. Different salts used to adjust osmotic potential have different levels of toxicity to *P.cinnamomi*, thus specific ions are more toxic than osmotic potential alone.

Production of sporangia in several species of *Phytophthora* does not occur in flooded soil; production is generally optimum in the range of -0.025 to -0.3 matric potential. Some sporangia have been reported at moisture levels as low as -3 to -4 bars (Sneh and McIntosh, 1974; Duniway, 1975). Sterne (1976) found that sporangium production was reduced by 50 per cent at -5 bars osmotic potential, and was prevented completely at -12 bars.

Aeration is an important factor in formation of sporangia. Mitchell and Zentmyer (1971) reported that production of sporangia is inhibited by low oxygen and also by high CO_2 concentrations. Duniway (1975) stressed the influence of aeration on formation of sporangia with *P.drechsleri*; sporangia were formed on the surface of saturated soil, but not at a depth of 5 mm in saturated (0 matric potential) or near-saturated (-0.01 bar) soils.

The microbial population of the soil also has a significant influence on production of sporangia, especially by species such as *P. cinnamomi*. *In vitro*, *P. cinnamomi* does not produce sporangia on sterile culture media as do many species. Production of sporangia occurs in a non-sterile soil extract (Mehrlich, 1935; Zentmyer and Marshall, 1959); bacteria, primarily pseudomonads, have been implicated as the primary microorganisms involved in this stimulation (Chee and Newhook, 1965; Marx and Haasis, 1965; Ayers and Zentmyer, 1971). Chen and Zentmyer (1970) found that sporangia also can be produced by *P. cinnamomi* under axenic conditions when young, vigorously growing hyphal mats are washed thoroughly to remove nutrients, and placed in a salt solution; this is essentially a process of nutrient depletion, and conceivably could be accomplished in the soil by the action of bacteria and other microorganisms.

Zoospore release from sporangia and movement in the soil are other vital parameters in relation to infection of roots by these propagules. Zoospore production and motility are obviously affected by matric potential. Indirect germination of sporangia, resulting in production of zoospores, generally occurs in high percentages at high matric potentials, from 0 to -.01 bars. Duniway (1976) found that 96, 81 and 78 per cent of sporangia produced zoospores at matric potentials of 0, -1 and -10 millibars (mb), respectively, in three different soils (sandy loams, silt loams and clay loams). Apparently some zoospores can be released from sporangia at matric potentials down to -300 or -400 mb (-0.3 to -0.4 bars), so saturated soil is not necessary but does facilitate zoospore release. Soil texture and water content, as well as matric potential, may affect indirect germination.

Once the zoospores are released from sporangia, their movement in soil is restricted to high matric potentials (between 0 and 0.01 bars); under these moisture conditions, soil pores sufficiently large to accommodate swimming zoospores are apparently filled with water. To a limited extent, zoospore movement at these high matric potentials is influenced by soil texture; at -10 mb, for example, there was no movement in a clay loam, and good movement in soil mix containing peat and fine sand (Duniway, 1976). In studies of infection, Duniway found that zoospores moved to all seedlings when the sporangia were on the surface of flooded soils, but there was much less movement and infection when sporangia were 3 to 8 mm below the surface of the flooded soils (clay loam, fine sandy loam and silt loam).

In our studies of root rot of *Persea indica* induced by *Phytophthora cinnamomi*, disease severity was a function of matric potential but not of osmotic potential. At matric potentials of 0, -0.05 and -0.10 bar, respectively, the percentage of diseased roots averaged 80 to 90%, 50 to 90% and 10 to 50% (Sterne, Zentmyer and Kaufman, 1977a). Infection was very slight at -0.25 bar in the sandy loam soil. Infection reached 50% at -0.25 bar in clay soil, however, and was severe at that potential when the sandy loam soil was amended with glucose and asparagine, indicating that at -0.25 bar in sandy loam soil, nutrient availability rather than matric potential *per se* limited disease development (Sterne, Zentmyer and Kaufman, 1977b). An unknown parameter in these studies also is the importance of the total amount of water in different soil types, as related to matric potential.

Following their release from sporangia, there is abundant evidence that *in vitro* (in soil extract or various solutions) zoospores are attracted to and accumulate on roots and in the vicinity of host plants. Chemotropism of germ tubes is evident in the case of zoospores germinating near the root. In my early studies of this phenomenon (Zentmyer, 1961a), zoospores were observed to be attracted in large numbers to the region of elongation on actively growing avocado roots, where they encysted, germinated, and soon infected the roots. Zoospores were shown to be attracted strongly by such amino acids as glutamic and aspartic acids (Zentmyer, 1966) that are present in root exudates.

Zoospores also may show chemotropism in response to substances released by the root, probably by exudation; spores settling down at a distance from the root will germinate with their germ tubes oriented toward the root. In addition, zoospores settle down at varying distances from roots as if in response to a gradient of chemicals released by the root (Table 1). Khew and Zentmyer (1973) found that zoospores of 5 species of *Phytophthora* were attracted to a wide range of chemicals, including vitamins, phenolic compounds, nitrogenous bases of nucleic acid, nucleotides, growth regulators, sugars, organic acids, and amino acids. Rovira (1965) showed that many different types of organic chemicals are exuded or released in other ways from roots. In further tests of amino acids, aspartic and glutamic acids were more active in chemotaxis than arginine and methionine.

TABLE 1

Attraction of zoospores of Phytophthora cinnamomi *to roots of avocado* (Persea americana)

Distance from Root (mm)	Average Number of Zoospores settling in Areas $0.5 mm^2$
0-0.5	34.0
0.5-1.0	14.7
1.0-1.5	11.1
1.5-2.0	8.7
2.0-2.5	5.0
2.5-3.0	4.2

Positively charged molecules were more attractive to zoospores than negatively charged molecules; further studies (Khew and Zentmyer, 1974) showed zoospores acted as negatively charged particles, and were attracted to the anode in electrotaxis studies. Thus, in attraction of zoospores to the root at the soil-root interface, release of particular types of chemicals is involved, as well as the charge on the molecule or on the root surface.

Most of the studies of chemotaxis have been done *in vitro*, but there are also several reports of accumulation of zoospores, in the initiation of the infection process, on roots in soil. Mehrotra (1970) demonstrated accumulation of zoospores on roots in soil, similar to that previously reported by other workers to occur *in vitro*; his studies were with *Phytophthora drechsleri* and *P. megasperma* var. *sojae*. The area of accumulation was in the region of elongation, and experiments with cellulose acetate membranes indicated that zoospores were attracted to the root exudate diffusing through the membrane.

Our early observations of root infection showed that following attraction of zoospores to the region of elongation in avocado roots, infection could be observed (in the form of brown lesions) in that area within 24 to 48 hours. Our recent studies (Ho and Zentmyer, 1977) showed that within one hour after zoospores accumulated on the root surface, germ tubes of the cysts had penetrated the epidermis. Thus, with a susceptible host in an environment favourable for the pathogen, rapid disease development and intensification can take place.

Other microbiological components of the soil environment - other fungi, bacteria, actinomycetes - are variously influenced by the physical and chemical factors in the soil. There are fewer data on the effects of matric and osmotic potentials of the soil on the saprophytic microflora than on soil pathogens. For example, many soil fungi will grow and survive at matric potentials much lower than those favouring growth and development of species of *Phytophthora*. Other microorganisms in the rhizosphere are undoubtedly influenced by the release of amino acids and other substances by the root. Responses of some of the bacteria and other soil fungi (e.g. *Pythium*, *Fusarium*) are well documented. There must be intense competition for these substances at the soil-root interface; under conditions of high matric potential, fungi such as *Phytophthora* would be expected to compete favourably.

Root invasion in the *Phytophthora* root rots, but not necessarily chemotaxis, can be influenced by at least two major protective mechanisms: host resistance, and by organic fungicides used either as soil drenches or as systemic chemicals applied to the foliage.

Host resistance

High resistance to *Phytophthora cinnamomi* has been found in a number of species of the genus *Persea* that are indigenous in the Americas; especially in the small-fruited species in the subgenus Eriodaphne (Zentmyer, 1961b). The avocado (*Persea americana* Mill.) is in the subgenus Persea that includes primarily large-fruited species.

Resistance apparently affects invasion of the root rather than chemotaxis or other interactions at the soil-root interface. There is some evidence of a reduction of chemotaxis to roots with high resistance, which may reflect smaller release of attracting chemicals from the resistant roots or release of substances somewhat antagonistic to *Phytophthora* zoospores. In our recent tests, however, attraction of zoospores of *P. cinnamomi* was similar in roots of susceptible (*Persea americana*) and resistant species (*P. borbonia*, *P. pachypoda*) of *Persea* (Ho and Zentmyer, 1977).

The principal difference was that subsequent lesion development and initiation of root rot was very restricted in the resistant species. It was concluded that, although there are some anatomical differences in roots of resistant and susceptible species of *Persea*, particularly in the endodermal layer, the major basis for resistance appears to be chemical.

We have isolated and identified a pre-formed fungitoxic chemical in resistant species of *Persea*, including *P.borbonia*, *P.caerulea* and *P.pachypoda*, that may be the primary basis for resistance (Zaki *et al.*, 1973). This has been named "borbonol", and is a long chain hydrocarbon with a lactone ring. Borbonol occurs in tissue of resistant species of *Persea* at concentrations up to 1000 μg/gram, and inhibits growth of *Phytophthora cinnamomi* at concentrations of 1 μg/ml in synthetic solution culture. Preliminary tests indicate that it is not released from roots and thus does not directly affect the pathogen, or other soil microorganisms, in the rhizosphere or in the soil. This agrees with the moderate to good chemotaxis of zoospores to roots of resistant species.

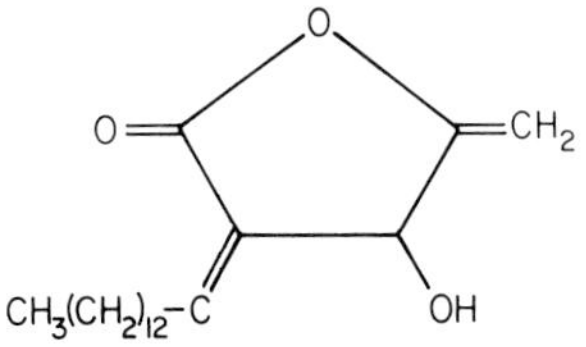

BORBONOL
(M.W. 308)

Fig. 1. Tentative structural formula for borbonol, based on data by J.A. Pettus, D. Kunert, J.J. Sims and G.A. Zentmyer.

Soil fungicides and systemics

Two unique types of chemicals, the acyl alanines (CIBA-Geigy Co. of Switzerland) (Schwinn, Staub and Urech, 1977; Urech, Schwinn and Staub, 1977), and the organic phosphites (Rhone-Poulenc Co. of France) can, we have found, greatly reduce root invasion and infection from *Phytophthora* when applied to the soil, or more significantly, from the standpoint of unusual systemic activity, when applied to the foliage. The most effective acyl alanine is CIBA-Geigy 48988 (Table 2); chemically this is [N-2,6-dimethylphenyl) -N-(methoxyacetyl)-alanine methyl ester] (Urech *et al.*,

TABLE 2

Control of root rot of Persea americana *with soil drenches of acyl alanine fungicide (CGA 48988)*

Treatment	Increase in Height (cm)	Per cent Healthy Roots	Weight of Roots (grams)	Weight of Tops (grams)
None	7.2	0	26.5	31.5
CGA 48988	167.1	58	291	646

Duration of experiment: 9 months. Avocado seedlings treated 3 times per week with 5 ppm solution. CGA 48988 = [N-(2,6-dimethylphenyl)-N-(methoxyacetyl)-alanine methyl ester].

1977). Rhone-Poulenc's LS74-783 (Aliette), aluminum ethyl phosphite, is the most effective phosphite (Bertrand *et al.*, 1977).

There have been a number of reports of the activity of systemic fungicides for plant disease control (Erwin, 1973). The great majority of these compounds, however, are systemic only in an upward direction. One exception is pyroxychlor [2-chloro-6-methyl-4-(trichloromethyl) pyridine]; this chemical has systemic activity downward against some species of *Phytophthora* and other fungi (Noveroske, 1973; Knauss, 1974).

Periodic soil drenches with the acyl alanines and organic phosphites effectively prevent *Phytophthora cinnamomi* from invading avocado roots and causing disease. The chemicals are taken up by the roots, and, based on inhibition of stem cankers, are translocated to some extent in an upward direction also. Based on *in vitro* tests in agar and solution culture, CGA 48988 could either prevent disease by inhibiting growth and sporulation of *Phytophthora* in the soil or by accumulating in the root in sufficient concentration to prevent or greatly reduce invasion. CGA 48988 reduces growth of *P. cinnamomi* by 95% at 2.5 μg/ml.

On the other hand, the organic phosphite LS74-783 is completely ineffective against *Phytophthora in vitro*, but is equally effective to the acyl alanines when applied to soil or to the plant. The phosphite thus must be preventing disease by accumulating in the root in a modified chemical form that is toxic to *Phytophthora* or by influencing the resistance of the host.

Indications that both chemicals are acting within the root to prevent disease are seen in that better disease control was obtained when plants were treated with several soil applications prior to transplanting into soil infested with *P. cinnamomi*.

Illustrating possible downward systemic activity, periodic foliar sprays with both chemicals also effectively prevented *P. cinnamomi* from invading avocado or *Persea indica* roots and from causing disease (Table 3 and Fig. 2). At present, the evidence is biological only; chemical tests are planned, or tests with labelled material, when these become available. The highly significant reduction in root rot as a result of foliar sprays indicates that the chemicals are acting within the root in a manner very similar to that resulting from soil application, or affecting the host so that the plant is more resistant to the pathogen.

Evidence is inconclusive at present as to whether these chemicals are affecting chemotaxis as well as root invasion. Some attraction of zoospores has occurred in preliminary tests, to roots of plants treated with foliar applications of the fungicides as well as to roots of untreated controls.

There is obviously a complex system of interactions in relation to root diseases incited by species of *Phytophthora* that is variously influenced by biological, physical, and chemical factors. Further identification of the precise relationships will provide valuable information in relation to disease development and control. The mechanisms involving resistance and fungicides are useful not only for direct disease control, but also in helping to identify the site of interaction between pathogen, other soil microorganisms, and possible protective substances.

TABLE 3

Control of Phytophthora root rot of Persea indica *with foliar sprays of acyl alanine (CGA 48988) and aluminum ethyl phosphite (LS74-783)*

Treatment	Increase in Height (cm)	Per cent Healthy Roots	Weight of Roots (grams)	Weight of Tops (grams)
None	0.68 a*	1 a	1.88 a	3.4 a
CGA 48988 spray 2000 ppm	7.26 b	95 b	10.6 b	14.8 b
LS74-783 spray 2000 ppm	5.5 b	93 b	8.4 b	12.9 b

Duration of experiment: 10 weeks; 5 replicates per treatment. Plants sprayed 3 times before transplanting into infested soil and weekly thereafter.
CGA 48988 = [N-(2,6-dimethylphenyl)-N-(methoxyacetyl)-alanine methyl ester]. LS74-783 = aluminum ethyl phosphite.
* Numbers followed by different letters are significantly different at the 1% level, by Duncan's Multiple Range Test.

Fig. 2. Control of Phytophthora root rot of *Persea indica* with foliar applications of systemic fungicides. 11 - non-infested control; 1 - weekly spray with aluminum ethyl phosphite (LS74-783); 3 - spray with LS74-783 prior to transplanting into infested soil; 6 - weekly spray with acyl alanine (CGA 48988); 10 - infested control. Treatments 1, 3, 6 and 10 are in soil infested with *Phytophthora cinnamomi*.

References

Ayers, W.A. and Zentmyer, G.A. (1971). Effect of soil solution and two soil pseudomonads on sporangium production by *Phytophthora cinnamomi*. *Phytopathology*, 61. 1188.

Bertrand, A., Ducret, J., Debourge, J.C. and Horriere, D. (1977). Etudes des propriétés d'une nouvelle famille de fongicides: les monoethyl phosphites metalliques, caracteristique physico-chimiques et propriétés biologiques. *Phytiat.-Phytopharm.*, 26.

Chee, K.H. and Newhook, F.J. (1965). Improved methods for use in studies on *Phytophthora cinnamomi* Rands and other *Phytophthora* species. *N.Z.J. Agr. Res.*, 8, 88.

Chen, D.W. and Zentmyer, G.A. (1970). Production of sporangia by *Phytophthora cinnamomi* in axenic culture. *Mycologia*, 62, 397.

Duniway, J.M. (1975). Formation of sporangia by *Phytophthora drechsleri* in soil at high matric potentials. *Can. J. Bot.*, 53, 1270.

Duniway, J.M. (1976). Movement of zoospores of *Phytophthora cryptogea* in soils of various textures and matric potentials. *Phytopathology*, 66, 877.

Erwin, D.C. (1973). Systemic fungicides: disease control, translocation and mode of action. *A. Rev. Phytopathol.*, 11, 389.

Ho, H.H. and Zentmyer, G.A. (1977). Infection of avocado and other species of *Persea* by *Phytophthora cinnamomi*. *Phytopathology*, 67, 1085.

Khew, K.L. and Zentmyer, G.A. (1973). Chemotactic response of zoospores of five species of *Phytophthora*. *Phytopathology*, 63, 1511.

Khew, K.L. and Zentmyer, G.A. (1974). Electrotactic response of zoospores of seven species of *Phytophthora*. *Phytopathology*, 64, 500.

Knauss, J.F. (1974). Pyroxychlor, a new and apparently systemic fungicide for control of *Phytophthora palmivora*. *Pl. Dis. Reptr.*, 58, 1100.

Marx, D.H. and Haasis, F.A. (1965). Induction of aseptic sporangial formation in *Phytophthora cinnamomi* by metabolic diffusates of soil micro-organisms. *Nature*, 206, 673.

Mehrotra, R.S. (1970). Techniques for demonstrating accumulation of zoospores of *Phytophthora* species on roots in soil. *Can. J. Bot.*, 48, 879.

Mehrlich, F.P. (1935). Non-sterile soil leachate stimulating to zoosporangia production by *Phytophthora* sp. *Phytopathology*, 25, 432.

Mitchell, D.J. and Zentmyer, G.A. (1971). Effects of oxygen and carbon dioxide tensions on sporangium and oospore formation by *Phytophthora* sp. *Phytopathology*, 61, 807.

Noveroske, R.L. (1973). Systemic activity of Dowco^{R} 269 for control of *Phytophthora parasitica* of tobacco. *2nd Int. Cong. Pl. Pathol.*, Minneapolis, Minn., Sept. 5-12, 1973 (Abstr. 0123).

Rovira, A.D. (1965). Plant root exudates and their influence upon soil microorganisms. *In* "Ecology of Soil-borne Plant Pathogens" (Eds. K.F. Baker and W.C. Snyder), p.170. Univ. Calif. Press, Berkeley.

Schwinn, F.J., Staub, T. and Urech, P.A. (1977). A new type of fungicide against diseases caused by Oomycetes. *Meded. Fac. Landbouww. Rijksuniv. Gent.*, 42, 1181.

Sneh, B. and McIntosh, D.L. (1974). Studies on the behavior and survival of *Phytophthora cactorum* in soil. *Can. J. Bot.*, 52, 795.

Sterne, R.E., Zentmyer, G.A. and Bingham, F.T. (1976). The effect of osmotic potential and specific ions on growth of *Phytophthora cinnamomi*. *Phytopathology*, 66, 1398.

Sterne, R.E., Zentmyer, G.A. and Kaufman, M.R. (1977a). The effect of matric and osmotic potential of soil on *Phytophthora* root disease of *Persea indica*. *Phytopathology*, 67, 1491.

Sterne, R.E., Zentmyer, G.A. and Kaufman, M.R. (1977b). The influence of matric potential, soil texture, and soil amendment on root disease caused by *Phytophthora cinnamomi*. *Phytopathology*, 67, 1495.

Urech, P.A., Schwinn, F. and Staub, T. (1977). A novel fungicide for the control of late blight, downy mildews, and related soil-borne diseases. *Proc. 1977 British Crop Protection Conf.*, p.623.

Zaki, A.K., Zentmyer, G.A., Keen, N.T. and Sims, J.J. (1973). Isolation of an antifungal substance from *Persea borbonia*. *2nd Int. Cong. Pl. Pathol.* Minneapolis, Minn., Sept. 5-12, 1973 (Abstr. 960).

Zentmyer, G.A. (1961a). Chemotaxis of zoospores for root exudates. *Science*, 133, 1595.

Zentmyer, G.A. (1961b). Resistance to *Phytophthora* root rot of avocado. *Proc. Carib. Reg. Am. Soc. Hortic. Sci.*, 5, 85.

Zentmyer, G.A. (1965). Bacterial stimulation of sporangium production in *Phytophthora cinnamomi*. *Science*, 150, 1178.

Zentmyer, G.A. (1966). Role of amino acids in chemotaxis of zoospores of three species of *Phytophthora*. *Phytopathology*, 56, 907 (Abstr.).

Zentmyer, G.A. (1978). *Phytophthora cinnamomi* and the diseases it causes. *Am. Phytopathol. Soc. Mono.*, 10, (in press).

Zentmyer, G.A. and Marshall, L.A. (1959). Factors affecting sporangial production by *Phytophthora cinnamomi*. *Phytopathology*, 49, 556 (Abstr.).

WHEAT RHIZOSPHERE pH AND THE BIOLOGICAL CONTROL OF TAKE-ALL

RICHARD W. SMILEY

Department of Plant Pathology, Cornell University, Ithaca, New York, U.S.A.

Summary

Take-all of wheat and lineal extension of *Gaeumannomyces graminis* var. *tritici* hyphae along roots of intact wheat plants were lower in soils treated with ammonium-nitrogen (NH_4-N) than with nitrate-nitrogen (NO_3-N). Differences in rhizosphere pH (pH_r) of up to 2.2 units were recorded for NH_4 vs. NO_3 fed plants in the glasshouse, and up to 1.2 units in the field. The bulk soil pH (pH_b) did not differ. Reduced disease was correlated with direct inhibition of the pathogen's growth at $pH_r < 5$ and, in some soils, with indirect inhibition at pH_r 5.0 - 6.5. In an alkaline soil that is suppressive to take-all the controlling influence of NH_4-N on growth of *G. graminis* was eliminated by soil fumigation or by treatment with aerated steam at 60°C, and reintroduced by additions of 1% nonsterile soil. Effects of fumigation on hyphal growth were absent in nonsuppressive soil, and in NO_3 treatments of the suppressive soil. Suppression of take-all by high NH_4/NO_3 ratios in the root zone has been attributed to direct inhibition of the pathogen's ectotrophic hyphal growth due to acidity in the rhizosphere of wheat growing in acid soils, and to a combined effect of acidity and specific antagonism by *Pseudomonas fluorescens* in some neutral to alkaline soils.

Introduction

Take-all of wheat (*Triticum aestivum* L.), caused by *Gaeumannomyces graminis* (Sacc.) Arx & Olivier var. *tritici* J. Walker, seriously limits grain production in many regions of the world (Asher and Shipton, 1979). Control of the root rot disease continues to depend upon manipulation of various crop management procedures because effective fungicides and resistant varieties are yet to be identified and exploited.

One field-proven control technique (Anon., 1941; Huber *et al.*, 1968) is to manage the dominant ionic form of nitrogen (N) in root zones of seedlings. Take-all can be controlled in the field with soil applications of ammonium sulphate, particularly when the fertilizer is applied in combination with 2-chloro-6-(trichloromethyl) pyridine [N-Serve] to slow nitrification. Ammonium-nitrogen (NH_4-N) has little or no effect on the disease if nitrification is mostly completed by the time of planting, and take-all is not reduced by soil applications of nitrate-nitrogen (NO_3-N) (Huber and Watson, 1974). Similar observations (reviewed by Hornby and Brown, 1977 and by Smiley and Cook, 1973) have also been made in other field and glasshouse investigations.

Suppression of take-all in NH_4-enriched soils was therefore well documented, but an explanation for the phenomenon was lacking. The NH_4/NO_3 balance in the root zone could affect take-all by influencing the nutrition of the host, the nutrition of the pathogen, the abiotic or biotic rhizosphere environment, or combinations of each. This paper reviews investigations (Smiley and Cook, 1973; Smiley, 1974, 1978a, 1978b, 1978c) which were conducted from 1969 to 1973 in the laboratories of Dr. R.J. Cook (USDA-ARS, Pullman, Washington) and of Dr. A.D. Rovira (CSIRO-Division of Soils, Adelaide, South Australia).

The Rhizosphere pH

Evidence in the literature suggested that absorption of predominantly NH_4-N by roots should reduce the rhizosphere pH (pH_r) whereas absorption of predominantly NO_3-N should increase it, even in the absence of changes in the bulk soil pH (pH_b). Wheat growing in slightly alkaline or acidic soils treated with NH_4-N or NO_3-N fertilizers (Smiley, 1974) exhibited pH_r differences of up to 2.2 units, and pH_r vs. pH_b differences of up to 1.2 units. Differences depended upon the forms, concentrations and placement of N in soil, upon the pH buffering properties of soils, and upon the variety and age of wheat. Reductions of pH_r in non-limed, base unsaturated, slightly acidic soils were directly proportional to the rate of fertilizer NH_4-N applied. No evidence was found for a transfer of the pH_r effect from NH_4-N absorption either up or down the root axis when fertilizers were supplied to only one segment of that root (Smiley and Cook, 1973). It is now possible, with the use of nitrification inhibitors, to maintain high NH_4/NO_3 ratios in field soils for extended time intervals, and therefore to alter the rhizosphere's chemical properties.

Soil pH and Growth of *G. graminis*

The sensitivity of *G. graminis* and take-all to acidic environments is well known, as is the pathogen's dependence upon ectotrophic runner hyphae during parasitic growth on wheat. When sterile straws were inoculated with the pathogen and buried in sterile soils adjusted to pH values of 3.5 to 8.2, no hyphal growth occurred at $pH_b < 4.8$ (Smiley and Cook, 1973). Inoculum of *G. graminis* was then placed as a point source beneath wheat seeds planted in nonsterile soil, and the lineal extension of runner hyphae was measured on roots 18 days later. Hyphal growth was retarded at $pH_r < 5.0$, regardless of the soil type, amount or form of N, or soil temperature.

Rhizosphere pH and Take-All

The severity of take-all in the glasshouse was studied in nonsterile soils which contained small amounts of randomly incorporated *G.graminis*-infested, chopped oat kernels. Take-all severity decreased below pH_r 6.8 - 7.0, regardless of other experimental variables. There was little effect of pH_r on numbers of primary infection sites, but the lesions in NH_4-treated soils were smaller than those in NO_3-treated and in nonfertilized soils. In methyl bromide-fumigated soils extreme acidity (pH_r < 4.9) was the only factor associated with reduced take-all severity. Disease control in the field was also associated with pH_r < 5.0. Control of take-all by NH_4-N in the field was mostly negated by additions of lime and by soil fumigation. Measurement of the pH within 1 mm of roots was the only reproducible method for assessing the influence of pH on parasitic activity of *G.graminis*. The pH necessary for control in the field and in fumigated soils (< 5.0) was far below that needed for control in the greenhouse (6.6), indicating a high variability that may depend upon the composition of the soil microflora, and on the environmental factors that affect that composition.

Smiley and Cook (1973) suggested that control of take-all by NH_4-N was the direct effect of acidity on *G.graminis* at pH_r < 5.0, and was indirect, through the aid of other microorganisms, at pH_r between 5.0 and 6.6. Growth of *G. graminis* was retarded as the pH dropped from 7 to 6, and this reduced vigour apparently reduced its inoculum potential and increased its vulnerability to antagonism by the resident rhizosphere microflora; provided that antagonists were present in that flora. Host resistance, as affected by changes in host nutrition, a mechanism favoured by Huber and Watson (1974), was not thought to be important since: 1) disease control was attained with NH_4-N in fumigated soil only at pH_r < 4.9; 2) lime negated the control achieved by NH_4-N; 3) NO_3-N was more favourable to wheat growth than was NH_4-N in noninoculated soils; 4) foliarly applied urea and placement of N at a point in soil above or below the infection court failed to reduce the incidence or severity of take-all; and 5) reductions in pH_r due to added fertilizer NH_4-N were clearly large enough to reduce the inoculum potential of *G.graminis*.

The Rhizosphere Microflora

All previously described work was conducted in the USA and it indicated that the suppression of take-all by NH_4-N was a result of indirect second- or third-order interactions, involving the pH and the microflora of the rhizosphere. These findings were duplicated in a study of eight Australian soils, of which two were selected for more critical study. An alkaline soil from a long-term wheat monoculture appeared to possess a microflora antagonistic toward *G.graminis*. A second soil had a variable cropping history, and did not suppress growth of *G.graminis*.

Lineal extension of *G.graminis* hyphae along roots of intact wheat plants growing in the Australian soils was less in the NH_4-treated than in the NO_3-treated suppressive soil (Smiley, 1978a). The inhibitory effect of NH_4-N was eliminated by soil fumigation and by treatment of the soil with aerated steam at 60°C, and was reintroduced by adding back 1% of the nonsterile soil. These responses were not found in the non-suppressive soil. A *G.graminis*-inhibitory, easily transferable, non-spore-forming microorganism therefore appeared to be active in rhizoplanes of NH_4-treated suppressive soil.

Fertilization of the soils with either NH_4-N or NO_3-N had little influence upon the total numbers of bacteria and streptomycetes isolated from the rhizoplane of wheat (Smiley, 1978a). Fumigation of the soils did not alter the streptomycete population on wheat rhizoplanes, but it did reduce the pseudomonad population by as much as 100 times, and increase the total bacteria counts by up to five times.

Rhizoplane isolates of bacteria and streptomycetes were then grown on plates of *G.graminis*-inoculated potato-dextrose-agar medium. Percentages of microorganisms antagonistic to *G.graminis* were much higher for the suppressive soil than for the nonsuppressive soil, and few antagonists could be isolated from fumigated soils. Of the microbial groups examined, the pseudomonads appeared to be the most important specific antagonists of *G.graminis*. Relatively high proportions of *Pseudomonas* isolates were antagonistic in the NH_4-treated suppressive soil. Very few or no antagonistic pseudomonads were isolated from NO_3-treated or fumigated suppressive soil, or from any treatments of the nonsuppressive soil.

Specific microbial groups were then added to the soils, and the pathogen growth rates on roots were measured (Smiley, 1978b). Young wheat roots, aseptically produced on sand in Petri plates, were inoculated with 20 randomly selected isolates of several rhizoplane microbial groups, and were then chopped and used as a microbially-colonized food base inoculum which was mixed into the fumigated soils that had or had not received additional treatment with 1% nonsterile soil. Wheat seed was then planted over point sources of *G.graminis* inocula, and the plants were later harvested for studies of the rhizoplane microflora, and for measurement of *G.graminis* growth.

Pseudomonas was consistently associated with the NH_4-induced suppression of *G.graminis* growth, whereas the relationships were poor with *Bacillus* species, the general microflora, the streptomycetes, and the fungi. Percentages of antagonists (*in vitro*) were about 50% for pseudomonads, and between 0 and 5% for the other groups. These studies have all strongly implicated the pseudomonads as specific antagonists of *G.graminis* in the suppressive soil.

Pseudomonas *species*

The identities of *Pseudomonas* species involved in the

biological control of take-all were then determined (Smiley, 1978c). Fluorescent pseudomonads dominated the total pseudomonad population in nonfumigated soil but were absent from the population in fumigated soil. The dominant N-form in the rhizosphere had no effect on the proportions of fluorescent and nonfluorescent pseudomonads. Percentages of antagonistic pseudomonads were much higher in the suppressive than in the nonsuppressive soil, and were also much higher in populations isolated from soil on agar media at pH 5.5 rather than at 7.0. Most of the antagonists were *P. fluorescens*, Biotype G, Set 30; very few were *P. putida*. Set 30 comprised 60 to 100% of the fluorescing isolates from the suppressive soil, and 0 to 55% of those from nonsuppressive soil. Furthermore, 83% of the Set 30 isolates from NH_4-treated suppressive soil were antagonists, compared to only 5% from the NO_3-treatments, and 0 to 14% from the nonsuppressive soils.

Fluorescent pseudomonads quickly occupy the ecological niche in which *G. graminis* must compete. They are selectively stimulated and are highly competitive in the rhizosphere, many are tolerant of antibiotics, and they are very tolerant of acidic environments. Experimental evidence therefore supports the hypothesis (Smiley, 1978b; Cook and Rovira, 1976) that control of take-all in NH_4-treated soil involves a biological control mechanism involving the pseudomonads in soil. The efficacy of this mechanism for controlling take-all will vary among soils due to the inherent differences in abiotic and biotic soil properties and due to difficulties in applying NH_4-N uniformly throughout the tilled soil layer.

Rhizoplane pH Measurements Needed

Although the inhibition of *G. graminis* by *P. fluorescens* appeared to be induced in some soils by a localized reduction in rhizosphere pH, the suppression of take-all by NH_4-N was demonstrated in one very highly buffered soil where changes in pH_r were not measurable (Smiley, 1978a,b). Since nutritional effects were apparently not responsible, I suggest that in this one exception the pH_r changes occurred so close to the rhizoplane that they were not measurable in the rhizosphere.

The pH_r values determined in these investigations for characterizing the parasitic growth of *G. graminis* on the rhizoplane were conveniently measured compromises between the easily measured pH_b, which has little effect on interactions at the soil-root interface, and the difficult-to-measure rhizoplane pH which is desired. The pH_r measurements were satisfactory in most soils, but for highly buffered soils a technique for routine measurement of rhizoplane pH is needed. Even in moderately buffered, NH_4-treated soils the pH and NH_4-N gradients from the rhizoplane to bulk soil were very steep in the first 3 to 5 mm (Smiley, unpublished). The gradients were measured in soils separated from roots by screens which allowed penetration only of root hairs, thus establishing two-dimensional rhizospheres. Replicated sampling boxes with root-hair screens allowed samples to be taken near roots of different ages, and allowed for adjacent comparisons to be made of different soils or soil treatments. A metered sampling system allowed successive 0.5 mm soil sections to be removed for studies of gradients in the root zone. Results of preliminary tests showed clearly that comparisons of soil chemical properties with the activities of root-infecting fungi are limited if both are not measured in the infection court.

Related Investigations

Hornby and Brown (1977) were unable to correlate take-all with reduced pH_r in field studies, however their results do not appear contrary to those reported here. The lowest pH_r they reported was 6.3, which is considerably higher than that found to be suppressive in my field studies, and the rhizosphere soil used in their studies was taken from up to 5 mm from the rhizoplane. Hornby and Brown also utilized Nitro-chalk (2:1 mixture of NH_4NO_3 and lime), and I too failed to induce control of take-all with NH_4NO_3 or $Ca(NO_3)_2$, as opposed to $(NH_4)_2SO_4$ or $NH_4H_2PO_4$.

Huber and Watson (1974) have questioned the significance of the pH_r relationships reviewed here. They consider the pH_r differences of up to 2.2 units in the glasshouse and 1.2 units in the field to be "insignificant relative to the level of control of take-all achieved". However, they failed to report contrary experimental evidence, or to report evidence supporting the favoured NH_4- vs. NO_3-N nutritional mechanism, which was unimportant in our studies.

Acknowledgements

The author wishes to thank the numerous individual cooperators and sponsors for these investigations; Washington State University, United States Department of Agriculture, North Atlantic Treaty Organization, Australian Commonwealth Scientific and Industrial Research Organization, Victorian Department of Agriculture, South Australian Department of Agriculture, and the Waite Agricultural Research Institute.

References

Anonomous (1941). Take-all or whiteheads of wheat and barley. *Advis. Leafl. Minist. Agric. Fish.*, 304, 1.

Asher, M.J.C. and Shipton, P.J. (1979). "Biology and Control of Take-All" (Eds. M.J.C. Asher and P.J. Shipton). Academic Press, London. (in preparation).

Cook, R.J. and Rovira, A.D. (1976). The role of bacteria in the biological control of *Gaeumannomyces graminis* by suppressive soils. *Soil Biol. Biochem.*, 8, 269.

Hornby, D. and Brown, M.E. (1977). Nitrate and ammonium in the rhizosphere of wheat crops and concurrent observations of take-all. *Plant Soil*, 48, 455.

Huber, D.M., Painter, C.G., McKay, H.C. and Peterson, D.L. (1968). Effect of nitrogen fertilization on take-all of winter wheat. *Phytopathology*, 58, 1470.

Huber, D.M. and Watson, R.D. (1974). Nitrogen form and plant disease. *Ann. Rev. Phytopathol.*, 12, 139.

Smiley, R.W. (1974). Rhizosphere pH as influenced by plants, soils, and nitrogen fertilizers. *Soil Sci. Soc. Amer. Proc.*, 38, 795.
Smiley, R.W. (1978a). Antagonists of *Gaeumannomyces graminis* from the rhizoplane of wheat in soils fertilized with ammonium- vs. nitrate-nitrogen. *Soil Biol. Biochem.*, 10, 169.
Smiley, R.W. (1978b). Colonization of wheat roots by *Gaeumannomyces graminis* inhibited by specific soils, microorganisms, and ammonium-nitrogen. *Soil Biol. Biochem.*, 10, 175.
Smiley, R.W. (1978c). Ammonium-nitrogen induction of take-all control by resident wheat-rhizoplane pseudomonads. *Phytopathology*, 68, in press.
Smiley, R.W. and Cook, R.J. (1973). Relationship between take-all of wheat and rhizosphere pH in soils fertilized with ammonium- vs. nitrate-nitrogen. *Phytopathology*, 63, 882.

SYSTEMIC ACTIVITY OF OXAMYL TO *MELOIDOGYNE JAVANICA* AT THE ROOT SURFACE OF TOMATO PLANTS

R.A. ATILANO and S.D. VAN GUNDY

Department of Nematology, University of California, Riverside, U.S.A.

Summary

The systemic activity of oxamyl was tested by applying it as a soil drench or to living leaves for downward translocation. The test plant was tomato, *Lycopersicon esculentum*, and the test organism the root knot nematode, *Meloidogyne javanica*. Applied to one part of the root system at a concentration of 10 ppm, oxamyl did not affect nematodes in another, separate part of the root system, but at 300 ppm the number of larvae entering the separate root system decreased, and fewer root tips were penetrated. At 300 ppm oxamyl also reduced nematode growth, development and egg production for up to 30 days. Observations of nematodes on agar plates suggested that the chemical was retained at the root surface, where it deterred feeding and penetration by the nematodes. The numbers of *Meloidogyne* males and larvae recovered from soil were unaffected but microbial feeding nematodes around roots of oxamyl-treated plants were significantly reduced. The protective effect of oxamyl appeared to be strongest two days after application after which it declined, but was still discernible up to 30 days after application of 300 ppm to roots. Hatching of eggs of *M. javanica*, and extraction of microbial feeders on the root surface was inhibited for up to four days in a mist extraction chamber, but thereafter the effect was lost suggesting loss of biological activity by leaching or by inactivation.

Introduction

Oxamyl (methyl N'N' - dimethyl - N -[(methyl carbamoyl) oxy]-1-thio-oxamimidate) has been used to control nematodes in the field as a soil treatment or as a foliar spray, but it has performed inconsistently. Radewald *et al.* (1970) demonstrated systemic activity against several nematodes in various host plants and Harrison (1971) reported that diffusates from potato plants sprayed with oxamyl lessened hatch of *H. rostochiensis* eggs compared with diffusates from unsprayed plants. Taylor and Alphey (1973) suggested a root exudate from plants sprayed with oxamyl was the cause of systemic activity in their experiments, but as yet direct evidence of systemic activity is lacking. The present study was undertaken to provide direct evidence of the systemic effect of oxamyl on *Meloidogyne javanica* (Treub) Chitwood inside and outside tomato roots.

Methods and Materials

Glasshouse studies

Glasshouse experiments were carried out with 10-14 day old tomato seedlings (*Lycopersicon esculentum* (Mill.) cv 'Pearson' and 'Rutgers') grown in sandy loam (75% sand, 24% silt, 1% clay) watered weekly with full strength Hoagland's solution. Temperatures were between 24 to 29°C, occasionally rising to 38°C. Experiments were completely randomized or in randomized blocks, and all nematode counts were transformed to logarithms for analysis.

To provide plants with separate root systems, tomato seedlings were transplanted into 170 g plastic food cups and the roots trained to grow into a second cup; these were called respectively the proximal and the distal cups (Fig. 1). 50 ml of oxamyl at 10 or 300 ppm were applied through the topmost part to the proximal cup, and untrea-

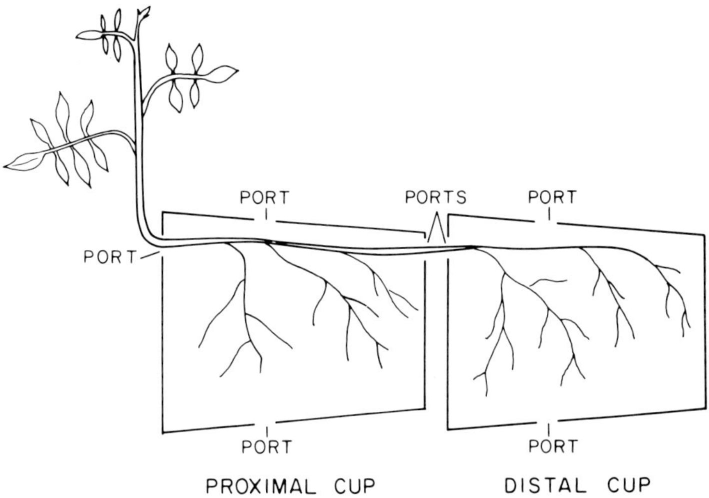

Fig. 1. Diagram of the double root system.

ted controls had water only, one and five days before and five days after infestation of each distal cup with 1,100 second stage larvae of *M.javanica*, collected from tomato roots between 24 and 48 hours after extracting in a mistifier (Southey, 1970).

72 hours after infestation, plants drenched one and five days before inoculation were harvested, roots stained in acid fuchsin/lactophenol (Southey, 1970) and nematodes counted. Thirty days after inoculation samples were taken from all treatments and stained. Eggs were obtained for counting by dispersing egg masses on roots in sodium hypochlorite and catching the eggs on sieves. Adult females were dissected from roots mounted in glycerine and their relative sizes estimated (Bird, 1959).

Distal roots of a second group of plants were inoculated and 27 days later transplanted into 905 gm cups, and three days later half treated with 500 ppm oxamyl drench, and half with water as a control. After a further 30 days roots were scrutinized for *M.javanica* larvae and free living nematodes.

Laboratory studies

Five day old tomato seedlings, grown on agar in petri dishes from seed soaked in 0.52% sodium hypochlorite and 10% ethanol were transferred singly to 100 x 15 mm petri dishes containing half strength Hoagland's agar with the hypocotyl emerging through a cotton wool sealed hole in the lid. These dishes were kept in a plastic basket lined with wet cheesecloth to prevent drying out of the agar, and incubated in a growth chamber at 28°C with 12 hour daily light. After establishing these cultures cotyledons or the stems were treated with oxamyl at 250, 500 and 1000 ppm in 1 cm^2 pieces of facial tissue or in lanolin paste for 24 to 120 hours, with controls having distilled water only. Two to five days later surface sterilized *M.javanica* larvae (Feldmesser, 1967) hatched from eggs obtained from roots (McClure *et al.*, 1973) using 0.05% sodium hypochlorite were placed on the agar near the growing seedlings, and their behaviour observed.

Leaf spray treatments

Foliar sprays of oxamyl at 500 and 10,000 ppm were applied 12 days after transplanting tomato seedlings into plastic tubes (7.5 x 3 cm) in the middle of 170 g styrofoam cups. Soil in the cups was protected from spray by absorbent paper. One to seven days later, 1000 to 1600 larvae were added per plant, and 48 to 72 hours after inoculation the number of larvae within roots was determined.

Results

Glasshouse studies

10 ppm oxamyl applied to the proximal roots did not decrease penetration of distal roots by *M.javanica* larvae, but 300 ppm applied one or five days before inoculation did (Table 1). The number of roots penetrated by larvae was also diminished (Table 2). Fewer galls were formed by *M. javanica* where 300 ppm oxamyl was applied to proximal roots

TABLE 1

The effects of oxamyl drench of proximal tomato roots on the number of M.javanica *larvae and female size within nontreated distal roots 72 hours and 30 days after infestation, respectively*

Treatment	Number of Larvae/ g of Roots		Female Size (cm^2)	
	10 ppm	300 ppm	10 ppm	300 ppm
Control	120a	212a	0.243a	0.223a
Oxamyl 5 days before infestation	103a	46b	0.234a	0.180b
Oxamyl 1 day before infestation	108a	89b	0.219a	0.154b
Oxamyl 5 days after infestation	-	-	0.181a+	0.123b

Mean within a concentration followed by the same letter are not significantly different from each other, Duncan's Multiple Range Test ($P \leq 0.05$).

\+ Treated with an additional 300 ppm oxamyl drench 11 days following infestation.

one and five days before and five days after infestation than following other treatments (Table 2). Although there were more galls on plants treated with oxamyl after inoculation than those treated before, the number of galls did not differ significantly between the three oxamyl treatments. Gall size appeared to be smaller where oxamyl was added, and root weight of plants inoculated with *M.javanica* was significantly more than those uninoculated, and greatest where no oxamyl had been applied.

The size of adult females of *M.javanica* was reduced where 300 ppm oxamyl were applied, the latest application five days after infestation, having the most effect (Table 1). Females that developed on plants treated with 300 ppm oxamyl also produced fewer eggs than those from untreated controls (Table 2).

At the end of 30 days, there was no significant effect of oxamyl on the number of larvae and males of *M.javanica* recovered from roots placed in the mist chamber or from soil around roots. By contrast, significantly more free-

TABLE 2

The effects of 300 ppm oxamyl drench of proximal tomato roots on the number of M.javanica *infected roots 72 hours after infestation and on the number of galls and number of eggs found 30 days after infestation of the distal untreated roots*

Treatment	No. Infected Roots / g of Roots	No. Galls / Root System	No. Eggs / Female
Control	75a	310a	240a
Oxamyl 5 days before infestation	14b	154b	98c
Oxamyl 1 day before infestation	28b	191b	145b
Oxamyl 5 days after infestation	-	231b	100c

Means followed by the same letter in each test are not significantly different from each other, Duncan's Multiple Range Test ($P \leq 0.05$).

living nematodes were collected from soil about the distal roots of untreated plants than from oxamyl treated plants (Table 3). More than twice as many nematodes were collected from untreated plants as from treated plants in the first day under the mist extraction equipment, but on subsequent days the difference became smaller suggesting that a factor contributing to the difference was being leached.

TABLE 3

The effects of oxamyl (300 ppm) drench of proximal tomato roots on the mean number of second-generation nematodes recovered from soil of the untreated distal root zone 30 days after infestation and treatment

Treatment	*M.javanica* Juveniles	*M.javanica* Males	Microbial Feeders
Control untreated	896	26	1458
Oxamyl treated	852	23	639
	NS	NS	($P \leq 0.01$)

Laboratory studies

M.javanica larvae placed on the agar culture surface about 2 cm from the nearest root showed no sign of attraction to it, but within five mm of the root their activity appeared to be influenced by it and they moved towards it. Most penetration of the root took place in a zone about 5 mm long beginning about 1 mm behind the root tip. Penetration was rapid, one larvae being observed to take two hours only to get entirely into the root, and eight hours later the root began to swell.

Oxamyl concentrations of 500 ppm and 1000 ppm were toxic to tomato plants, but plants grew and nematode galls were produced at 250 ppm. The initial movement of larvae towards roots of oxamyl treated plants was the same as that in untreated plants, but once they reached the roots fewer remained and tried to penetrate, and these few took longer to penetrate than they did in untreated roots. In two instances, complete penetration took 6 and 8 hours, and in both cases the cell walls around the penetration site temporarily appeared damaged or discoloured, a feature not seen in nematode penetration of untreated roots. Within a few hours, with continued root growth, the cell structure where the nematode was returned to a normal appearance.

Leaf spray experiments

Tests were carried out to determine the effects of Triton B-1956$^{(R)}$ glycerol and sucrose in enhancing the effectiveness of leaf sprays. Sucrose appeared to enhance the effectiveness of oxamyl more than other adjuvants when applied two days, but not four days, before infestation (Table 4). All oxamyl treatments reduced the number of larvae entering roots and the percentage of roots entered during the 72 hour period, but treatments differed little in the number of larvae entering each infected root. The same adjuvants were used in a second test with oxamyl sprays of 500 and 10,000 ppm applied two days before infestation. The lower rate of oxamyl was ineffective with all adjuvants, but the higher rate gave similar reductions in the number of larvae penetrating the roots and the per-

TABLE 4

The effect of 10,000 ppm oxamyl foliar sprays containing different adjuvants on the number of M. javanica *second-stage larvae within tomato roots 72 hours following infestation*

Treatments	No. Larvae / g Root	No. Larvae / Infected Root	% Infected Roots
Control. Triton B-1956 @ 2 ml/l	340a	5.2a	25.6a
Control. Sucrose @ .6M	278a	6.0a	25.7a
Control. Glycerol @ .1M	139ab	2.1ab	13.0b
Oxamyl + Triton B-1956 @ 4 days before infestation	129b	3.8a	19.7bc
Oxamyl + Triton B-1956 @ 2 days before infestation	71b	1.7ab	11.4b
Oxamyl + Sucrose @ 4 days before infestation	55bc	0.9b	4.6bc
Oxamyl + Sucrose @ 2 days before infestation	7c	3.2ab	1.5c
Oxamyl + Glycerol @ 4 days before infestation	37bc	2.6ab	3.9bc
Oxamyl + Glycerol @ 2 days before infestation	93b	4.4a	4.9bc
Oxamyl + Glycerol @ 1 day before infestation	112b	4.1a	8.5bc

Means followed by a common letter in each test are not significantly different from each other, Duncan's Multiple Range Test ($P \leq 0.05$).

centage of roots entered by larvae, with no significant differences between adjuvants.

Discussion

These studies support the hypothesis that oxamyl and/or its toxic metabolites are transported systemically from the sprayed foliage to roots and from drenched roots distally to roots not in direct contact with the chemical. In our tests it took roughly thirty times as much chemical in the foliage sprays as in soil drenches to produce consistent and similar responses in nematode penetration and development. These concentrations, 10,000 ppm and 300 ppm respectively, are considerably higher than those ordinarily recommended for field use and may account for some of the ineffective results in field experiments where low dosages were used for testing.

Various hypotheses for the mode of action of organophosphate and organocarbamate nematicides include direct kill of nematodes, temporary incapacitation of sensory or musculature systems, alteration of root qualities or exudates that repel nematodes. The latter has gained considerable support although the results have been variable. Our observations on nematode behaviour on agar plates would suggest that root exudates containing oxamyl and/or its toxic metabolites were not a factor in reducing the number of *M.javanica* larvae entering root tips and producing galls. It seems more likely that the systemic action more nearly resembles the temporary incapacitation of the sensory or musculature systems of the larvae at the root surface and their feeding and penetration was slowed and reduced at the root tip. This resulted in fewer larvae per g-root, fewer galls per root system and smaller, less developed galls in the treated plants than in the untreated controls. These observations are supported by Volvas and Lamberti (1974) who tested several chemicals in a split-root system for the control of *M.incognita*. They found that oxamyl (10 kg/ha ai) retarded penetration of the untreated split roots for 17 days and female development for 30 days. Our studies clearly demonstrate systemic activity on the development of the nematode life cycle. Oxamyl applications up to five days after soil infestation significantly reduced female size, altered female morphology and reduced the production of eggs of *M.javanica* in tomato roots.

There is also some indirect evidence that oxamyl and/or its metabolites was translocated to the root surface and exhibited some biological activity there. There was a significant reduction of the microbial-feeding nematodes in the soil around roots of oxamyl-treated plants. These results could be taken to suggest that the growth of microbial flora at the root surface was reduced and provided less food for the microbial-feeding nematodes or that they suffered from temporary incapacitation of the sensory or musculature system. The toxic effects on root microflora is supported by two studies on nitrogen transformations in soil treated with oxamyl. Elliot, Marks and Tu (1977) found reductions of *Nitrosomonas* spp. but *not Nitrobacter* spp. in soil around tobacco plants transplanted with oxamyl inhibited nitrification and decreased the rate of denitrification in soil. They also found a harmful effect on nitrogenase activity in treated soil. These data suggest that systemic nematicides may have pronounced effects on the microflora and microfauna around soil roots. Biological activity in this zone is further supported by nematode extractions from roots in the mist chamber where hatching of eggs of *M.javanica* and microbial-feeding nematodes was inhibited up to four days after removal of the plant from the soil. There were no differences at six, eight and twelve days suggesting that the active portion of the chemical was either inactivated or leached from the roots.

One of the major obstacles to foliar chemical sprays and their uptake by plants is retention and uniform coverage at the plant surface. Adjuvants or stickers are sometimes beneficial. Starr and Mai (1975) did not find any improvement in oxamyl effectiveness with several adjuvants. We tested sucrose, glycerol and Triton B-1956 as possible adjuvants and found some improvement with sucrose when applied two days prior to infestation. The use of adjuvants in combination with a study of the environmental factors that affect such activity in plants such as day length, temperature, light intensity and water deficits might be worthy of further study.

References

Bird, A.F. (1959). Development of the root-knot nematodes *Meloidogyne javanica* (Treub) and *Meloidogyne hapla* (Chitwood) in tomato. *Nematologica*, 4, 31.

Elliot, J.M., Marks, C.F. and Tu, C.M. (1977). Effects of certain nematicides on soil nitrogen, soil nitrifiers and populations of *Pratylenchus penetrans* in flue-cured tobacco. *Can. J. Plant Sci.*, 57, 143.

Feldmesser, J. (1967). An *in vitro* technique for nematode inoculation and culture of plants. *Nematologica*, 13, 141.

Harrison, M.B. (1971). Evidence of the mode of action of a systemic nematicide. *J. Nematol.*, 3, 311 (Abstr.).

McClure, M.A., Kruk, T.H. and Misagh, I. (1973). A method for obtaining quantities of clean *Meloidogyne* eggs. *J. Nematol.*, 5, 230.

Radewald, J.D., Shibuya, F., Nelson, J. and Bivens, J. (1970). Nematode control with 1410, an experimental nematicide-insecticide. *Plant Dis. Rep.*, 54, 187.

Southey, J.R. (Ed.) (1970). Laboratory methods for work with plant and soil nematodes. Ministry Agric. Fish. Food Tech. Bull. No. 2.

Starr, J.L. and Mai, W.F. (1975). Effects of adjuvants on the efficacy of oxamyl. *Plant Dis. Rep.*, 59, 510.

Taylor, C.E. and Alphey, T.J.W. (1973). Aspects of the systemic nematicidal potential of DuPont 1410 in the control of *Longidorus* and *Xiphinema* virus vector nematodes. *Ann. Appl. Biol.*, 75, 464.

Vlassak, K. and Livens, J. (1975). Effect of some pesticides on nitrogen transformations in soil. *Sci. Total Environ.*, 3, 363.

Volvas, N. and Lamberti, F. (1974). Studies of the systemic action of some chemicals in the control of the root-knot nematode. XII. Simposio Internacional de Nematologica. *Soc. Eu. Nematol.*, p.107.

THE INTERFACE IN RELATION TO ROOT FUNCTION AND GROWTH

J.F. LONERAGAN

School of Environmental and Life Sciences, Murdoch University, Perth, Western Australia

Summary

This paper briefly reviews the relationship of the chemical environment to root growth and nutrient absorption.

Low concentrations of O_2 and of Ca^{2+} and borate ions and high concentrations of Al^{3+} and H^+ ions inhibit root growth, and interfere with nutrient absorption. The concentration of nutrient at the root surface is also an important determinant of nutrient flux into roots but the relationship is modified by other factors. The flux of K^+ into roots of plants growing in nutrient solution where the concentration of K^+ is similar to that occurring in soils is modified strongly by the K-status of the plant. The flux of Zn^{2+} and possibly of other divalent ions responds strongly to the ionic environment and seems to be independent of the Zn-status of the plant. It increases with Zn^{2+} concentration but decreases with increasing concentrations of other cations. Flux of phosphate and other anions increases with increasing concentrations of both the anion and other salts. Flux of the univalent phosphate ion is much faster than that of the divalent ion and is modified to some extent by the P-status of the plant.

To understand the effect of the chemical environment on root growth and function, more information is needed on those factors which affect the formation, development, and function of mycorrhizae. More information is also needed on the activities of particular regions of the root and of the micro-environment produced at the soil-root interface by those activities.

The Chemical Environment and Root Growth

Root growth involves several processes - extension of the main axis by cell division and elongation, differentiation of the structure of the main axis, formation of root hairs, initiation and development of laterals, and formation of specialized structures in association with micro-organisms - root nodules, mycorrhizae, and proteoid roots. Each of these processes may respond somewhat differently to the chemical environment as has been shown for example for cell division compared with cell elongation (Burstrom, 1952) and for the development of nodules which sometimes fail to form under conditions favourable for growth of the host root (Munns, 1968; Loneragan, 1972).

Some components of the chemical environment act directly on root growth. Others act indirectly by affecting the nutritional status of the plant as a whole.

Oxygen

Root elongation may cease within 2 or 3 minutes of removing oxygen from the root environment. The concentrations of O_2 required at the root interface for maximal elongation vary greatly with species, with root temperature, and with the aerial environment (Greenwood, 1969; Eavis, Taylor and Huck, 1971; Luxmoore and Stolzy, 1972).

Calcium and borate ions

Roots must have minimal concentrations of Ca^{2+} and borate ions in their immediate environment to grow. No other nutrient is required in the immediate environment since they may move from other plant parts to any part of the root system which has no external supply. But even supplying luxury quantities of Ca^{2+} and borate ions to one part of a split root system did not induce any growth in those roots bathed in a solution without them (Haynes and Robbins, 1948).

The concentrations of Ca^{2+} ions required to maintain root structure and to permit some growth range from 0.01 μM for wheat (Burstrom, 1952) to 0.3 μM for some herbs and legumes (Loneragan, Snowball and Simmons, 1968). Concentrations of 10 μM or more are required for maximal root elongation. However, the relationship of Ca^{2+} ion concentrations to physiological processes is likely to vary with the concentrations of H^+ and other cations (Jones and Lunt, 1967).

At borate concentrations of 0.5 to 1 μM, roots of beans, tomatoes, and flax develop normally. Maize roots develop at even lower concentrations (Neales, 1960, 1964).

Hydrogen and aluminium ions

In soil, the effects of acidity are sometimes complicated by a high concentration of Al^{3+} in the soil solution but, even in its absence, acidity can suppress root growth. Thus at pH 4.0 wheat roots stopped elongating: Burstrom (1952) attributed this effect to the action of H^+ ions in suppressing cell division in root meristems. In his experiments, cell numbers in root tips responded strongly to pH from 4 to 6 but did not respond at all to changing concentrations of Ca^{2+} ions from 0.01 to 1 μM. By contrast, at pH 4.4 the fine structure of meristematic cells of maize roots became deranged within 2 to 3 hours in the absence of Ca^{2+} but was not affected in the presence of 10 μM Ca^{2+} (Marschner, Handley and Overstreet, 1966).

Aluminium ion concentrations of 1000 μM or more completely suppressed cell division and elongation in onion roots (Clarkson, 1965). Root growth of sensitive legumes was also depressed by 20 μM Al^{3+} in solution and stopped by 70-80 μM (Munns, 1965; Andrew, Johnson and Sandland, 1973;

Gunatilaka, 1977). The effects of Al^{3+} on root growth vary considerably with species, pH, temperature, and concentration of other cations especially Ca^{2+} and Mg^{2+} (Foy, 1974; Rhue and Grogan, 1976). Many studies now attribute poor growth of roots on acid soils to high Al^{3+} concentrations in the soil solution. For example, a concentration of 33 μM Al^{3+} at pH 5.0 was thought responsible for poor growth of cotton roots (Adams and Pearson, 1970).

Other nutrients

For some time, it has been known that application of nutrients to local areas of the root zone may promote branching and growth of roots there (Wiersum, 1957; Duncan and Ohlrogge, 1958; Alston, 1976). Recent work has shown that N and P are particularly effective (Hackett, 1972; Drew, 1975; Drew and Saker, 1975a,b). The development of roots in the fertilized zone seems to be at the expense of root development elsewhere (Miller, 1974) and probably only occurs in plants suffering from a general deficiency of the nutrient concerned (Drew and Saker, 1978).

The Chemical Environment and Ion Absorption

Those factors of the chemical environment which directly influence root growth may also drastically affect ion absorption. In the absence of oxygen, absorption ceases. The deterioration of membrane structures by low Ca^{2+} or high H^+ ions also results in inhibition of ion absorption and leakage of ions from the root. High concentrations of Al^{3+} and other polyvalent ions inhibit the absorption of Ca^{2+} (Clarkson and Sanderson, 1971). Recent research has also shown that in the absence of borate ions, maize roots stop absorbing other nutrients even before they stop growing (Pollard *et al.*, 1977).

Other factors of the chemical environment may be less drastic in their action. Of these the concentration of the absorbing ion is perhaps the most important but the relationship of nutrient concentration at the soil-root interface to nutrient flux into the root may vary with the form of the nutrient, the concentration and nature of other ions in solution, the nutrient status of the plant, and with plant species. Reductive activity at the root surface is likely to be important for the absorption of Fe and possibly also Mn (Chaney *et al.*, 1972). The presence of micro-organisms and the infection of roots with mycorrhizal fungi may also strongly modify the nature of the interface and the response of ion absorption to the chemical environment there (see Rovira, this volume).

Concentration-flux relationship

In short term studies with 'low salt' plant materials (i.e. plants with a low level of the nutrient being studied) the relationship of external concentration of a nutrient to the flux of that nutrient into roots follows a dual pattern as illustrated for K^+ flux into excised barley roots in Figure 2. This relationship has characterized the behaviour of a wide range of ions in a wide range of plant materials. It has been taken to indicate the existence of at least 2 discrete mechanisms of absorption - a mechanism 1 characteristic of low concentration ranges (1 mM) and mechanism 2 or a series of mechanisms operative at high concentrations (1-50 mM). Many features of absorption differ between the 2 proposed mechanisms lending support to their separate identities (Epstein, 1976). The evidence suggests that mechanism 1 is active and selective whereas mechanism 2 may simply result from passive, non-selective diffusion from high external concentrations of ions into the root (Barber, 1972).

In contrast to the dual isotherms of nutrient flux into 'low salt' roots, nutrient flux in long term experiments with whole plants shows only a single isotherm. Many of the relevant experiments have been summarized by Pitman (1976) and are presented in Figure 1. In Figure 1, net flux of each nutrient into roots shows an initial steep response with increasing nutrient concentration followed by a marked attenuation or plateau. In the region of steep response plants were generally deficient in the nutrient under study. The concentration-flux relationships of nutrients also tended to fall into 2 groups, univalent ions whether positive or negative being absorbed at rates 10 to 100 times those of divalent ions at equivalent molar concentrations.

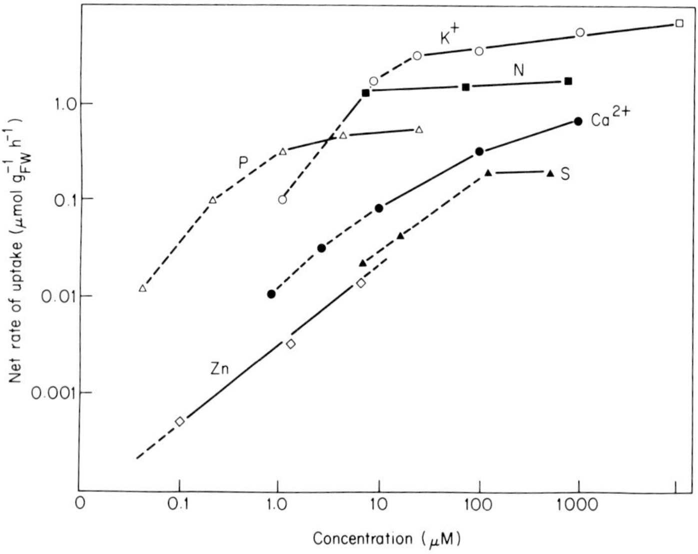

Fig. 1. Rates of net uptake of various elements calculated from change in external concentration or internal content. Dashed lines show where growth limitation was observed. Note this is a log/log plot to increase the range of data given. (○) K^+, data from Asher and Ozanne, 1967, assuming R_W = 0.13 d^{-1}, mean of 14 spp; (□) Pitman, 1972, for same R_W for barley; (△) P from Asher and Loneragan, 1967, Loneragan and Asher, 1967, mean of 8 spp; (■) N from Clement *et al.*, 1974, for *Lolium perenne*; (●) Ca^{2+} from Loneragan and Snowball, 1969a,b, mean of 30 spp; (▲) S from Bouma, 1967a,b,c *Trifolium subterraneum*; (●) Zn^{2+} from Carroll and Loneragan, 1968, 1969, mean of 8 spp. (Reproduced with permission from Pitman, 1976.)

Two factors have contributed to the striking differences between concentration-flux isotherms of the two types of study. Firstly, many short term studies have used very high salt concentrations whereas long term experiments have generally used concentrations of nutrients which occur in soil solutions and most, but not all, of these are within the mechanism 1 range of concentrations. Secondly, plants have feed-back mechanisms which depress nutrient fluxes of some ions preventing nutrient concentrations from rising to excessively high values: these mechanisms do not operate in 'low salt' plants but become important with increasing nutrient status. The importance of each of these factors in modifying the concentration-flux relationships varies with each nutrient. Univalent cations, divalent cations, and anions show 3 distinct patterns of behaviour.

Monovalent cations

Pitman (1976) has pointed out that many studies have shown that the concentration of Na^+ and K^+ in mature plant roots are relatively constant both with time and with distance along the root, provided that the culture solution does not become depleted. Moreover, the concentration of K^+ within the plant responds only weakly to increasing K^+ concentrations in solutions which produce large responses to K^+ flux in low salt roots (Fig. 2).

It thus appears that, once the older tissues of the plant have reached their characteristic K^+ concentrations, net K^+ flux into them ceases and net K^+ flux into roots supplies only the new growth of roots and of shoot. To a large extent then, K^+ flux into roots is determined by the growth rate of the plant (Pitman, 1972; Pitman *et al.*, 1974). This control of K^+-flux is achieved by some feed-back mechanism which results in the plant adjusting its K^+-influx in response to the K^+-content of its tissues (Johansen *et al.*, 1970). At a very low concentration of K^+, tissues do not reach a satisfactory level of K^+ and K^+-flux continues at high rates relative to the external concentration. As a result, the relationship of K^+-flux to K^+-concentration in long term experiments in flowing culture solutions is asymptotic contrasting markedly with the re-

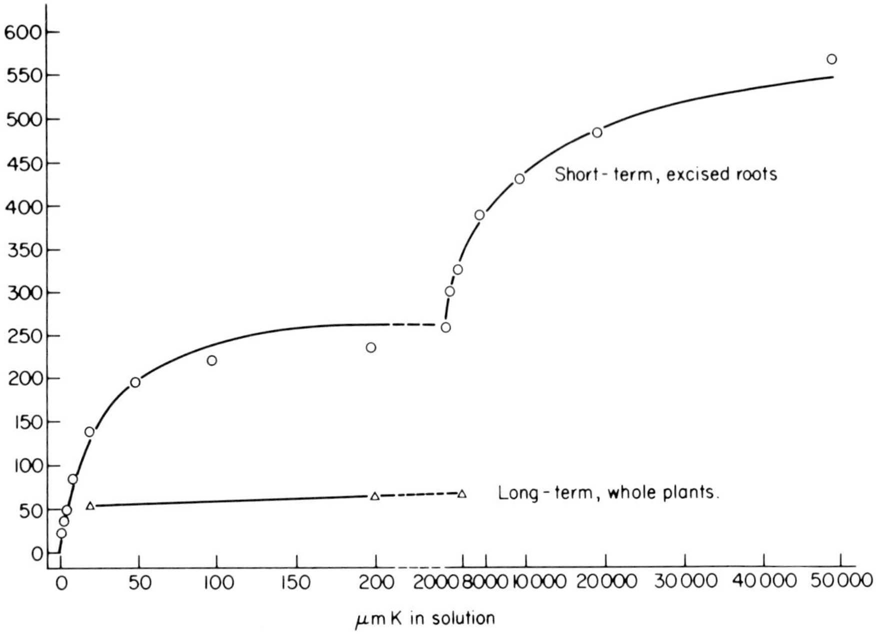

Fig. 2. Effect of K^+ concentration in solution on rates of K^+ absorption in (a) short-term study with excised roots of barley (data of Epstein *et al.*, 1963) and (b) long-term study with whole plants of barley (mean of Ca_{250} and Ca_{2500} over 20 days). (Reproduced with permission from Johansen *et al.*, 1968.)

lationship in low salt roots (Asher and Ozanne, 1967; Asher, 1964).

Divalent cations

By contrast with K^+, the fluxes of divalent cations into roots seem to be largely governed by the environment of the root and, so far as the limited data permit a judgement, seem to be independent of the quantities of cations previously absorbed.

For example, Zn^{2+} flux into several species was approximately linear with increasing concentrations of Zn^{2+} in solution to values far higher than those usually encountered in soil solution (Carroll and Loneragan, 1968, 1969: Fig. 1). In wheat, Zn^{2+} flux was not affected by Zn-status but was markedly depressed by increasing concentrations of H^+, Cu^{2+}, and all alkali and alkaline earth cations (Fig. 3; Chaudhry and Loneragan, 1972a,b,c.)

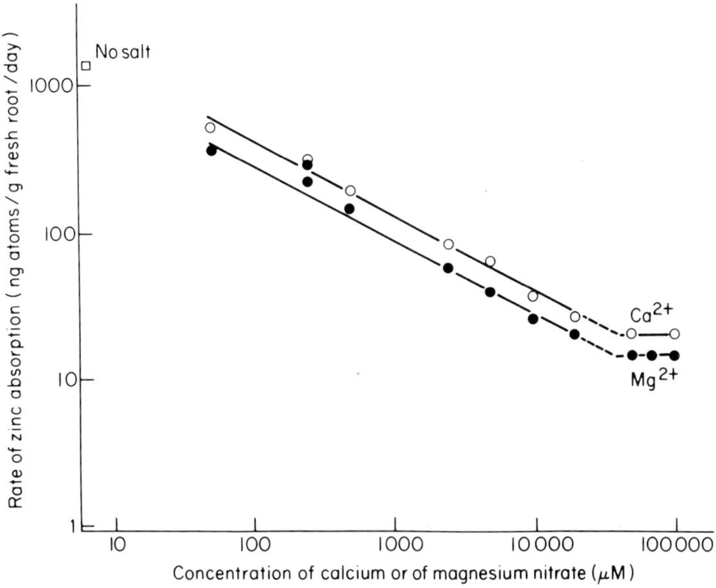

Fig. 3. The effects of Ca^{2+} and Mg^{2+} nitrate concentrations in solution on the flux of Zn^{2+} into wheat roots in short term experiments. (Reproduced with permission from Chaudhry and Loneragan, 1972c.)

The chelation of Zn^{2+} also depresses its absorption by plants. This effect seems to result from a decrease in the concentration of free Zn^{2+} ions at the root surface since either increasing the concentration of chelating agent or substituting agents of increasing avidity for Zn^{2+} decreases Zn^{2+} absorption (DeKock and Mitchell, 1957; Guinn and Joham, 1962). The situation in soils is far more complex since chelating agents may increase the concentration of metal in solution and increase substantially its movement through the soil to the absorption interface (Wilkinson, 1972).

The flux of Mn^{2+} into roots shows many similarities to that of Zn^{2+} being strongly inhibited by H^+ and Ca^{2+} (Robson and Loneragan, 1970). Fluxes of Mg^{2+} and Ca^{2+} into roots are also depressed by other ions in solution. Thus increasing concentrations of K^+ depressed flux of both Mg^{2+} and Ca^{2+} into roots of several species grown in flowing culture solutions (Johansen *et al.*, 1968a; Spear *et al.*, 1978). The depression of Ca^{2+} flux by K^+ contrasts markedly with the lack of any effect of Ca^{2+} on K^+ flux into the same plants (Johansen *et al.*, 1968b).

The uptake of Ca^{2+} ions by roots is also sensitive to H^+ (Arnon *et al.*, 1942) and varies with species being generally much higher in many legumes and some herbs than in grasses (Loneragan and Snowball, 1969a,b).

Anions

The behaviour of phosphate suggests that, under some conditions, the ionic environment of the root dominates absorption but that under other conditions the P-status of the plant dominates. Its behaviour is complicated further by the fact that it may be present as either a monovalent or a divalent ion.

In the experiment of Figure 1, phosphate was present primarily as a univalent ion since the solution was maintained at pH 5.4. Increasing pH increases the proportion of phosphate present in divalent form and progressively decreases the rate of uptake from a given concentration. This effect has been attributed to various differences in the affinity of phosphate absorption sites for the monovalent and divalent ions (van den Honert, 1937; Hagen and Hopkins, 1955; Hagen, Leggett and Jackson, 1957; Carter and Lathwell, 1967; Edwards, 1970). But perhaps the effects of pH may result simply from differences in the behaviour of univalent and divalent anions towards electronegative charges of the cell wall or the plasmalemma.

This latter explanation is consistent with the effect of ionic strength of the absorbing solution on the flux of anions into roots. Thus salts of divalent and, to a lesser extent, monovalent cations enhanced the flux of Br^- and phosphate ions into roots (Hooymans, 1964; Hyde, 1966; Franklin, 1969; Robson *et al.*, 1970). For both anions, Ca^{2+} salts had an immediate action which was particularly effective at low concentrations of the anions. In one species, increasing the $CaSO_4$ concentration from 250 to 2500 μM more than doubled the phosphate flux into plants growing in flowing culture solutions which contained a low but constant phosphate concentration (0.2 μM: Robson, Edwards and Loneragan, 1970). Ionic strength at the interface may therefore be an important determinant of the phosphate flux into roots, especially at low phosphate concentrations.

Differences in ionic strength can partly account for apparent discrepancies in data from experiments with excised roots and whole plants (Fig. 4: Loneragan and Asher, 1967; Edwards, 1970). The ionic strength of the external solution has been shown to correlate with the electrical potential difference between the solution and the cells of roots and may, therefore, influence the driving forces operating on phosphate ions entering the cell (Tang and Laudelot, 1971).

The P-status of the plant also appears to be an important determinant of phosphate flux into roots. Apparently under some conditions, as P-status of the plant increases, a feed-back mechanism slows down net phosphate flux (Drew and Saker, 1978; Clarkson *et al.*, 1978; de Jager, p.391, this volume). But, unlike the feed-back mechanism for K^+, the mechanism for phosphate appears to operate at a rather coarse level of control since several species have been shown to accumulate toxic concentrations of P in their leaves from very low phosphate concentrations in solution (Asher and Loneragan, 1967; Robson, Edwards and Loneragan, 1970).

The Influence of the Root on its Chemical Environment

The chemical environment at the soil-root interface may

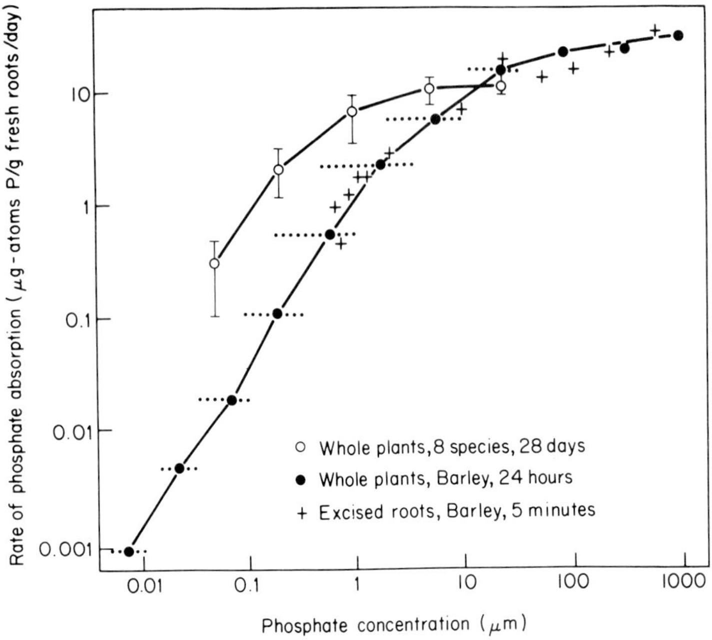

Fig. 4. Effect of phosphate concentration in the nutrient solution on P flux into roots: (o) Whole plants of 8 species in flowing culture solutions at pH 5.4 for 28 days (data of Loneragan and Asher, 1967); (•) Whole plants of barley in KH_2PO_4 solutions at pH 5.5-6.5 for 24 hours (data of Russell and Martin, 1953); (+) Excised roots of barley in KH_2PO_4 solutions at pH 4.0 for 5 minutes (data of Hagen *et al.*, 1957). (Reproduced with permission from Loneragan and Asher, 1967.)

be modified by plant activity. Transpiration induces mass flow of soil water to the root surface where the accompanying ions accumulate, react, or are absorbed (Barber, 1962). Nutrient absorption itself modifies the environment. Roots excrete "mucigel", other organic compounds, HCO_3^-, and H^+ and also display reducing activity.

The extent and significance of plant activity in modifying the soil-root interface can not yet be assessed in a satisfactory quantitative way. For example, it is known that the plant excretes either HCO_3^- or H^+ to balance any difference in the absorption of cations and anions (Jackson and Adams, 1963). It is also known that the absorption of cations and anions such as Ca^{2+} and phosphate may vary independently along the length of the root (Clarkson *et al.*, 1968; Harrison-Murray and Clarkson, 1973). However, the pH in the micro-environment of the soil-root interface during absorption has not been measured. Moreover, although the quantities of HCO_3^- or H^+ excreted by the whole root can be calculated, the location of absorption of cations and anions is not known sufficiently well to allow satisfactory prediction of micro-environmental pH changes which are likely to occur. Such pH changes could have profound effects both on the release of ions from the soil and on their absorption by the plant.

References

Adams, F. and Pearson, R.W. (1970). Differential response of cotton and peanuts to subsoil acidity. *J. Agron.*, 62, 9.

Alston, A.M. (1976). Effects of depth of fertilizer placement on wheat grown under three water regimes. *Aust. J. agric. Res.*, 27, 1.

Andrew, C.S., Johnson, A.D. and Sandland, R.L. (1973). Effect of aluminium on the growth and chemical composition of some tropical and temperate pasture legumes. *Aust. J. agric. Res.*, 24, 325.

Arnon, E.I., Fratzke, W.E. and Johnson, C.M. (1942). Hydrogen ion concentration in relation to absorption of inorganic nutrients by higher plants. *Pl. Physiol.*, 17, 515.

Asher, C.J. (1964). Potassium uptake by pasture plants. Ph.D. Thesis, Univ. W. Australia.

Asher, C.J. and Loneragan, J.F. (1967). Response of plants to phosphate concentration in solution culture: I. Growth and phosphorus content. *Soil Sci.*, 103, 225.

Asher, C.J. and Ozanne, P.G. (1967). Growth and potassium content of plants in solution cultures maintained at constant potassium concentrations. *Soil Sci.*, 103, 155.

Barber, D.A. (1972). "Dual isotherms" for the absorption of ions by plant tissue. *New Phytol.*, 71, 255.

Barber, S.A. (1962). A diffusion and mass-flow concept of soil nutrient availability. *Soil Sci.*, 93, 39.

Bouma, D. (1967a). Growth changes of subterranean clover during recovery from phosphorus and sulphur stresses. *Aust. J. biol. Sci.*, 20, 51.

Bouma, D. (1967b). Nutrient uptake and distribution in subterranean clover during recovery from nutritional stresses. I. Experiments with phosphorus. *Aust. J. biol. Sci.*, 20, 601.

Bouma, D. (1967c). Nutrient uptake and distribution in subterranean clover during recovery from nutritional stresses. II. Experiments with sulphur. *Aust. J. biol. Sci.*, 20, 613.

Burstrum, H. (1952). Studies on growth and metabolism of roots. VIII. Calcium as a growth factor. *Physiol. Plant*, 5, 391.

Carroll, M.D. and Loneragan, J.F. (1968). Response of plant species to concentrations of zinc in solution. I. Growth and zinc content of plants. *Aust. J. agric. Res.*, 19, 859.

Carroll, M.D. and Loneragan, J.F. (1969). Response of plant species to concentrations of zinc in solution. II. Rates of zinc absorption and their relation to growth. *Aust. J. agric. Res.*, 20, 457.

Carter, O.G. and Lathwell, D.J. (1967). Effect of chloride on phosphorus uptake by corn roots. *Agron. J.*, 59, 250.

Chaney, R.L., Brown, J.C. and Tiffin, L.O. (1972). Obligatory reduction of ferric chelates in iron uptake by soybeans. *Pl. Physiol.*, 50, 208.

Chaudhry, F.M. and Loneragan, J.F. (1972a). Zinc absorption by wheat seedlings. I. Inhibition by macronutrient ions in short-term experiments and its relevance to long-term zinc nutrition. *Soil Sci. Soc. Amer. Proc.*, 36, 323.

Chaudhry, F.M. and Loneragan, J.F. (1972b). Zinc absorption by wheat seedlings. II. Inhibition by hydrogen ions and by micronutrient cations. *Soil Sci. Soc. Amer. Proc.*, 35, 327.

Chaudhry, F.M. and Loneragan, J.F. (1972c). Zinc absorption by wheat seedlings and the nature of its inhibition by alkaline earth cations. *J. exp. Bot.*, 23, 552.

Clarkson, D.T. (1965). The effect of aluminium and other trivalent metal cations on cell division in the root apices of *Allium cepa*. *Ann. Bot.*, 29, 309.

Clarkson, D.T. and Sanderson, J. (1971). Inhibition of the uptake and long distance transport of calcium by aluminium and other polyvalent cations. *J. exp. Bot.*, 22, 837.

Clarkson, D.T., Sanderson, J. and Russell, R.S. (1968). Ion uptake and root age. *Nature*, 220, 805.

Clarkson, D.T., Sanderson, J. and Scattergood, C.B. (1978). Influence of phosphate-stress on phosphate absorption and translocation by various parts of the root system of *Hordeum vulgare* L. (barley). *Planta*, 139, 47.

Clement, C.R., Hopper, M.J., Canaway, R.J. and Jones, L.H.P. (1974). A system of measuring the uptake of ions by plants from flowing solutions of controlled composition. *J. exp. Bot.*, 25, 81.

DeKock, P.C. and Mitchell, R.L. (1957). Uptake of chelated metals by plants. *Soil Sci.*, 84, 55.

Drew, M.C. (1975). Comparison of the effects of a localised supply of phosphate, nitrate, ammonium, and potassium on the growth of the seminal root system and the shoot in barley. *New Phytol.*, 75, 479.

Drew, M.C. and Saker, L.R. (1975a). Nutrient supply and the growth of the seminal root system in barley. II. Localized compensatory increases in lateral root growth and rates of nitrate uptake. *J. exp. Bot.*, 26, 79.

Drew, M.C. and Saker, L.R. (1975b). Further studies on the modification to root growth and ion uptake caused by localized enrichment of phosphate in the rooting zone of barley. *Ann. Report. ARC, Letcombe Lab.*, 1974, 8.

Drew, M.C. and Saker, L.R. (1978). Nutrient supply and growth of the seminal root system in barley. III. Compensatory increases in growth of lateral roots and in rates of phosphate uptake, in response to a localized supply of phosphate. *J. exp. Bot.*, 29 (in press).

Duncan, W.G. and Ohlrogge, A.J. (1958). Principles of nutrient uptake from fertilizer bands. II. *Agron. J.*, 50, 605.
Eavis, B.W., Taylor, H.M. and Huck, M.G. (1971). Radicle elongation of pea seedlings as affected by oxygen concentration and gradients between shoot and root. *Agron. J.*, 63, 770.
Edwards, D.G. (1970). Phosphate absorption and long-distance transport in wheat seedlings. *Aust. J. biol. Sci.*, 23, 255.
Epstein, E. (1976). Kinetics of ion transport and the carrier concept. *In* "Transport in Plants. II. B. Tissues and Organs" (Eds. U. Lutte and M.G. Pitman), p.70. Enc. Pl. Physiol. N.S. 2. Springer Verlag, Berlin.
Epstein, E., Rains, D.W. and Elzam, O.E. (1963). Resolution of dual mechanisms of potassium absorption by barley roots. *Proc. natl. Acad. Sci.*, 49, 684.
Foy, C.D. (1974). Effects of aluminium on plant growth. *In* "The Plant Root and its Environment" (Ed. E.W. Carson), p.601. University Press, Virginia, Charlottesville, USA.
Franklin, R.E. (1969). Effect of adsorbed cations on phosphorus uptake by excised roots. *Pl. Physiol.*, 44, 697.
Greenwood, D.J. (1969). Effect of oxygen distribution in the soil on plant growth. *In* "Root Growth" (Ed. W.J. Whittington), p. 202. 15th Easter School, Nottingham, London, Butterworths.
Guinn, G. and Joham, H.E. (1962). Effects of two chelating agents on absorption and translocation of Fe, Cu, Mn, and Zn by the cotton plant. *Soil Sci.*, 94, 220.
Gunatilaka, A. (1977). "Effects of Aluminium Concentration on the Growth of Maize, Soybean and four Tropical Root Crops". M. Ag. Sci. Thesis, Univ. Queensland.
Hackett, C. (1972). A method of applying nutrients locally to roots under controlled conditions and some morphological effects of locally applied nitrate on the branching of wheat roots. *Aust. J. biol. Sci.*, 25, 1169.
Hagen, C.E. and Hopkins, H.T. (1955). Ionic species in orthophosphate absorption by barley roots. *Pl. Physiol.*, 30, 193.
Hagen, C.E., Leggett, J.E. and Jackson, P.C. (1957). The sites of orthophosphate uptake by barley roots. *Proc. natl. Acad. Sci.*, 43, 496.
Harrison-Murray, R.S. and Clarkson, D.T. (1973). Relationship between structural development and absorption of ions by the root system of *Cucurbita pepo*. *Planta (Berl.)*, 114, 1.
Haynes, J.L. and Robbins, W.R. (1948). Calcium and boron as essential factors in the root environment. *J. Amer. Soc. Agron.*, 40, 707.
Honert, T.H. van den (1933). Onderzoekingen over de voedingsphysiologie van het suikerriet: II. Mededeel Proefsta. *Java-Suikerind.*, 23, 1120.
Hooymans, J.J.M. (1964). The role of calcium in the absorption of anions and cations by excised barley roots. *Acta bot. neerl.*, 13, 507.
Hyde, A.H. (1966). Nature of the calcium effect in phosphate uptake by barley roots. *Pl. Soil*, 24, 328.
Johansen, C., Edwards, D.G. and Loneragan, J.F. (1968a). Interactions between potassium and calcium in their absorption by intact barley plants. I. Effects of potassium on calcium absorption. *Pl.*

Physiol., 43, 1717.
Jackson, P.C. and Adams, H.R. (1963). Cation-anion balance during potassium and sodium absorption by barley roots. *J. gen. Physiol.*, 46, 369.
Johansen, C., Edwards, D.G. and Loneragan, J.F. (1968b). Interactions between potassium and calcium in their absorption by intact barley plants. II. Effects of calcium and potassium concentration on potassium absorption. *Pl. Physiol.*, 43, 1722.
Johansen, C., Edwards, D.G. and Loneragan, J.F. (1970). Potassium fluxes during potassium absorption by intact barley plants of increasing potassium content. *Pl. Physiol.*, 45, 601.
Jones, R.G.W. and Lunt, O.R. (1967). The function of calcium in plants. *Bot. Rev.*, 33, 407.
Loneragan, J.F. (1972). The soil chemical environment in relation to symbiotic nitrogen fixation. *In* "Use of Isotopes for Study of Fertilizer Utilization by Legume Crops", p.17. Int. At. Energy Agency, Vienna.
Loneragan, J.F. and Asher, C.J. (1967). Response of plants to phosphate concentration in solution culture. II. Rate of phosphate absorption and its relation to growth. *Soil Sci.*, 103, 311.
Loneragan, J.F. and Snowball, K. (1969a). Calcium requirements of plants. *Aust. J. agric. Res.*, 20, 465.
Loneragan, J.F. and Snowball, K. (1969b). Rate of calcium absorption by plant roots and its relation to growth. *Aust. J. agric. Res.*, 20, 479.
Loneragan, J.F., Snowball, K. and Simmons, W.J. (1968). Response of plants to calcium concentration in solution culture. *Aust. J. agric. Res.*, 19, 845.
Luxmoore, R.J. and Stolzy, L.H. (1972). Oxygen diffusion in the soil-plant system. IV. Oxygen concentration and temperature effects on oxygen relations predicted for maize roots. *J. Agron.*, 64, 720.
Marschner, H., Handley, R. and Overstreet, R. (1966). Potassium loss and changes in fine structure of corn root tips induced by H^+-ion. *Pl. Physiol.*, 41, 1725.
Miller, M.H. (1974). Effects of nitrogen on phosphorus absorption by plants. *In* "The Plant Root and its Environment" (Ed. E.W. Carson), p.643. University Virginia Press, Charlottesville.
Munns, D.N. (1965). Soil acidity and growth of a legume. II. Reactions of aluminium and phosphate in solution and effects of aluminium, phosphate, calcium and pH on *Medicago sativa* L. and *Trifolium subterraneum* L. in solution culture. *Aust. J. agric. Res.*, 16, 743.
Munns, D.N. (1968). Nodulation of *Medicago sativa* in solution culture. I. Acid-sensitive steps. *Pl. Soil*, 28, 129.
Neales, T.F. (1960). Some effects of boron on root growth. *Aust. J. biol. Sci.*, 13, 232.
Neales, T.F. (1964). A comparison of the boron requirements of intact tomato plants and excised tomato roots grown in sterile culture. *J. exp. Bot.*, 15, 647.
Pitman, M.G. (1972). Uptake and transport of ions in barley seedlings. III. Correlation of potassium transport to the shoot with plant growth. *Aust. J. biol. Sci.*, 25, 243.
Pitman, M.G. (1976). Ion uptake by plant roots. *In* "Transport in Plants. II. B. Tissues and Organs" (Eds. U. Lüttge and M.G. Pit-

man), p.95. Enc. Pl. Physiol. N.S.2. Springer Verlag, Berlin.
Pitman, M.G., Lüttge, U., Läuchli, A. and Ball, E. (1974). Ion uptake to slices of barley leaves and regulation of K content in cells of the leaves. *Z. Pflanzenphysiol.*, 72, 75.
Pollard, A.S., Parr, A.J. and Loughman, B.C. (1977). Boron in relation to membrane function in higher plants. *J. exp. Bot.*, 28, 831.
Rhue, R.D. and Grogan, C.O. (1976). Screening corn for aluminium tolerance. *In* "Proceedings of Workshop on Plant Adaptation to Mineral Stress in Problem Soils" (Ed. M.J. Wright), p.297. Cornell Univ. Ag. Exp. Stn., Ithaca.
Robson, A.D., Edwards, D.G. and Loneragan, J.F. (1970). Calcium stimulation of phosphate absorption by annual legumes. *Aust. J. agric. Res.*, 21, 601.
Robson, A.D. and Loneragan, J.F. (1970). Sensitivity of annual *Medicago* species to manganese toxicity as affected by calcium and pH. *Aust. J. agric. Res.*, 21, 223.
Russell, R.S. and Martin, R.P. (1953). A study of the absorption and utilization of phosphate by young barley plants. I. *J. exp. Bot.*, 4, 108.
Spear, S.N., Edwards, D.G. and Asher, C.J. (1978). Response of cassava (Manihot esculenta Crantz) to potassium concentration in solution. III. Interactions between potassium, calcium and magnesium. *Field Crops Res.*, (in press).
Tang van Hai and Laudelot, H. (1971). Phosphate uptake and root electrical potential. *J. exp. Bot.*, 22, 830.
Wiersum, L.K. (1957). The relationship of the size and structural rigidity of pores to their penetration by roots. *Pl. Soil*, 9, 75.
Wilkinson, H.F. (1972). Movement of micronutrients to plant roots. *In* "Micronutrients in Agriculture" (Eds. J.J. Mortvedt, P.M. Giordano and W.L. Lindsay), p.139. Soil. Sci. Soc. Amer. Madison, Wisconsin.

MEASUREMENT OF ROOTING PATTERNS FOR DETERMINATE AND INDETERMINATE SOYBEAN GENOTYPES WITH A FIBER-OPTIC SCOPE

J.L. SANDERS and D.A. BROWN

Arkansas Agricultural Experiment Station, Fayetteville, Arkansas, U.S.A.

Summary

A fiber-optic scope technique was utilized to investigate the rooting habits of indeterminate and determinate soybean genotypes. Although these genotypes were not isolines, they were similar lines. There were no significant differences in the total root densities produced by either genotype throughout the growing season, but there were differences in root distribution. Indeterminate soybeans tended to produce larger root densities under the row than determinate ones. However, during the reproductive stages of soybean development, determinate soybeans produced significantly larger root densities in the interrow spaces. Both genotypes maintained at least 50 per cent of their root system in the Ap horizon throughout the growing season. From the seasonal data, four stages of root development were deduced and correlated with stages of soybean development (shoot portion). The data indicate that the root distribution may be more important than root density in influencing soybean yields. The performance of the fiber-optic scope technique proved to be a valuable tool for determining soybean root densities and distributions. Particular care should be taken in the installation of the observation tube to insure uniform contact between the soil and the surface of the tube. In soils where swelling or shrinkage occur, preliminary tests should be made to ascertain the effect on the uniformity of the soil-tube interface contact.

Introduction

Research programmes designed to develop a better understanding of plant root activity and its relation to the total plant performance have increased exponentially during the last few years. Over twenty papers, covering many aspects of soil-plant relationships, were presented before the crop physiology section of the American Society of Agronomy meetings in 1977. The research data presented covered many diverse areas: a survey of methods used for sampling and characterizing plant roots by Böhm (1977); in situ root characterization of plant roots using neutron radiography by Willit *et al.* (1977); soil-plant root water relations by Heatherly *et al.* (1977), Klepper and Taylor (1977) and Unger (1977). Arkin (1977) reported on a calculator method for estimating root length, and Barber (1977) summarized the effect of various plant root parameters on the nutrient status of plants. A new fiber-optic technique for measuring root growth of soybeans under field conditions was reported by Sanders and Brown (1978). This paper described the experimental procedure involved in the use of a modified fiber-optic duodenoscope for observing and photographing root development within the soil profile.

Briefly, the method utilizes a medical duodenoscope. The instrument is composed of a light source, a flexible optical tube (125 cm long) and an objective lens that transmits the subject image to a focusing eye-piece or to a 35 mm camera. The scope is equipped with a focus adjustment, a right angle objective lens, and a camera adapter. The field of view can be varied by increasing or decreasing the distance of the objective lens from the subject; therefore the experimentor can photograph either detailed or gross root-morphology. Figure 1 illustrates the essen-

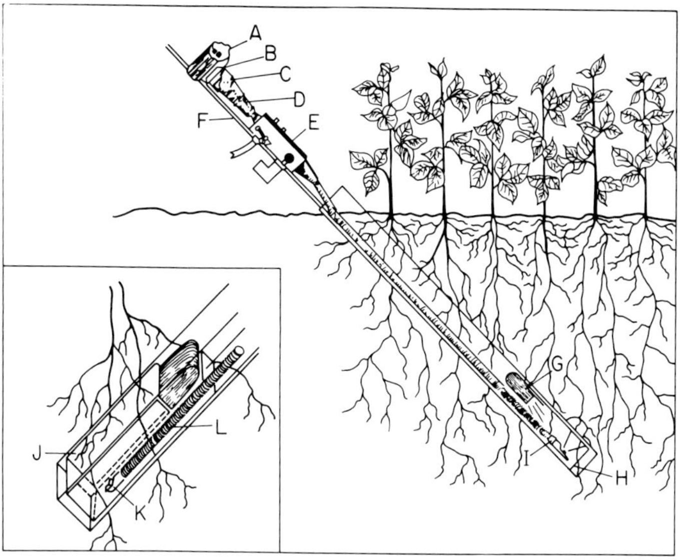

Fig. 1. Use of the fiber-optic scope and observation tube for photographing root growth of field grown soybeans; (A) 35 mm camera, (B) no-lens couples, (C) camera adapter, (D) focus adjust, (E) model ACMI, F5-A Duodenoscope, (F) slide support for scope, (G) photoflash, (H) observation tube, (I) ring holder, (J) light absorbing plate, (K) objective lens, (L) optic fibres.

tial features of the duodenoscope and how it is used inside a transparent observation tube for photographing soybean roots *in situ*. In this study the objective lens was set at 3.5 cm from the soil viewing surface. A flash unit is attached 15 cm behind the objective lens. The flash unit is synchronized with the camera (35 mm Pentax) by a five-foot connector. A 3X tele-extender is attached to the camera so that the subject would fill the entire frame. The scope is mounted on a black, plexiglass slide measuring 6.7 x 1.8 x 165 cm designed to slide in and out of the observation tube. The best results have been obtained with a Kodak High Speed Ectachrome EH135 (ASA 400), the camera being set for a one-sixtieth of a second exposure. Root monitoring is conducted by inserting the unit into a transparent plexiglass tube imbedded within the root zone. The observation tube is made of 0.3 cm thick plexiglass and measures 7.6 x 7.6 x 109 cm. The observation tube is installed downward into the root zone at a 45 degree angle and positioned so that the two upper sides were set at 45 degree angles to the horizontal. This arrangement allows the two faces of the tube to be available for root interception and prevents roots from elongating down the outer surface. The tube is installed by preparing a hole with a truck-mounted hydraulic soil-probe (Giddings; model GSRP-S).

The soil probe measures 7.6 x 7.6 x 101.6 cm (3 x 3 x 40 in.). It is slotted on one side and the open end was tapered in order to leave the surrounding soil with as little compaction as possible. After the soil core is removed with the probe, the observation tube is installed to a depth of 100 cm at a 45 degree angle which made the bottom end of the tube approximately 72 cm in horizontal-depth below the soil surface. The open end of the tube is then sealed to prevent the entry of light and the collection of moisture on the inside of the tube.

The soil was wet when the tube was inserted and allowed to form a firm contact with the faces of the observation tube over a two-month period. The soil-tube contact should be checked to be certain no large voids exist along the interface. Observation and characterization of the root system was made throughout the entire growing season.

Soybean roots (*Glycine max* (L.) Merr.) (Lee-68) at the R-7 stage of maturity were photographed and the root length was measured by two methods; the duodenoscope and the "Newman-core" method (see Sanders and Brown, 1978). Four depth-intervals (0-18, 18-36, 36-54 and 54-72 cm) were evaluated. The plants were grown in a Peridge silt loam soil (Typic Paleudalf; fine silty, mixed, mesic). Total root lengths as estimated by the "Newman-core" method for the 0-72 cm depth interval was 56.36 m, while the estimate by the scope was 52.84 m. Both methods gave similar results, however, the scope tended to estimate a greater root length than did the soil core method at all depths except the 0-18 cm. Both methods showed that the majority of the root system was in the 0-18 cm depth; 68% for the soil cores and 47% for the scope. Root density (cm/cm^3) decreased with soil depth from 10.58 to 1.52 by the soil core method and from 6.77 to 2.83 for the scope measurement. The coefficient of variation was 52.64 and 14.10 per cent for the core and scope measurements, respectively.

This study represents an extension of the previous work using the fiber-optic scope to characterize the rooting patterns of determinate and indeterminate soybeans.

Methods and Materials

In 1977 a field study of the distribution of determinate and indeterminate soybean roots was conducted. The field soil was a Peridge silt loam (Typic Paleudalf, fine silty, mixed, mesic). The fiber-optic scope technique was used to evaluate the rooting patterns of Lee 68 (determinate) and R72-346 (indeterminate) soybeans. These genotypes were used because in the previous grafting experiment, significant differences in nutrient uptake, growth, and yield were noted as being due to the root genotype. Soil moisture was maintained in the range of < 1 bars tension by means of furrow irrigation. Root distribution was not only monitored with depth under the row but between adjacent rows. Observation tubes were installed at a 45 degree angle at distances of 0, 25.4 and 50.8 cm from the centre of the row. This arrangement of tubes was used across three rows and thereby allowed each tube an equal opportunity to intercept the plant roots. Four replications at each location between the rows were used. Photographs were taken in an alternating manner on both sides of the observation tube and reported as the midpoint of the tube at 0, 25.4 and 50.8 cm from the row centre. Five stages of soybean development (as described by Fehr *et al.* (1971) were monitored for changing root density: V-4, V-8, R-2, R-3 (indeterminate) and R-4 (determinate) and R-7. The 35 mm transparencies were projected onto a Newman-type grid (Newman, 1965); root lengths were then converted to root density (cm of root/cm^3 of soil) by dividing the root length shown on a photograph by the representative volume of soil it represented (4.2 cm length x 2.75 cm width x 0.3 cm viewing depth). Root densities were weighed and reported according to the following soil horizons: Ap (0 to 20.5 cm); B1t (20.5 to 45.1 cm) and B12t (45.1 to 69.7 cm). The greatest depth at which roots were photographed was at 69.7 cm vertically from the soil surface.

The root densities were analyzed on the basis of a split-plot-design, each variety constituting a main plot.

Results

The root density and distribution patterns for indeterminate and determinate soybeans were found to differ with stage of plant growth and soil depth. The data given in Table 1 show that root densities were different for each soil horizon during both the vegetative and reproductive stages of growth. The Ap horizon had significantly higher root densities throughout the growing season. The B1t and B12t were generally not significantly different; however, later in the growing season the B12t tended to have higher root densities. Except at the R-7 stage of soybean development, root density between the rows was significantly different throughout the growing season. Both indeterminate and determinate root densities increased under the planted row until the R-3, R-4 stage; following these stages the densities decreased (Figure 2a). At the 25.4 and 50.8 cm positions root densities for indeterminate and determinate increased from the V-4 to the R-7 stage of soybean development (Figs. 2b, 2c, respectively).

Observable root growth under the row ceased after the R-3, R-4 stage (Fig. 2a) but roots in the 25.4 and 50.8 cm

TABLE 1

Average root densities and distribution patterns for indeterminate and determinate soybean plants

Stage of Soybean Development	Soil Horizons	Indeterminate			Determinate		
		Distance from the Row Centre (cm)					
		0	25.4	50.8	0	25.4	50.8
		Root Density (cm/cm^3)					
V-4	Ap	1.80	.31	0	1.50	.06	0
	B1t	.50	.02	0	.11	.01	0
	B12t	.01	0	0	.05	.01	0
	CV = 126%						
V-8	Ap	2.37	.58	0	1.85	.61	0
	B1t	.56	.14	0	.52	.05	0
	B12t	1.21	.46	0	.42	.04	0
	CV = 91%						
R-2	Ap	4.43	3.68	1.77	4.18	4.12	3.14
	B1t	2.25	1.10	.83	1.45	1.70	1.17
	B12t	3.11	2.13	1.48	1.21	2.09	1.25
	CV = 33%						
R-3, R-4	Ap	6.05	5.22	3.09	8.02	7.01	4.02
	B1t	3.35	2.14	1.15	3.14	3.05	2.12
	B12t	4.02	2.55	2.18	2.16	3.14	4.23
	CV = 31%						
R-7	Ap	5.12	6.30	3.93	5.61	6.60	4.50
	B1t	2.64	2.25	.96	1.96	3.43	2.40
	B12t	3.66	2.49	2.31	2.35	3.53	4.97
	CV = 22%						

positions indicated that growth had not ceased until later in the growing season (Figs. 2b and 2c). It was possible that root growth under the row was still occurring after the R-3, R-4 stage, but that the decomposition of roots occurred at a faster rate. Therefore, overall root growth for indeterminate and determinate soybeans continued beyond the R-3, R-4 stage of soybean development. Kaspar and Stanley (1977) and Mitchell and Russell (1971) have observed a continuation of root growth during the latter stages of soybean development.

The total root densities (averaged over depth and spacings) of indeterminate and determinate soybeans were not significantly different during the growing season, but there was a difference in their individual root distribu-

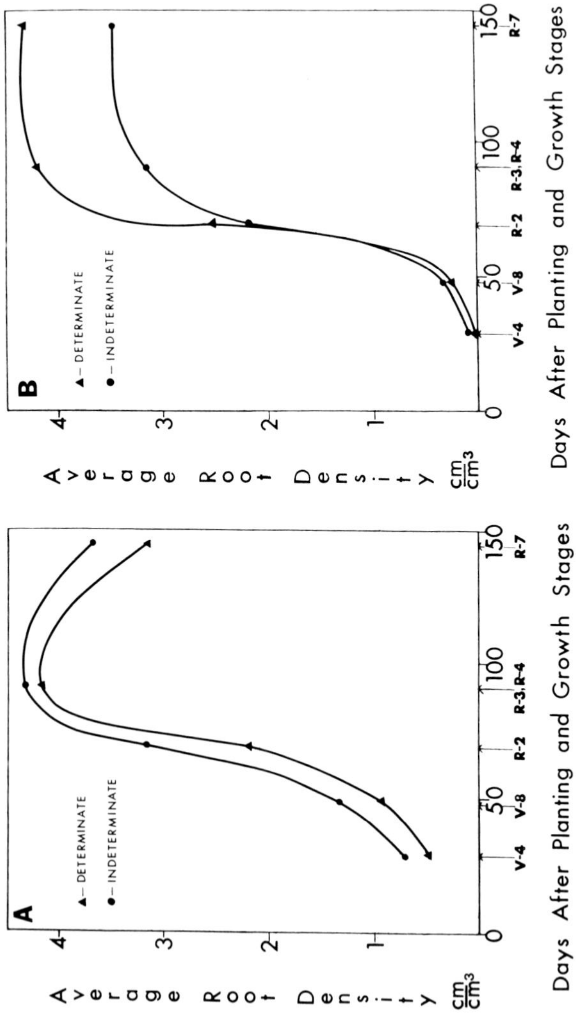
A
▲— DETERMINATE
●— INDETERMINATE
B
▲— DETERMINATE
●— INDETERMINATE
Average Root Density $\frac{cm}{cm^3}$
0
1
2
3
4
50
100
150
V-4
V-8
R-2
R-3, R-4
R-7
Days After Planting and Growth Stages

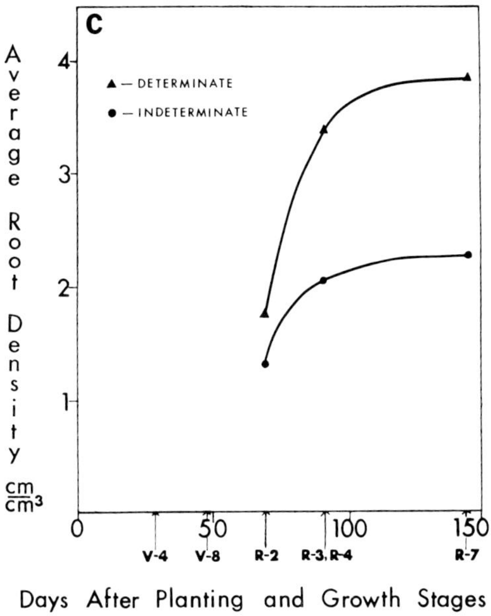

Fig. 2. (A, B, opposite; C, above)
Average root density at 0, 25.4 and 50.8 cm (A, B, C, respectively) distances from the row centre for indeterminate and determinate soybeans at different stages of growth.

tion throughout the soil profile. Generally, indeterminate soybeans accumulated higher root densities under the row throughout the growing season. Even though indeterminate and determinate inter-row root densities were similar during the early vegetative stages of soybean development, the determinate soybean root densities were higher during the reproductive stages.

Mitchell and Russell (1977) described three stages of soybean root growth using a soil monolith technique. In this study four distinct stages of root growth and development can be formulated from the seasonal indeterminate and determinate soybean root densities and distributions. Stage one corresponded to vegetative shoot growth (emergence to the V-8 stage of soybean development) and was characterized by root proliferation under the row. Stage two consisted of vegetative shoot growth and flowering (V-8 to the R-2 stage of soybean development) periods during which time root densities increased rapidly in the interrow spacings and especially in the Ap horizon. Stage three represented the pod growth and development period (R-2 to the R-3, R-4 stages of soybean development) in which root densities increased throughout the soil profile, but more so in the Ap horizon. Stage four represented the period of seed development and physiological maturity (R-3, R-4 to the R-7 stage of soybean maturity), in which root densities decreased under the row, apparently due to root decomposition. Very little new root growth elsewhere in the profile was evident.

Determinate soybeans yielded significantly more than indeterminate soybeans (3078 and 2171 kg/ha, respectively). However, the only significant difference in root distribution was the higher interrow densities in determinate varieties during the R-4 and R-7 stages of soybean development. This data indicates that root distribution or activity may be more important than total root density in influencing soybean yields.

The fiber-optic scope technique proved to be a valuable tool in differentiating soybean root density and distribution. Even though both genotypes were similar in genetic character and were grown under optimum conditions, the scope technique provided data which enabled us to distinguish significant differences between indeterminate and determinate soybean rooting patterns. Particular care should be taken in the installation of the observation tube to insure secure, uniform contact between the soil and the surfaces of the tube. In soils where swelling and shrinking occur, preliminary tests should be made to ascertain the effect on the soil-tube interface contact. With these precautions, the technique should be particularly useful in root studies where treatment effects are expected to be larger than those encountered in this experiment.

Acknowledgements

Published with the approval of the Director of the Arkansas Agricultural Experiment Station, Fayetteville. This project was supported in part by a grant from the Arkansas Soybean Promotion Board which is gratefully acknowledged.

References

Arkin, G.F., Chaison, R. and Jordan, W.R. (1977). Using your desktop calculator to estimate root length. *Agronomy Abstracts 1977*, 79, Amer. Soc. Agron., Madison.

Barber, Stanley A. (1977). Effect of various plant root and soil parameters on plant nutrient status. *Agronomy Abstracts 1977*, 80, Amer. Soc. Agron., Madison.

Böhm, W. (1977). Methods of characterizing root distribution in soil. *Agronomy Abstracts 1977*, 81, Amer. Soc. Agron., Madison.

Fehr, W.R., Caviness, C.E., Burmood, D.T. and Pennington, J.S. (1971). Stage of development descriptions for soybeans (*Glycine max* (L.) Merr.). *Crop Sci.*, 11, 929.

Heatherly, Larry G., Russell, W.J. and Ginn, L.H. (1977). Effect of tillage on water extraction, rooting density, and yield of soybeans. *Agronomy Abstracts 1977*, 86, Amer. Soc. Agron., Madison.

Kaspar, T.C. and Stanley, C.D. (1977). Soybean root growth during the reproductive stages of development. *Agronomy Abstracts 1977*, 87, Amer. Soc. Agron., Madison.

Klepper, Betty and Taylor, H.M. (1977). Seasonal and diurnal changes in root water uptake. *Agronomy Abstracts 1977*, 87, Amer. Soc. Agron., Madison.

Mitchell, R.L. and Russell, W.J. (1971). Root development and rooting patterns of soybean (*Glycine max* (L.) Merr.) evaluated under field conditions. *Agron. J.*, 63, 313.

Newman, E.I. (1965). A method of estimating total length of root in a sample. *J. Appl. Ecol.*, 2, 139.

Sanders, J.L. and Brown, D.A. (1978). A new fiber-optic technique for measuring root growth of soybeans under field conditions. *Agron. J.*, 70 (in press).

Unger, P.W. (1977). Effects of deep tillage and profile modification on soil physical properties, root growth and yield. *Agronomy Abstracts 1977*, 94, Amer. Soc. Agron., Madison.

Willatt, S.T. (1977). In situ root studies using neutron radiography. *Agron. J.*, 70, 581.

CALCULATION OF THE ROOT DENSITY REQUIRED FOR GROWTH IN SOILS OF DIFFERENT P-STATUS

M. VAN NOORDWIJK and P. DE WILLIGEN

Institute for Soil Fertility, Haren (Gr.), The Netherlands

Summary

Root density determines the physical availability of mobile soil phosphorus to the plant. Calculations about diffusion of phosphorus to the roots were made, with the assumption that the rate of uptake is determined by the requirements of the crop in the linear growth phase. The poorer the soil, the higher must be the root density to be able to extract sufficient phosphorus from the soil. This relation is quantified analytically for a linear adsorption process and through numerical simulation for the curvilinear adsorption under agricultural conditions. An agricultural evaluation of plant-available soil phosphorus by extraction with water is possible, but the best volume ratio of soil to water to be used depends on the root density.

Introduction

A gap still exists between the experimental basis of fertilizer recommendations, and an explanation of the processes involved through knowledge of soil physical chemistry and plant behaviour. The explanations usually are only qualitative. The first bridges across this gap have only recently been build by quantitative models of solute movement in the soil-root system (Nye and Tinker, 1977). Our approach is a little more practical and applied than that of other models, starting with the phosphorus requirements of a crop at its maximum rate of growth, as determined by factors other than phosphorus shortage.

The main problem is to determine how extensive a root system must be to extract enough phosphorus to sustain this growth rate in soils of different P-status. The agricultural approach is the reverse: how high must the P-status of a soil be to give maximum growth of crops of different rooting densities. Closely related to this is the question how the desired P-status can best be characterized. Most literature on soil extraction methods deals with chemical aspects of extraction with water or any of a wide range of acids (Ansiaux, 1977), to simulate extraction by roots. But since phosphorus must diffuse to the roots and since this process is strongly affected by adsorptive phenomena in the soil, there are important physical problems too in characterizing the availability to roots. The distance a phosphate ion must traverse from its position in the soil to the nearest active root surface is an important parameter, which depends on root density (i.e. length of roots per unit volume of soil). Our approach is to simplify the problem very roughly, to obtain results that only indicate orders of magnitude, but which can be easily generalized. Our model is based on a number of simplifying assumptions.

Assumptions

Physical chemistry of phosphorus in the soil

We assume that only two fractions of phosphorus are important to the plant: inorganic phosphorus in the aqueous phase and that adsorbed to the solid phase, together forming the mobile phosphorus. The two fractions are related by an adsorption isotherm, which is a characteristic of a given soil. We assume this adsorption relation to be valid for the desorption too (i.e. no hysteresis) and that local equilibrium is maintained at all times. In the Netherlands phosphorus recommendations for arable crops are based on the Pw-value, the amount of phosphorus extracted with water, with a volume ratio of soil to water of 1:60 (Van der Paauw, 1971). This desorption can be approximately calculated for this and any other soil to water ratio from the adsorption isotherm (De Willigen and Van Noordwijk, 1978).

Geometry of the soil-root system

We assume a homogeneous distribution of roots of uniform thickness (radius 0.025 cm) in the phosphorus-containing upper layer of the soil (20 cm in our calculation). As we assume the uptake rate of all roots to be equal, an equivalent cylinder of soil is attributed to every root, without transport across its outer boundary. We do not consider mycorrhiza or root hairs and in our first approximation we assume a stationary root system without growth.

Uptake of phosphorus by the root

We assume the uptake rate by the roots to be regulated completely by the requirements of the plant, defined by growth and phosphorus content of the dry matter. This implies that the uptake of individual roots has to be correspondingly higher in a sparse root system than in a dense system. We assume that all roots have the same possibilities for uptake, independent of age (Clarkson *et al.*, 1977). The required uptake cannot be realized when the concentration at the root surface falls below some limiting concentration determined by the relation between concentration and uptake rate, with a maximum uptake rate (which determines the minimum amount of roots for sufficient uptake) of $0.5\ 10^{-12}$ Mol cm^{-2} s^{-1} and in the concentration dependent phase a root absorbing power of 2.10^{-4} cm s^{-1} (Nye, 1977). Thus far we only considered crops in the linear phase of growth with a growth rate of 200 kg ha^{-1} day^{-1} (Sibma, 1968) and a constant P-content of 2.2 mg P g^{-1} dry matter.

We express the possibilities for uptake of a soil-root

system by t_c, the time during which a root system may realize its required uptake rate. After t_c the plant cannot keep its phosphorus content at the most favourable level and production will be affected. Increasing the root density would of course be the best strategy for the plant, but this diminishes the dry matter available for shoot growth.

Analytical Solution in Case of a Linear Adsorption Isotherm

For circumstances where the adsorption isotherm is approximately linear, an analytical solution for the concentration profile around the root could be derived. The variables are combined into the following dimensionless groups:

u = concentration (actual concentration divided by the original)

x = distance to the centre of the root (divided by Ro, the radius of the root)

ρ = radius of soil cylinder (divided by Ro)

τ = time (multiplied by "apparent diffusivity" - the adsorption reduces the diffusivity in water by a factor of 10 to 2000 - and divided by Ro^2)

ϕ = demand-supply relation (apparent diffusivity times amount of mobile phosphorus initially present divided by uptake per unit surface area of soil per day times Ro)

η = depth of P-concentration zone (divided by Ro).

The solution for the concentration in space and time is:

$$(1)\quad u(x,\tau) = 1 + \frac{\rho^2}{2\eta\phi}\Big[\frac{2\tau}{1-\rho^2} + \frac{x^2-\rho^2}{2(1-\rho^2)} + \frac{\rho^2}{1-\rho^2}\ln\left(\frac{\rho}{x}\right) - \frac{\rho^2}{(1-\rho^2)^2}\ln\rho - \frac{\rho^2+1}{4(1-\rho^2)} - \pi\sum_{n=1}^{\infty}\frac{e^{-\alpha_n^2\tau}}{\alpha_n}\,\frac{J1(\rho\alpha_n)\{J0(x\alpha_n)\,Y1(\alpha_n) - Y0(x\alpha_n)\,J1(\alpha_n)\}}{J1^2(\alpha_n) - J1^2(\rho\alpha_n)}\Big]$$

where J0, Y0, J1 and Y1 are Bessel functions of the first

(J) and second (Y) kind and of order 0 and 1. The α_n are the roots of $J1(\alpha)\ Y1(\rho\alpha) - J1(\rho\alpha)\ Y1(\alpha) = 0$. The series part of this solution was given by Carslaw and Jaeger (1959) in section 13.4.

The first part of this formula does not contain products of distance and time, while the infinite sum does. This means that if the second part can be neglected, the depletion of the soil has the same rate at every distance from the root and is constant in time, i.e. is a steady rate process. After some time, determined by the demand-supply relation and root density, the second part may in fact be neglected as time appears in the negative exponential term only. The depletion converges to a steady rate as shown in Figure 1.

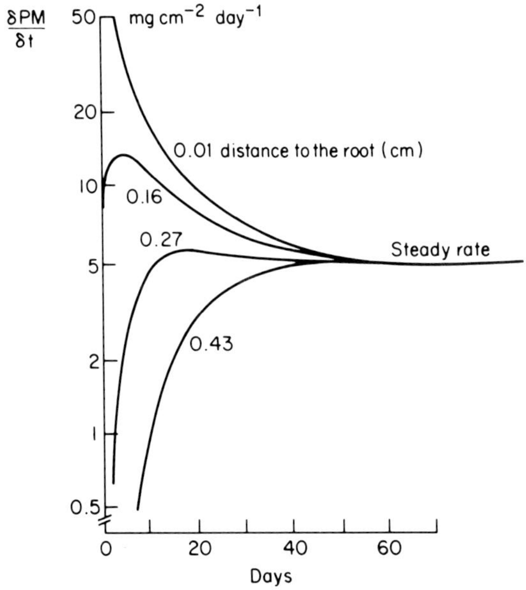

Fig. 1. Depletion of the mobile phosphorus (Pm) in the soil at different distances from the root surface. After some time the rate of depletion becomes independent of the distance ("steady rate").

The time t_c can be found by solving for the concentration at the root surface equal to the limiting concentration. When the steady rate is reached it is a simple matter to solve explicitly:

$$(2)\quad \tau_c = \phi'\eta\,\frac{(\rho^2-1)}{\rho^2} - \left(\frac{\rho^2}{2}\ln\rho - \frac{3\rho^2}{8} - \frac{1}{8\rho^2} + \frac{1}{2}\right)\left(\frac{p^2}{p^2-1}\right)$$

with ϕ' differing from ϕ in that the amount of phosphorus at the limiting concentration is substracted from the amount of mobile phosphorus at t_o.

Solution by Simulation in the Case of a Non-linear Adsorption Isotherm

The adsorption isotherms, determined for five different soils, are distinctly non-linear in the agriculturally important concentration range. They may be described by two Langmuir terms (Holford and Mattingley, 1976a). No analytical solution could be found, so a computer simulation model was used, tested with the linear adsorption solution. For the five soils at three different Pw-values (mg P_2O_5 dm^{-3} soil), viz., 10, 30 and 50 (agricultural evaluation poor, sufficient and high) t_c was calculated (for details: De Willigen and Van Noordwijk, 1978). The results in Figure 2 show that the poorer the soil, the higher the root density has to be for good growth. An agricultural norm of t_c = 100 days, may be reached by a root density of 1 cm $^{-2}$ at Pw 50, a root density 2 to 3 at Pw 30 and a much higher density, depending on the soil, for Pw 10. We conclude that the value of Pw is a good measure of possibilities for uptake in a certain range, but for higher root densities the Pw value does not have the same meaning for different soils.

The diameter of roots and the presence of root hairs likewise are important, as may be seen in Figure 3. With thinner roots a larger root density is required for the same uptake. But the same biomass per unit volume of soil is more efficient when distributed over a larger number of thinner roots. Provided that the distribution of the roots stays homogeneous and the physiological abilities for uptake and transport along the root remain constant, there is no optimum root diameter for uptake, as the thinnest are best.

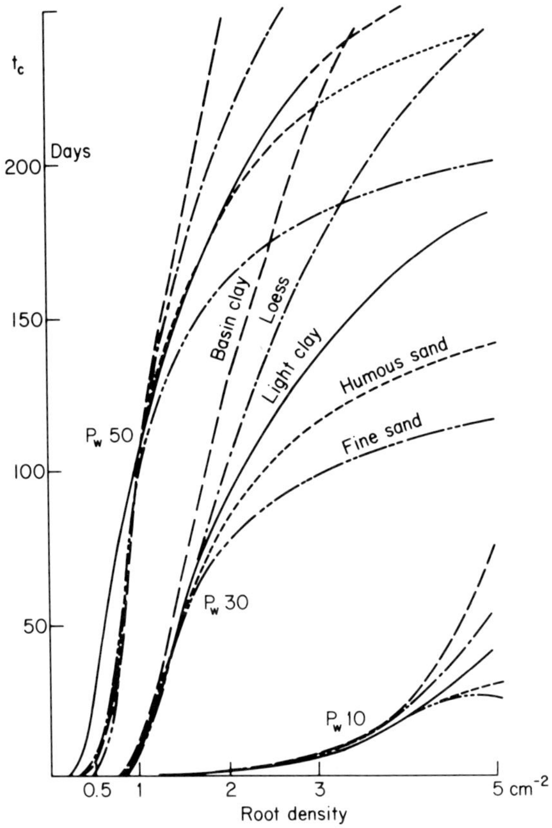

Fig. 2. Simulation results with a non-linear adsorption isotherm. The time during which the uptake is sufficient for plant requirements (t_c) is given as a function of root density (cm of roots cm^{-3} of soil) and P-status.

Some remarks can be made on Pw as a method of soil analysis. The optimum soil/water ratio to be used in this method was derived by means of pot experiments with young potato plants (Van der Paaauw, 1971). Our results suggest that the optimum ratio of soil to water for characterizing availability depends on the root density, as shown in Figure 4. Different values of Pw(v) according to the volume ratio (v) water to soil were calculated from the adsorption isotherms and compared with t_c for our model calculations and actual uptake data of pot experiments of

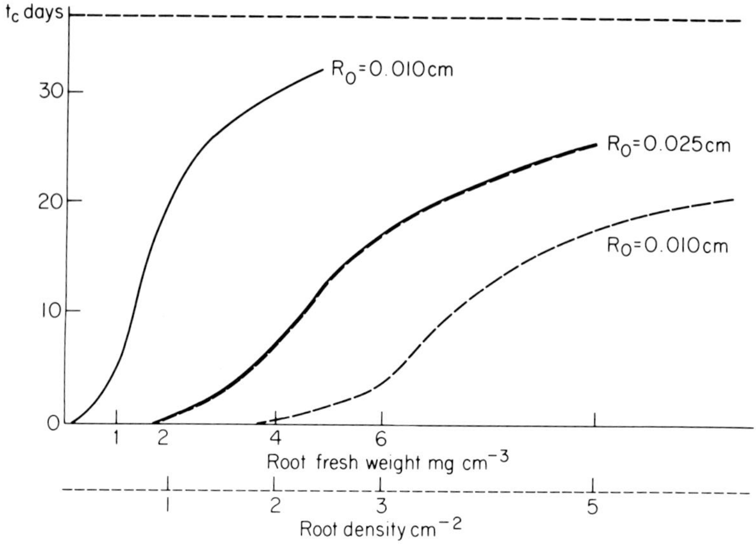

Fig. 3. Effect of root diameter on possibilities for uptake. For the fine sand of Pw 10 of Figure 2, a smaller root diameter is compared both on the basis of root density and on the basis of root biomass.

Holford and Mattingly (1976b) with *Lolium perenne* (with unknown, but presumably high root density estimated to be over 10 cm^{-2}, Mattingly pers. comm.). The higher the root density, the larger the fraction of mobile phosphorus ("chemically available") that can be transported to the root at the required rate ("physically available"). To simulate this in soils with different adsorption capacities more water has to be used in the extraction procedure. The agricultural experience that Pw (60) is a good measure for arable crops but much worse for permanent grassland with a much higher root density may well be interpreted along these lines.

Acknowledgements

Thanks are due to Dr. H.A. Sissingh for measuring the adsorption isotherms and to Dr. N.N. Temme (Mathematisch Centrum, Amsterdam) and Dr. P.A.C. Raats for mathematical help.

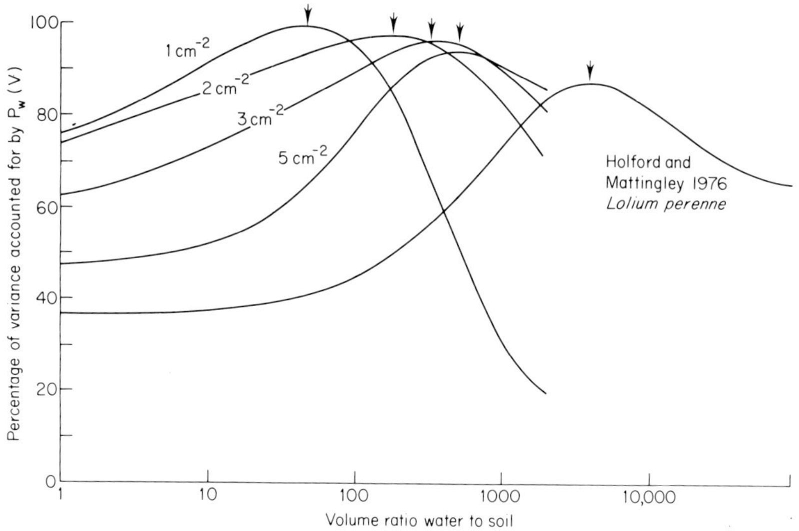

Fig. 4. Percentage of the variance in plant performance on different soils accounted for by Pw(v), i.e. calculated values for a water extract of the soil at different volume ratios of soil to water (v). The upper four lines represent the correlation of t_c (the length of the period of undisturbed uptake, calculated in our model for different soils and P-status) with calculated values of Pw(v), for each root density. In the lower line the actual uptake of phosphorus after 40 days by *Lolium perenne* in pot experiments on 3 calcareous soils, was correlated with Pw(v) values calculated from adsorption isotherms (data of Holford and Mattingly, 1976). The optimal extraction ratio increases with root density.

References

Ansiaux, J.R. (1977). The level of soil phosphorus status. *Phosphorus in Agriculture*, 70, 1.

Carslaw, H.S. and Jaeger, J.C. (1959). "Conduction of Heat in Solids", 2nd Edition. Clarendon, Oxford, 510 pp.

Clarkson, D.T., Sanderson, J. and Scattergood, C.B. (1977). The modification of phosphate transport of barley by phosphorus deficiency. *Agric. Res. C. Letcombe Lab. Ann. Rep. 1976*, 9.

Holford, I.C.R. and Mattingly, G.E.G. (1976a). A model for the behaviour of labile phosphate in soil. *Plant and Soil*, 44, 219.

Holford, I.C.R. and Mattingly, G.E.G. (1976b). Phosphate adsorption and availability [to the] plant of phosphate. *Plant and Soil*, 44, 377.

Nye, P.H. and Tinker, P.B. (1977). "Solute Movement in the Soil-Root

System". Blackwell, Oxford, 342 pp.

Nye, P.H. (1977). The rate limiting step in plant nutrient adsorption from soil. *Soil Sci.*, 123, 292.

Paauw, F. van der (1971). An effective water extraction method for the determination of plant-available soil phosphorus. *Plant and Soil*, 34, 467.

Sibma, L. (1968). Growth of closed green crop surfaces in the Netherlands. *Neth. J. Agric. Sci.*, 16, 211.

Willigen, P. de, and Noordwijk, M. van (1978). Berekeningen over fosfaat-opname afhankelijk van worteldichtheid en fosfaatrijkdom. *Report, Institute for Soil Fertility*, (in prep.).

LOCALIZED STIMULATION OF ROOT GROWTH AND PHOSPHATE UPTAKE IN *ZEA MAYS* L. RESULTING FROM RESTRICTED PHOSPHATE SUPPLY

A. DE JAGER

Botanical Laboratory, University of Utrecht, Utrecht, The Netherlands

Summary

The effect of restricted phosphate supply on root growth and phosphate uptake was studied in plants of *Zea mays*, with four crown roots on the first whirl. Following a P-deficient growth period the time course of reactions was compared, during 6, 9 or 14 days, when either all the roots received a complete nutrient solution or three of them a P-deficient one. The dry matter production was the same in both. This is accounted for by two compensatory reactions, both appearing after about 3 days, (a) stimulated root growth and (b) increased P-uptake by the one root in the complete medium. The latter effect is P-specific, since the uptake of nitrate and potassium was not stimulated simultaneously. The stimulated growth can be explained as the result of competition between +P and -P roots for assimilates, the relative strength of the sinks being determined by their P-concentration.

The specificity of the increased P-uptake casts doubt upon explanations depending on enhanced assimilate supply or hormonal influences.

Introduction

Roots seldom exhibit maximum rates of growth and ion uptake; this is demonstrated by the temporary increase that occurs in the rate of these processes after balanced growth has been disturbed by, for instance, mineral deficiency or root pruning. Restriction of the supply of one nutrient to part of the root system (localization) is another way of influencing the rates of root growth and ion uptake. This method has the advantage that the relative growth rate (RGR) of the plant and the shoot-root ratio may be normal (Drew and Saker, 1975). By such means Goedewaagen (1942), for instance, observed a stimulation of growth and P-uptake by those roots to which P-supply was restricted.

Drew, Saker and Ashley (1973) state that 'the mechanisms by which different parts of the root system appropriate dry matter remain obscure'. Suggested explanations for the stimulation of the P-uptake vary from enhanced assimilate supply to the absorption sites (Louwerse, 1967) or hormonal control of nutrient absorption (Pitman, 1972) to regulation by internal concentration of the nutrient to be absorbed (Glass, 1975). In this article a possible explanation of the regulation of root growth and P-uptake is given.

Materials and Methods

Zea mays cv. *caldera* was sown in sand. After 7 days the seedlings were rinsed, most of the seminal root system was removed, and the plants were grown in a P-deficient nutrient solution. Four days later the seeds and the remaining seminal roots were removed, leaving only crown roots of about 5 cm. Experiments began after another 3 or 4 days when the crown roots were about 20 cm long. The nutrient solution described by Steiner (1968) was used at half strength. In the P-deficient solution KH_2PO_4, 0.5 mM, was replaced by K_2SO_4, 0.25 mM. The pH was adjusted to 4.0 by adding H_2SO_4. From day 7 until the start of the experiments conditions were as described under Figure 1.

At the start of the experiment plants were mounted on containers, 1.5 x 1.5 x 12.0 dm^3, or on sets of four glass tubes, one root of a single plant being placed in each. Control plants were supplied with a single solution to all roots, treated plants had one root in a P-containing solution and three in a P-deficient one. The nutrient solutions were aerated continuously and renewed regularly according to the rate of depletion.

In the experiments using containers, plants were harvested daily, and in the other experiments only at the end. At harvest shoots and roots were weighed fresh and then either dried at 90°C to determine dry weight, or fixed to determine root length (Tennant, 1975) and root diameter. From these latter data root surface area was calculated. The dried plant parts were analysed for P-concentration by colorimetry after dry ashing at 550°C and dissolving the ash in 3N HCl. The uptake of phosphate was either determined directly by measurement of the depletion of the nutrient solution, as was the uptake of potassium and nitrate, or calculated according to Williams (1948) and Drew and Saker (1975), using dry weight and phosphate concentration data. Potassium and nitrate concentrations in the nutrient solution were determined by atomic absorption and potentiometry respectively.

Results

Localization of supply of phosphate to one of the four crown roots does not affect plant growth (Fig. 1a). RGR-values ($g\ g^{-1}d^{-1}$) for the +P localized and the +P control plant are .22 and .23 respectively (all values given here are calculated from day 2 to day 9). The +P root of the +P localized plant ($+R_1$) shows an enhanced growth compared with the +P roots of the +P control plant ($+R_c$) (Fig. 1b). The RGR-values are .20 and .15 respectively. The -P roots of the +P localized plant ($-R_1$) show a reduced RGR of .12. The ratio of the dry weights of the $+R_1$ and the $-R_1$ (Fig. 1c) shows no further rise after six days indicating that the RGR-values have become equal. The rate of P-uptake by

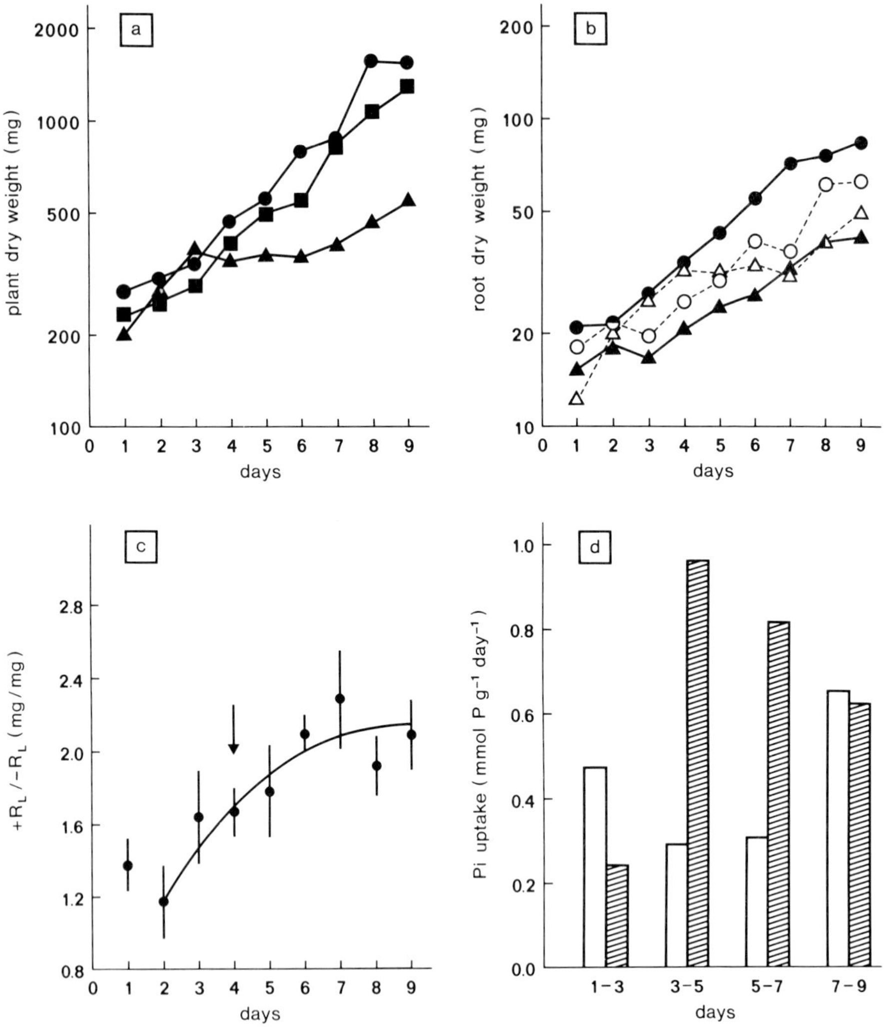

Fig. 1. Effect of localization of P-supply (0.5 mM) on growth and P-uptake of plants with a low P concentration at the start of the experiment. Control plants had all their roots in a complete or a -P nutrient solution. Treated plants had one root in a complete solution and three roots in a -P solution. Conditions: illumination by fluorescent light at an intensity of 60 W/m^2; 16 hours light at 30°C and 30% RH; 8 hours dark at 24°C and 45% RH.

(a) plant dry weight; ●-● , +P control; ▲-▲ , -P control; ■-■ , +P localized plant.

(b) root dry weight per one root; O-O, +P root control ($+R_c$); Δ-Δ, -P root control ($-R_c$); ●-● , +P root localized ($+R_l$); ▲-▲ , -P root localized ($-R_l$).

(c) ratio of the dry weights of the $+R_l$ and the $-R_l$. The arrow indi-

the $+R_l$ is higher than that of the $+R_c$ (Fig. 1d). This difference appears about the third day and disappears at the end of the experiment.

Increase in light intensity during the experiment enhances the growth rate of the +P control and the +P localized plant almost equally (.41 and .37) (Fig. 2a). Figure 2b demonstrates that the $+R_l$ has an enhanced RGR as compared to the $+R_c$ (.43 and .39) considering that the +P localized plant shows a slightly lower RGR. As in the first experiment the $-R_l$ shows a reduced RGR (.24) compared with both the $+R_c$ and the $+R_l$. The difference in the RGR-values between the $+R_l$ and the $-R_l$ remains throughout the experiment, giving a steady rise of $+R_l/-R_l$ (Fig. 2c). The rate of P-uptake by the $+R_l$ (Fig. 2d) was stimulated as at the lower light intensity, but now the stimulation occurs throughout the experiment and the rate of uptake is greater in both types of P-absorbing roots.

The different ways in which the $+R_l$ and the $-R_l$ respond to the treatment might be related to their P-concentration. P-concentration of shoots and roots is shown in Figure 3. Only results for the lower light intensity are given, those at the higher light intensity being essentially the same. It is not surprising that the P-concentration of the $+R_l$ is much higher than that of the $-R_l$ (Fig. 3b). On the other hand the P-concentration of the $+R_l$ and the $+R_c$ are quite similar although different amounts of phosphate pass through them. After some time the P-concentration of the $-R_l$ shows an increase which is preceded by the increase in P-concentration of the shoot (Fig. 3b and 3a respectively)

Fig. 1 (opposite) (cont'd)
cates the time where the weight of the $+R_l$ has become significantly greater than that of the $-R_l$ ($p < 0.05$).
(d) rate of P-uptake by the $+R_c$ (white) and the $+R_l$ (hatched); the bars represent calculated values over the indicated two-day periods.
Values are the means of 6 (P-localized plants), 4 (+P-control plants) and 2 (-P-control plants) determinations.

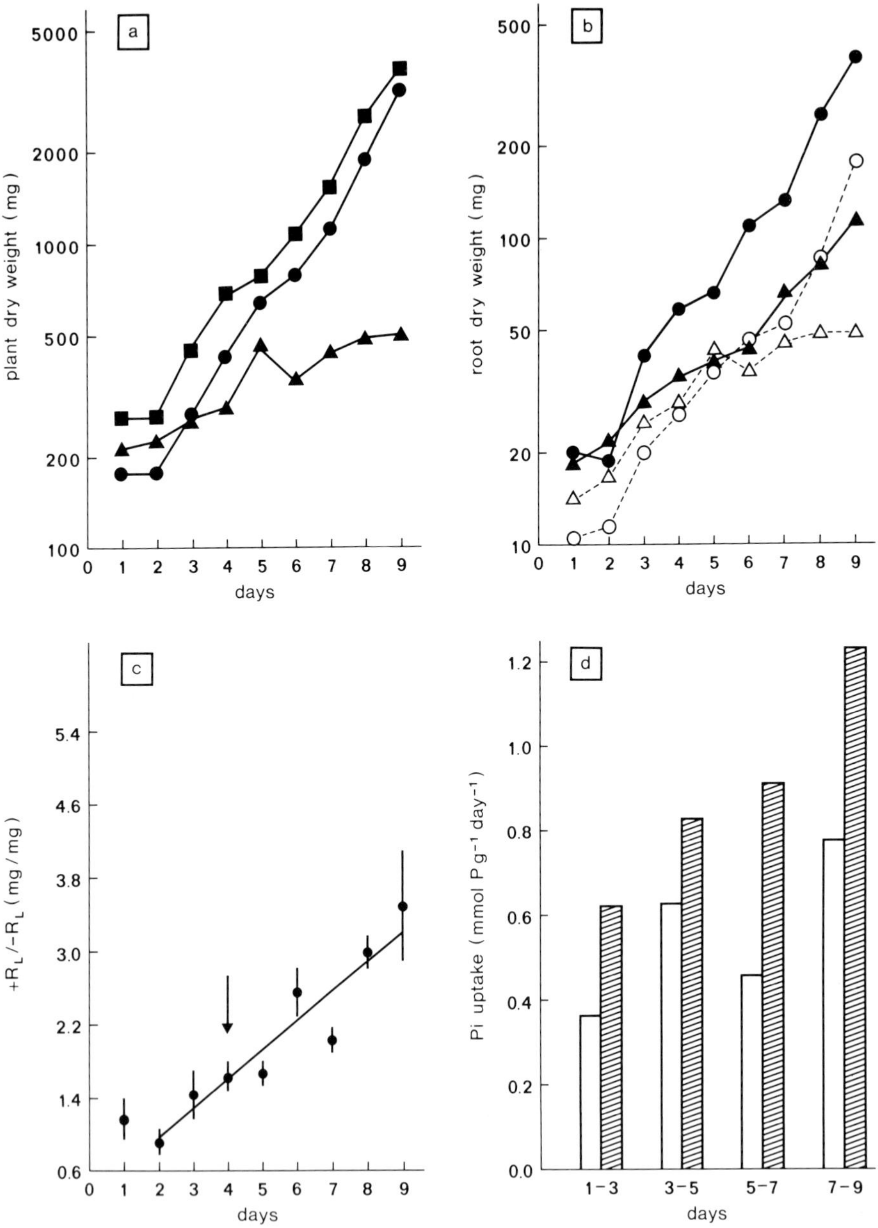

Fig. 2. Localization of P-supply at a high light intensity. As this experiment was similar to experiment 1 (except for the conditions, given below) the reader is referred to the caption of Figure 1. Conditions: illumination by high pressure mercury lamps

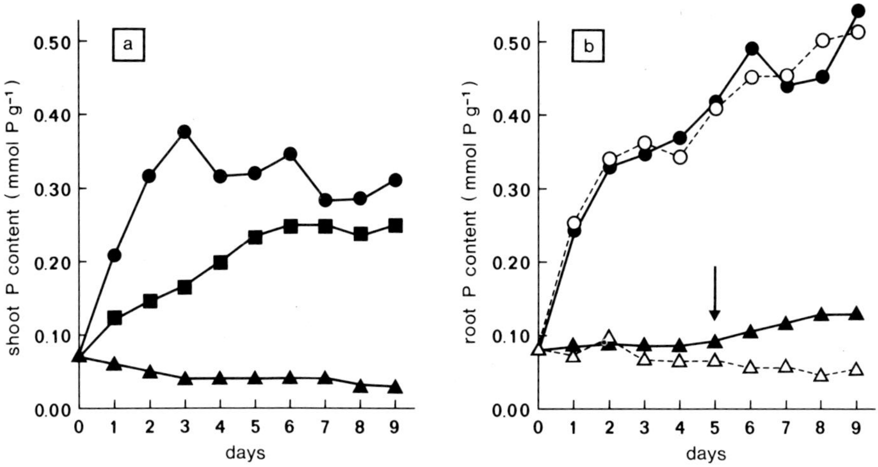

Fig. 3. Effect of localization of P-supply at 60 W/m^2 on P-content of shoots and roots of plants with a low P-content at the start of the experiment. Conditions, and n, were as described under Figure 1.
(a) shoots; ●-● , +P control; ▲-▲ , -P control; ■-■ , +P localized plants.
(b) roots; O-O, $+R_c$; Δ-Δ, $-R_c$; ●-●, $+R_1$; ▲-▲ , $-R_1$.
The arrow indicates the time where the P-content of the R_1 has become significantly higher than that of the $-R_c$ ($P < 0.05$).

of the +P localized plant.

The enhanced P-uptake by the $+R_1$ (Fig. 1c) after 3 days is not the result of different root morphology since, in a further experiment, it was shown by calculations on the basis of fresh weight, or length, or surface area of the root (Table 1). As Table 1 shows K-uptake is the same for both types of root, so the stimulation of P-uptake seems to be P-specific. In another experiment the uptake of phosphate, potassium and nitrate was determined simultaneously during 14 days. Since the ratio of P- and K-uptake (Fig. 4a) and the ratio of P- and NO_3-uptake (Fig. 4b) are much higher for the $+R_1$ than for the $+R_c$ we can conclude, as

---

Fig. 2 (opposite) (cont'd)
at an intensity of 120 W/m^2; 16 hours light at 34°C and 30% RH; 8 hours dark at 26°C and 45% RH.

from Table 1, that the stimulation of P-uptake is P-specific. By relating the uptake of K and NO_3 at the end of the experiment to the root weight we were able to show that the implicit assumption of similar uptake rates for these ions by both types of root was correct.

TABLE 1

Uptake of phosphate by the $+R_l$ and the $+R_c$ as related to fresh weight ($\mu mol\ g^{-1}d^{-1}$), length ($\mu mol\ cm^{-1}d^{-1}$) and surface ($\mu mol\ cm^{-2}d^{-1}$) of the root. Measurements, as depletion of the nutrient solution, were made on the third and on the sixth day after the start of the supply of phosphate to plants with a low P-content. On day 6 potassium uptake is given as well. Conditions: illumination by fluorescent light giving 90 W/m²; 16 hours light at 32°C and 30% RH; 8 hours dark at 26°C and 45% RH

Day of Harvest	Type of Root	n	Uptake per Gram fresh		Uptake per Unit length		Uptake per Unit Surface	
			P	K	P	K	P	K
3	$+R_c$	3	26.8		0.034		0.32	
	$+R_1$	4	24.8		0.036		0.35	
6	$+R_c$	3	23.3	102.2	0.031	0.133	0.34	1.46
	$+R_1$	3	31.4[a]	103.4	0.041[b]	0.135	0.44[c]	1.44

Probability by chance (t-test) a: $p = 0.026$, b: $p = 0.043$, c: $p = 0.069$.

Discussion

Our experiments show that restriction of the supply of phosphate to only a part of the root system stimulates both the RGR and the P-uptake of the P-absorbing root. These results are in accordance with those of Gile and Carrero (1917), Goedewaagen (1942) and Stryker, Gilliam and Jackson (1974) and other authors.

As a result of the stimulation of both processes the plants with restricted P-supply showed, in our experiments, a P-content of 75-80% of that of +P control plants and grew at the same rate. The enhanced growth rate of the root to which P-supply had been localized coincided with a reduced growth rate of the remaining roots to the effect

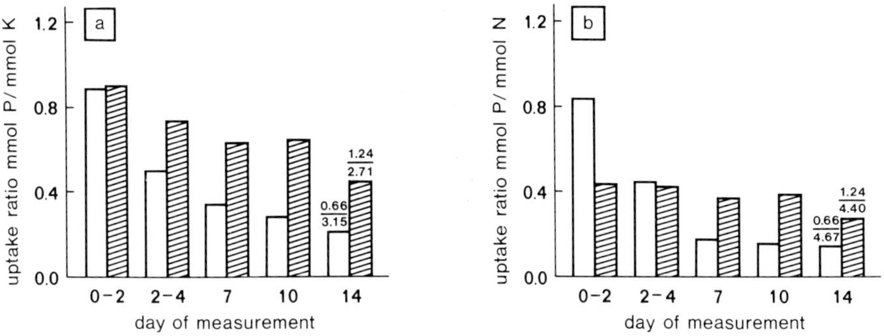

Fig. 4. Ratio of the uptake of phosphate to the uptake of potassium (a) or nitrate (b) by the $+R_1$ (white) and the $+R_c$ (hatched) at different times after the start of P-supply to one or all of the roots of plants with a low P-content. The uptake was measured as depletion of the nutrient solution. The figures given on day 14 represent the actual rates of uptake (mmol g dry^{-1}d^{-1}) of phosphate (numerator) and potassium or nitrate (denominator). Conditions as described in Table 1. The concentration of the nutrient solution was halved except for P. The concentration of the three nutrients was then as follows: P 0.5 mmol, K 1.8 mmol, NO_3 3.0 mmol.

that the shoot-root ratio of the +P localized plant was the same as in the +P control plant. Thus P-localization causes a change in the distribution of dry matter within the root system possibly as a result of competition for carbohydrates. The dry matter distribution between the roots might be regulated by their P-concentration. The P-supplied roots, which had a high P-concentration, grew faster at first than the P-devoid roots. Later on, at least at the lower light intensity, the growth rate of the P-devoid roots showed a relative recovery which was correlated with an increase in their P-concentration. Since in P-deficient plants the concentration of free sugars is high no competition for these substances will exist. This would explain why the roots of the -P control plants showed, at first a higher growth rate than the -P roots of the +P localized plants even though the P-concentration of the former was equal or lower than that of the latter (Figs. 1b and 3b). Drew *et al.* (1973) also mention competition for carbohydrates as a possible mechanism whereby roots appropriate dry matter.

As regards the P-devoid roots of the +P localized plants a number of facts point to considerable activity, as a consequence apparently of P-redistribution. These are, (1) the appearance of second order laterals, that did not occur on roots of -P control plants, (2) quite normal rates of uptake of K and NO_3 (results unpublished), (3) the restored growth rate (Fig. 1b) and (4) the increased P-concentration (Fig. 3b). Our conclusion is thus at variance with that of Stryker *et al.* (1974) that roots lack the ability to use assimilates whenever there is no external P-supply. Redistribution of P to the P-devoid roots may occur either by mass flow, dependent on the import of assimilates, as suggested by Marshal and Wardlaw (1973) or independent of the flow of assimilates and preferentially to the P-devoid roots, as suggested by results of Moore (1949). In both cases the P-concentration of the shoot will determine the amount of descending P. Indeed the increase in the P-concentration of the P-devoid roots is preceded by an increase in the P-concentration of the shoot of the +P localized plant (Figs. 3b and 3a).

A number of possible explanations for the enhanced P-uptake by the P-supplied root of the +P localized plant, such as the higher RGR or an altered morphology cannot apply since the effect is specific for phosphate (Table 1 and Fig. 4), the uptake rate of K and NO_3 remaining unaffected. For the same reason explanations based on enhanced metabolism put forward by Louwerse (1967) and Pitman, Mowat and Nair (1971), and those based on hormonal action as suggested, for example, by Pitman (1972) do not apply either. Recently, substantial evidence has been presented that the uptake of certain inorganic ions such as nitrate (Cram, 1973), potassium (Glass, 1975) and phosphate (Blasco, Ducet and Azoulay, 1976) is primarily regulated by the internal concentration of this element. This could also be an acceptable explanation for the observed specificity of the stimulated P-uptake.

It is the concentration of inorganic P which would regulate the rate of P-uptake. Thus the fact that there was no measurable difference in the total-P-concentration between both types of P-absorbing root is not necessarily contradictory to the assumption.

The question is now whether the lower P-concentration of the shoot can lead to a lower inorganic phosphate concentration of the absorbing cortical cells. Pitman (1972) argues that the redistribution of potassium from the shoot to the roots is quantitatively too unimportant a process to account for the wide range of uptake rates. Indeed, changes in the concentration of the whole cell would be negligible. Changes in the concentration of the cytoplasm might, however, be considerable (Prichodko, 1974) since the cytoplasm forms a comparatively small pool. Bieleski (1973) estimates the amount of cytoplasmic inorganic phosphate to be 1-5% of the total inorganic phosphate of the cell. From our data it can be calculated that redistribution of P to the P-devoid roots of the +P localized plants amounts to 5% of the concomitant P-uptake by the P-supplied root. Per day this is about twice the total amount of cytoplasmic inorganic phosphate of the whole +P root.

Summarizing, we propose that the regulation of the P-uptake by roots is exerted by the inorganic phosphate concentration of the root cells, especially of the cytoplasm, and that this in turn is controlled by the process of P-redistribution, reflecting the P-status of the shoot. To test this hypothesis we need more information about the correlation between redistribution on the one hand and the cytoplasmic inorganic phosphate concentration and the rate of P-uptake on the other hand.

Acknowledgements

I am indebted to Professor Dr. R. Brouwer for his interest and encouragement throughout. I like to thank Dr. O.

M. van Andel and Dr. H. Konings for critically reading the manuscript and Mrs. S.M. McNab for revising the English text.

I also wish to thank Miss C.C. van Ditmarsch and Mrs. I. Bosboom-Hellendoorn for their technical assistance.

References

Bieleski, R.L. (1973). Phosphate pools, phosphate transport and phosphate availability. *Ann. Rev. Plant Physiol.*, 24, 225.

Blasco, F., Ducet, G. and Azoulay, E. (1976). Mise en evidence de deux systemes de transport du phosphate chez *Candida tropicales*. *Biochimie*, 58, 351.

Brouwer, R., Jenneskens, P.J. and Borggreve, G.J. (1961). Growth responses of shoots and roots to interruptions of the nitrogen supply. *Jaarboek I.B.S. 1961*, 29.

Brouwer, R. (1963). Some aspects of the equilibrium between overground and underground plant parts. *Jaarboek I.B.S. 1963*, 31.

Cram, W.J. (1973). Internal factors regulating nitrate and chloride influx in plant cells. *J. exp. Bot.*, 24, 328.

Drew, M.C., Saker, L.R. and Ashley, T.W. (1973). Nutrient supply and the growth of the seminal root system in barley. I. The effect of nitrate concentration on the growth of axes and laterals. *J. exp. Bot.*, 24, 1189.

Drew, M.C. and Saker, L.R. (1975). Nutrient supply and the growth of the seminal root system in barley. II. Localized compensatory increases in lateral root growth and rates of nitrate uptake when nitrate supply is restricted to only part of the root system. *J. exp. Bot.*, 26, 79.

Gile, P.L. and Carrero, J.O. (1917). Absorption of nutrients as effected by the number of roots supplied with the nutrient. *J. Agr. Res.*, 9, 73.

Glass, A. (1975). The regulation of potassium absorption in barley roots. *Plant Physiol.*, 56, 377.

Goedewaagen, M.A.J. (1942). "Het Wortelstelsel der Landbouwgewassen" (Ed. Ministerie van Landbouw en Visserij). Algemeene Landsdrukkerij, Den Haag.

Jungk, A. and Barber, S.A. (1975). Plant age and the phosphorus uptake characteristics of trimmed and untrimmed corn root systems. *Plant and Soil*, 42, 227.

Louwerse, W. (1967). The influence of the plant nutritional status on bleeding and salt uptake. *Acta Bot. Neerl.*, 16, 42.

Marshall, C. and Wardlaw, I.F. (1973). A comparative study of the distribution and speed of movement of ^{14}C assimilates and foliar-applied ^{32}P-labelled phosphate in wheat. *Austr. J. Biol. Sci.*, 26, 1.

Moore, R.F. (1949). Downward translocation of phosphorus in separated maize roots. *Am. J. Bot.*, 36, 166.

Pitman, M.G., Mowat, J. and Nair, H. (1971). Interactions of processes for accumulation of salt and sugar in barley plants. *Austr. J. Biol. Sci.*, 24, 619.

Pitman, M.G. (1972). Uptake and transport of ions in barley seedlings. III. Correlations between transport to the shoot and relative growth rate. *Austr. J. Biol. Sci.*, 25, 905.

Prichodko, N.V. (1974). The uptake of nutrients by plant roots and possible mechanisms of regulation. *Physiology and Biochemistry of cultivated Plants*, 6, 605 (in Russian).

Steiner, A.A. (1968). "Soilless Culture". Proc. 6th Coll. Int. Potash Inst. Florence, 324.

Stryker, R.B., Gilliam, J.W. and Jackson, W.A. (1974). Nonuniform phosphorus distribution in the root zone of corn: growth and Phosphorus uptake. *Proc. Am. Soc. Soil Sci.*, 38, 334.

Tennant, D. (1975). A test of a modified line intersect method of estimating root length. *J. Ecol.*, 63, 995.

Williams, R.F. (1948). The effect of phosphorus supply on the rates of intake of phosphorus and nitrogen, and upon certain aspects of phosphorus metabolism in gramineous plants. *Austr. J. Sci. Res. Ser. B.1.*, 333.

FINE REGULATION OF IRON UPTAKE BY THE Fe-EFFICIENT PLANT *HELIANTHUS ANNUUS*

V. RÖMHELD and H. MARSCHNER

Institut für Pflanzenernährung, Universität Hohenheim, Stuttgart, West Germany

Summary

Differences in susceptibility of plant species to iron chlorosis are well known. Under iron stress sunflower (*Helianthus annuus* L.), as an Fe-efficient species, rapidly increases the reducing capacity on the root surface and lowers the pH of the nutrient solution (substrate). Simultaneously the rate of uptake and translocation of iron increases up to 100-fold.

Morphological changes of the roots also occur under iron deficiency. The growth of the main root is stunted and many lateral roots appear just behind the main root tip. The root tips are swollen as result of the enlarged cortex cells and additional cell division of rhizodermis cells. Each cell of the peripheral rhizodermis layer in this swollen tip region has developed a long root hair.

Results are presented which show that the root tip region is exclusively responsible for the iron stress reactions. Even before symptoms of chlorosis are visible, root tips respond to insufficient iron supply, i.e. to latent iron deficiency. The changes in morphology and physiology of the root tip region of sunflower plants occur periodically and are an expression of a very effective mechanism for fine regulation of iron uptake. It can be assumed that this fine regulation mechanism plays an important role in iron mobilization in the rhizosphere especially in soils with high pH values.

Introduction

Inorganic iron-III-compounds are sparingly soluble in well aerated soils with high pH values (calcareous soils/ arid climate). The concentration of inorganic iron in the soil solution predicted by theory is by far too low to cover the demand of plants to allow a normal growth rate (Lindsay, 1974). Therefore, production of chelators for iron from soil organic matter and microbial activity seems to be essential to maintain iron nutrition of plants in such soils. However, as already demonstrated by Weiss (1943), in crop plants genetical differences in susceptibility to iron chlorosis ("lime chlorosis") and in iron uptake also exist. So-called Fe-efficient plant species are adapted to these soils with low iron solubility by their capacity to mobilize iron in the rhizosphere. Reactions involved in this special adaption to increase iron availability under iron stress are increases in reducing capacity at the surface of root cells in general (Chaney *et al.*, 1972) and decreases in pH of the substrate (Oertli and Jacobson, 1960), caused by decreased anion uptake (Venkat Raju *et al.*, 1972).

In this paper the time courses of morphological and physiological changes of the root system of sunflower plants under iron stress are studied. The experiments were performed mainly in water culture with iron chelates. For comparison, in some experiments a Fe-inefficient species (*Zea mays* L.) was also used.

Abbreviations: BPDS: 4.7-di(4-phenylsulphonate)-1.10 phenanthroline, bathophenanthroline disulphonic acid;
EDDHA: ethylenediaminedi (o-hydroxyphenylacetic acid);
EDTA: ethylenediamine-tetraacetic acid.

Material and Methods

Sunflower (*Helianthus annuus* L., cv. Sobrid) and maize (*Zea mays* L., cv. Velox) were precultured in a complete nutrient solution with FeEDTA (10^{-4}M). To induce iron deficiency, sunflower plants were grown for a further 6 days and maize plants for a further 10 days without iron.

The composition of the nutrient solution during preculture and treatment periods was as follows: (M): $Ca(NO_3)_2$ 2.0×10^{-3}; K_2SO_4 0.75×10^{-3}; KH_2PO_4 0.50×10^{-3}; $MgSO_4$ 0.65×10^{-3}; H_3BO_3 1.0×10^{-5}; $MnSO_4$ 1.0×10^{-6}; $(NH_4)_6Mo_7O_{24}$ 5.0×10^{-8}; $CuSO_4$ 5.0×10^{-7}; $ZnSO_4$ 5.0×10^{-7}.

Reducing capacity

Using a modification of the method of Chaney *et al.* (1972), the reducing capacity of isolated roots (0.5 g samples) was measured with BPDS reagent (3×10^{-4}M) and Fe-III-EDTA (1×10^{-4}M) as the Fe-III-source in a nutrient solution without micronutrients (pH 6.2 - 6.5) for 20 h.

Uptake and translocation of ^{59}Fe

^{59}Fe-III-hydroxide: ^{59}Fe-III-hydroxide was prepared by using $Fe(NO_3)_3$ labelled with ^{59}Fe (specific activity 1 μC/μmole Fe) and precipitated with ammonia before centrifugion. After three-fold washing of the precipitate with distilled water ^{59}Fe-III-hydroxide was thoroughly suspended in water by mixing with a "Ultraturrax" (Azarabadi and Marschner, 1978).

^{59}FeEDDHA: labelled ^{59}Fe EDDHA was prepared two days before the experiment by mixing $FeCl_3$ with KEDDHA (specific activity 1 μC/μmole Fe) and adding KOH to pH 6.2. Prior to the uptake experiments, ^{59}FeEDDHA solution was filtered (membrane filter 0.01 μ) to ensure the removal of colloids of ^{59}Fe-III-hydroxide.

Results

Requirement for FeEDDHA for optimal growth rate

Sunflower and maize plants responded quite differently to increasing concentrations of FeEDDHA (Table 1). For optimal growth the requirement of the Fe-inefficient maize for FeEDDHA was nearly 50 times higher than that of the Fe-efficient sunflower. For maximal chlorophyll formation, 5×10^{-4}M FeEDDHA was necessary for sunflower, whereas in maize the maximum was at 5×10^{-3} or higher. High chelate concentrations were much more toxic for sunflower than for maize. In this experiment [the pH of] the nutrient solution was kept between pH 6.5 - 7.5, i.e. the higher Fe-efficiency of sunflower did not necessarily require lowering the pH of the solution.

Capacity for iron reduction, pH decrease and iron uptake

The reducing capacity of the roots was similar in both plant species at normal iron nutrition (Table 2) but increased nearly 3-fold under iron stress in sunflower. The corresponding increase in maize roots was very small.

TABLE 1

Effect of increasing FeEDDHA concentration on growth and chlorophyll content of sunflower and maize plants

FeEDDHA-conc. (M)	Dry weight mg/shoot		Chlorophyll mg/g D.W.	
	sunflower	maize	sunflower	maize
- Fe	155	153	0.9	2.2
1×10^{-7}	225	151	3.3	2.3
1×10^{-6}	422	154	6.1	2.2
1×10^{-5}	<u>882</u>	189	7.9	2.5
1×10^{-4}	860	239	8.3	5.1
5×10^{-4}	676	<u>523</u>	9.0	8.0
1×10^{-3}	321	501	7.7	9.4
5×10^{-3}	78	364	6.4	10.9

Growth period: 12 days; pH 6.5 - 7.5

TABLE 2

Effect of iron preculture on the reducing capacity of roots of sunflower and maize plants

Preculture		Reducing capacity (μmole Fe^{II}/g F.W.)
Sunflower	+ Fe	1.9
	- Fe	5.8
Maize	+ Fe	1.8
	- Fe	2.1

TABLE 3

Effect of iron preculture and pH on translocation of iron into the shoots of sunflower and maize plants (^{59}Fe was supplied as ^{59}Fe-III-hydroxide)

pH	preculture	^{59}Fe content (nmole/g D.W.) sunflower	maize
6.5	+ Fe	3.2	5.2
6.5	- Fe	5.7	3.8
4.0	+ Fe	43.8	48.6
4.0	- Fe	212.7	56.7

^{59}Fe-III-hydroxide 10^{-4}M; uptake period 24 h

This increase in the reducing capacity of sunflower roots occurred at latent iron deficiency prior to the decrease in the pH of the culture solution. This was brought about by further increase in iron deficiency (Fig. 1). In contrast, maize was not able to lower the pH even when the symptoms of iron chlorosis became severe.

The increase in reducing capacity of sunflower roots under iron stress was also reflected in the somewhat better utilization and translocation of iron into the shoot even at pH 6.5 and with Fe-III-hydroxide as the source of iron (Table 3).

Artificial lowering of the pH of the culture solution to pH 4.0 dramatically increased the translocation of iron into the shoot in both plant species, independent of the

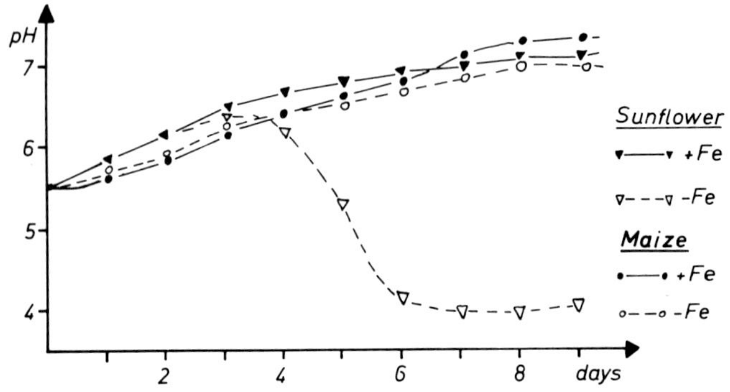

Fig. 1. Effect of iron supply to sunflower and maize plants on the pH changes in the nutrient solution.

iron nutritional status (Table 3). In these circumstances there was a further 5-fold increase in iron translocation in response to iron stress in sunflower but hardly any further increase in maize.

Morphological changes of the root system under iron stress

In maize, apart from lower growth rates under iron stress there were no other obvious changes in root morphology. In contrast to maize, in sunflower under iron stress the growth of the main root was stunted and many short lateral roots emerged just behind the main root tip. Furthermore, the tips of all roots were swollen and yellow due to accumulation of riboflavin. Light microscopic investigations of transverse sections from these swollen root tips showed no changes in the stele but enlargement of the cortical cells and additional cell division in the rhizodermis, resulting in 2 - 4 times more cell layers in this area [and therefore a corresponding increase in root diameter]. Moreover, each cell of the peripheral rhizodermis layer in this tip region developed a long root hair. When the iron supply in water culture was sufficient for normal growth, root hairs were almost absent in sunflower roots. The morphological changes of the sunflower root system under iron stress are summarized in Table 4 and Figure 2.

TABLE 4

Effect of iron nutritional status on the morphology of sunflower roots

Parameter	Iron preculture + Fe	Iron preculture - Fe	Relative increase (+ Fe = 100%)
Length of adventitious roots (cm)	13	6	46
diameter of root (mm) (2 mm behind root tip)	0.54	0.81	150
diameter of stele (mm)	0.20	0.22	110
thickness of cortex (mm)	0.17	0.30	176
cell layers			
cortex (without rhizodermis)	7	7	0
rhizodermis	1	3	300
number of root hairs per circumference	12	144	1200
length of root hairs (mm)	n.d.	0.1-0.2	-

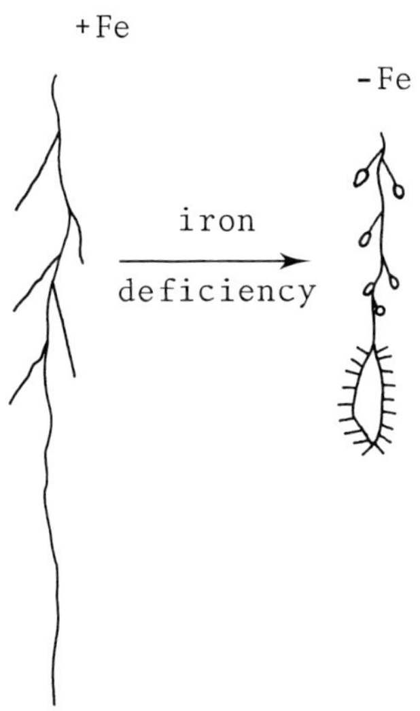

a) Increase in size of cortex and in rate of cell division in the rhizodermis
b) Promoted growth of root hairs
c) Accumulation of reducing substances and chelators on the root surface
d) Decrease of the pH of the substrate

a-d ⟶ increased mobilization of sparingly soluble iron-III-compounds in the rhizosphere

Fig. 2. Changes in morphology and physiology of roots of sunflower plants under iron deficiency.

Correlation between morphology and physiology of sunflower roots under iron stress

Using a modification of the technique of Clarkson and Sanderson (1973) the capacity of various root zones of chlorotic sunflower plants to lower the pH and to reduce iron was determined (Table 5). It was obvious that these physiological expressions of iron stress were restricted to the root tip region where the morphological changes also occur.

In transverse sections of the swollen tip zones the outer 2 - 3 cell layers of the cortex (rhizodermis) contained dark coloured substances. There are indications that these substances were phenols, i.e. compounds capable of both reducing and complexing iron. These changes in the physiology of the root tip of sunflower under iron stress were largely responsible for the steep increase in iron uptake from FeEDDHA by sunflower roots after preculture without iron (Table 6). The 8-fold increase in iron uptake of the iron deficient root system was primarily the result of the steep increase in iron uptake by the root tip zones (70-fold). Compared to this the increase in the basal zones was relatively small. It also seems probable that root tip zones were primarily responsible for the steep increase in iron translocation into the shoot of sunflower plants under iron stress, already shown in Table 3.

TABLE 5

Capacity of various root zones of chlorotic sunflower plants to lower the pH and to reduce iron on the root surface

	Root zone (cm)			Entire root system
	0-1.5 (root tip)	1.5-3.5	3.5-5.5	
pH value* (after 10 h)	4.1	5.8	5.7	4.5
reducing capacity (μmole Fe^{II}/g F.W.)	6.7	0.9	0.6	5.6

* pH value at the beginning of the experiment 6.0

TABLE 6

Effect of iron preculture on iron accumulation in various root zones of sunflower plants

Root zone (cm)	^{59}Fe content (nmole/g D.W.) preculture + Fe	- Fe	Relative increase (+ Fe = 100%)
entire root system	570	5026	882
5.5 - 3.5	583	1861	319
3.5 - 1.5	595	2462	414
1.5 - 0 (root tips)	319	21774	6825

^{59}FeEDDHA 2 x 10^{-5}M; uptake period 10 h; pH 6.2 - 6.4

TABLE 7

Effect of FeEDDHA concentration during the pretreatment (0-6 days) on ^{59}Fe accumulation and translocation rate in roots and shoots of sunflower plants

FeEDDHA conc. (M) (pretreatment)		^{59}Fe content (nmole/g D.W.) day of pretreatment: 1	2	3	4	5	6
2 x 10^{-6}	shoots	6	636	1329	6	45	786
	roots	266	2502	1491	261	1294	1759
5 x 10^{-4}	shoots	5	3	3	3	12	3
	roots	353	265	282	280	293	267

^{59}FeEDDHA 2 x 10^{-5}M; uptake period 10 h; pH 6.2 - 6.5

Iron stress responses at latent iron deficiency

Often one more swollen yellowish zone occurs along the axes of roots of sunflower plants grown in nutrient solutions with FeEDDHA-concentrations between 10^{-6} - 10^{-5}M, i.e. near the optimal concentration for growth (Table 1). This could be an indication that, during the growth, temporarily latent iron deficiency existed during which the root tips responded with the above mentioned morphological and physiological changes. This assumption was checked in the following experiment.

Sunflower plants were pretreated with 2 levels of non-labelled FeEDDHA (2×10^{-6}M and 5×10^{-4}M) and after various periods of pretreatment (1 - 6 days) the plants were supplied with ^{59}FeEDDHA (2×10^{-5}M) for 10 h. Table 7 demonstrates that dramatic fluctuations occurred in rate of iron accumulation and translocation in the plants pretreated with the low concentrations of FeEDDHA. During the entire experiment no visual changes in chlorophyll content occurred. In contrast to the conditions under latent iron deficiency, in the plants with the high iron supply the rate of uptake and translocation was much lower, but constant over the time.

In a preliminary experiment with sunflower plants grown in a calcareous soil a similar trend (although less dramatic) in the time course of ^{59}Fe transport into the shoot has been observed as in the plants grown in nutrient solution with the low iron pretreatment.

Discussion

The typical iron stress reactions of an Fe-efficient plant species, namely increase in reducing capacity of the roots and lowering of the substrate pH (Brown and Ambler, 1970), are confirmed with sunflower plants. These physiological changes are probably causally related to morphological changes, both restricted to the root tips. The root tips are responsible for the pH decrease under iron stress, brought about by inhibited anion uptake (Venkat Raju *et al.* 1972). Accumulation of riboflavin under iron stress as result of disturbance in purine metabolism (Schlee *et al.*, 1968) also occurs primarily in the root tip (Venkat Raju *et al.*, 1972).

Morphological changes of the root tips similar to these under iron stress in sunflower can be induced by auxin treatment of the roots (Ekdahl, 1957; Burström and Svensson, 1972). One might speculate, therefore, that, under iron stress, either the auxin transport from the shoot into the root is intensified or that the decomposition rate of auxins is impeded as a result of accumulation of phenolic substances in the root tips under iron stress (Römheld, 1978). As demonstrated by Pingel (1976) phenolic substances decrease the decomposition rate of IAA by inhibiting the IAA-oxidase activity. Higher IAA concentrations in the iron stressed root tips might induce pH decrease in the substrate as a result of an IAA-activated proton efflux pump (Hager *et al.*, 1971).

In agreement with results of Ambler *et al.* (1971) with soybeans the increased reducing capacity was restricted to certain root zones; in sunflower it is restricted to the rhizodermal layers of the root tips where additional cell divisions occur under iron stress. Presumably, accumulation of phenolic substances is primarily responsible for this increase in reducing capacity. However, increased dehydrogenase activity might be also involved, since it has been shown that the activity of the enzyme is closely correlated with the rate of cell division (Karpov and Potapov, 1975).

In addition to changes in root morphology, increased reducing capacity of certain root zones is the first step in regulating the plant response to iron stress. As a result, soluble iron-III-compounds (Fe-III-chelates e.g.) are reduced at higher rates at the root surface, leading to increased rates of uptake and translocation of iron. The inability of maize to increase the reducing capacity under iron stress seems to be the main factor responsible for the much higher demand for Fe-III-chelates for optimal growth (Table 1).

If the first step of this regulation mechanism is insufficient to overcome the iron stress of sunflower plants, a second step occurs whereby proton efflux increases. The corresponding pH decrease in the substrate not only increases the solubility of inorganic iron-III-compounds (Venkat Raju and Marschner, 1972), but might also increase the membrane permeability and thus the release of reducing substances from the roots (Chaney *et al.*, 1972; Marschner *et al.*, 1974).

These morphological and physiological changes in sunflower roots occur periodically in substrates with low iron availability. The rapid and effective response of this fine regulation mechanism is reflected in the enormous fluctuations in uptake and translocation of iron at latent iron deficiency (Table 7). After a period of increased iron uptake, the iron stress is overcome within 1 day, with simultaneous loss of the ability for increased iron uptake (Marschner *et al.*, 1974). Rapid loss in ability for iron translocation to shoots during the period of recovery from iron stress has been reported by Clarkson and Sanderson (1973) using barley plants.

Based on preliminary results with sunflower plants grown in calcareous soil it can be assumed that this fine regulation mechanism of iron uptake plays an important role also in soils with low iron availability.

Acknowledgements

These studies were supported by the Deutsche Forschungsgemeinschaft.

References

Ambler, J.E., Brown, J.C. and Gauch, H.B. (1971). Sites of iron reduction in soybean plants. *Agron. J.*, 63, 95.

Azarabadi, S. and Marschner, H. (1978). Role of the rhizosphere in utilization of inorganic iron-III-compounds by corn plants. *Z. f. Pflanzenernähr. u. Bodenkd.*, **141**, (in press).

Brown, J.C. and Ambler, J.E. (1970). Further characterisation of iron uptake in two genotypes of corn. *Soil Sci. Soc. Amer. Proc.*, **34**, 249.

Burström, H.G. and Svensson, S.-B. (1972). Hormonal regulation of root growth and development. *In* "Hormonal Regulation and Development". Proc. Adv. Study Inst. Izmir, Verlag Chemie, Weinheim.

Chaney, R.L., Brown, J.C. and Tiffin, L.O. (1972). Obligatory reduction of ferric chelates in iron uptake by soybeans. *Plant Physiol.*, **50**, 208.

Clarkson, D.T. and Sanderson, J. (1973). Iron nutrition of barley plants. *Agric. Research Council Letcombe Laboratory, Annual Report 1972*, 3.

Ekdahl, I. (1957). On the growth mechanism of root hairs. *Physiol. Plant*, 10, 798.
Hager, A., Menzel, H. and Krauss, A. (1971). Versuche und Hypothesen zur Primärwirkung des Auxins beim Streckungswachstum. *Planta*, 100, 47.
Karpov, E.A. and Potapov, N.G. (1975). Reducing activity of the root surface in corn plants in connection with its differentiation and absorption capacity. *Sov. Plant Physiol.*, 22, 298.
Lindsay, W.L. (1974). Role of chelation in micronutrient availability. *In* "The Plant Root and its Environment" (Ed. E.W. Carson), p.507. University Press of Virginia, Charlottesville, 1974.
Marschner, H., Kalisch, A. and Römheld, V. (1974). "Mechanism of Iron Uptake in different Plant Species". Proc. 7th Intern. Colloqu. Plant Analysis and Fertilizer Problems, Hannover.
Oertli, J.J. and Jacobson, L. (1960). Some quantitative considerations in iron nutrition of higher plants. *Plant Physiol.*, 35, 683.
Pingel, U. (1976). Der Einfluss phenolischer Aktivatoren und Inhibitoren der IES-Oxidase-Aktivität auf die Adventivbewurzelung bei Tradescantia albiflora. *Z. Pflanzenphysiol.*, 79, 109.
Römheld, V. (1978). "Mechanismus der Aufnahme und Verlagerung von Eisenchelaten bei höheren Pflanzen." Dissertation Techn. Univ. Berlin.
Schlee, D., Reinbothe, H. and Fritsche, W. (1968). Der Einfluss von Eisen auf den Purinstoffwechsel und die Riboflavinbildung von Candida guillermondii. *Z. Allg. Mikrobiologie*, 8, 127.
Venkat Raju, K. and Marschner, H. (1972). Regulation of iron uptake from relatively insoluble iron compounds by sunflower plants. *Z. Pflanzenernähr. u. Bodenkd.*, 133, 227.
Venkat Raju, K., Marschner, H. and Römheld, V. (1972). Effect of iron nutritional status on iron uptake substrate pH and production and release of organic acids and riboflavin by sunflower plants. *Z. Pflanzenernährung u. Bodenkunde*, 132, 117.
Weiss, M.G. (1943). Inheritance and physiology of efficiency in iron utilization in soybeans. *Genetics*, 28, 253.

ABSTRACTS

1. EFFECT OF NO_3^- CONCENTRATION IN SOLUTION ON RATE OF UPTAKE BY *LOLIUM PERENNE*

C.R. CLEMENT, M.J. HOPPER and L.H.P. JONES

Grassland Research Institute, Hurley, Berkshire, UK

Plant roots may encounter a wide range of NO_3^- concentrations in the soil solution, i.e. from 20,000 down to 0.02 mg N l^{-1} (1,430 mM to 1.43 µM). Studies based on flowing solution culture have shown that the growth and rate of NO_3^- uptake by *Lolium perenne* are practically independent of concentration in solution over the range 200 - 0.2 mg N l^{-1}. However, as absorbing roots in the soil will produce depletion zones, the question arises as to whether the rate of NO_3^- uptake is affected when the concentration declines to below 0.2 mg N l^{-1}. This has been examined in further studies with *Lolium perenne* grown in flowing solution culture. Although the concentration of NO_3^- in solution declined from 0.13 to below 0.01 mg N l^{-1} the rate of NO_3^- uptake was only slightly reduced (15 per cent) and still exceeded the equivalent of 10 kg N $ha^{-1}d^{-1}$. These results indicate that in the soil the rate of uptake will be limited by transport processes to the root rather than by its ability to absorb from very low concentrations.

2. SOIL WATER STRESS AND AIR HUMIDITY EFFECTS ON THE ROOT SYSTEM OF SORGHUM

GONZALO CRUZ-ROMERO and CARLOS RAMOS

U.S. Salinity Laboratory, Riverside, California, USA

Sorghum (*Sorghum bicolor* L., Moench) was grown in two sunlit chambers with controlled air temperature at relative humidities (R.H.) of 80% and 20%.

In the humid air chamber the following soil water treatments were imposed:

a) *No water stress*: irrigation with tap water (EC = 0.06 S/m) at matric potentials (ψ_m) of -0.02 to -0.03 MPa.
b) *Largely matric stress*: irrigation with tap water at ψ_m values of -1.0 to -1.5 MPa.
c) *Osmotic stress*: two levels were obtained by irrigation with saline waters of EC = 0.4 and 0.8 S/m and SAR = 2, at ψ_m values of -0.02 to -0.03 MPa.

In the dry air chamber plants were irrigated as in a).

Hydraulic resistance of entire root systems (R_t) was calculated for all treatments 58-68 days after sowing (16th to 18th leaf stage) as follows: $R_t = (\psi_s - \psi_x)/F$.

ψ_s is the soil water potential in MPa; ψ_x is the xylem sap water potential in the lower part of the stem, in MPa; and F is the water loss rate of the entire plants, in $cm^3 s^{-1}$. At the time of measurements ψ_m was between -0.01 and -0.02 MPa. From R_t, hydraulic root resistance per unit root length (R_l), unit root weight (R_w), or unit leaf area (R_a) were obtained.

In all treatments a significant linear decrease of R_a, R_l or R_w was found with increasing transpiration rate (T). The best correlation with T was obtained when using R_a. R_a ranged from 2×10^5 to 10^6 MPa s cm^{-1} and T from 0.8×10^{-6} to 2.6×10^{-6} cm s^{-1}.

The top/root ratio (dry wt./dry wt.) of plants that had experienced high soil osmotic stress or air humidity stress (R.H. = 20%) were significantly higher than in the other treatments.

At low values of T (about 10^{-6} cm s^{-1}), R_a was nearly twice in the osmotically stress plants compared to the non-saline, well irigated plants. T values of 2×10^{-6} - 2.5×10^{-6} cm s^{-1} gave R_a values in all treatments that were not significantly different.

3. EFFECTS OF CALCIUM, PHOSPHORUS AND POTASSIUM ON MYCORRHIZAL PLANTS

MELVIN J. DAFT

Department of Biological Sciences, The University of Dundee

In 1904 Hiltner defined the rhizosphere as that portion of the soil that is influenced by the root system. Now that it is generally accepted that most plants are mycorrhizal this definition should perhaps be modified to include mycorrhizas and the various prokaryotes that live within the root tissues. Plant roots provide an environment where the soil microbial populations exist in an unstable equilibrium. The densities of the populations decrease with distance from the root and are dependent on many environmental factors such as light, temperature, aeration, pH, soil water and nutrient availability. In Dundee we have shown some of the effects of light, pH, nutrient availability on mycorrhizal development and how mycorrhizal infection influences nodulation by species of *Rhizobium* and the free living bacteria within the rhizosphere.

The supply of photosynthate to the mycorrhiza is probably an important factor. Defoliation, reduction in day length and irradiance reduces the mycorrhizal growth response and the development of the endophyte. Mycorrhizal development is much lower in media with low pH values compared to growth media with high pH values that also contain adequate levels of calcium. Growth responses in mycorrhizal plants have frequently been attributed to a greater uptake of phosphorus from the growth medium. However, this response depends on the host plant and the type of growth media. In tripartite associations of plant, mycorrhizal fungus and a nodulating bacterium there are responses to calcium, phosphorus and potassium. These responses can be detected as increased rates of nitrogen fixation, changes in carbon/nitrogen ratios and increased growth rates. Mycorrhizal roots support a much larger and diverse population of free living bacteria. This has been found using agar plate counts and direct light microscopy.

4. THE ROLE OF ABSCISIC ACID IN THE METABOLISM OF PLANT ROOTS

P.C. DEKOCK

Macaulay Institute for Soil Research, Craigiebuckler, Aberdeen, Scotland

The plant growth inhibitor, Abscisic Acid (ABA) induces changes in the mineral composition of the shoots of plants which resemble ageing changes: there is a reduction in the potassium and phosphorus concentrations in the tissues and an increase in the amounts of calcium and iron. ABA also causes starch to accumulate. Benzyl-adenine (BA) reverses these effects when added together with ABA, causing growth to continue normally with the starch concentrations lowered and the ratios of potassium to calcium and phosphorus to iron greater than those typical of growing tissues. ABA inhibits root growth, inducing the same changes in the composition of the root as of shoots. BA reverses these changes.

5. EFFECT OF IRRIGATION FREQUENCY, AND WATER SALINITY AND QUANTITY ON ROOT WATER UPTAKE DISTRIBUTION AND LEAF WATER POTENTIALS OF ALFALFA

CHRISTIAAN DIRKSEN and MICHAEL J. HUBER

U.S. Salinity Laboratory, Riverside, California, USA

Alfalfa was grown in six laboratory soil columns 1.05 m deep for more than two years and cut at about 24 day intervals. All columns were irrigated first daily, and then at 4 to 12 day intervals. By selecting leaching fractions (LF) and salinities (EC) of the water, a wide variety of osmotic (ψ_o) and matric (ψ_m) soil water potential distributions were established. Detailed root water uptake (λ) distributions were obtained by measuring soil water content (θ) with an automatic gamma ray attenuation scanner. One column received only tap water.

Under daily irrigation, water uptake occurred mainly in the daily replenished zone. EC increased rapidly with depth within that zone and little below it. At low LF this zone was only about 0.30 m deep. Observed λ distributions agreed well with those derived theoretically from steady EC profiles. With water available elsewhere in the profile water uptake ceased around EC = 1.3 S m^{-1}. Under stress, alfalfa concentrated the soil water up to EC = 3.0 S m^{-1}. One column, sampled at the end of the daily irrigations, showed a maximum root density of 11.3 cm cm^{-3} at 0.10 m depth and a nearly uniform density of 3.0 cm cm^{-3} below 0.35 m; the other columns showed qualitatively similar weight distributions.

Under infrequent irrigation, λ shifted with time. The EC profile of the 4 day frequency was similar to that under daily irrigation. Decreasing frequency of irrigation resulted in deeper leaching and lower EC, and also much lower ψ_m and reduced transpiration. General conclusions on the relative effect of ψ_m and ψ_o (irrigation frequency) on dry matter production and water use efficiency cannot be made.

Leaf water potentials ψ_L generally varied between -0.8 and -2.0 MPa during light and were about 0.4 MPa higher during darkness. In the nonsaline column ψ_L decreased more rapidly than ψ_m, suggesting increased soil-root contact resistance with decreasing θ. In the saline columns there was no correlation between ψ_L and θ, but good correlation between ψ_L and spatially-averaged EC for individual columns. Uptake-weighted average EC showed reasonably good correlation with ψ_L for all columns combined.

6. A KINETIC MODEL FOR THE NET FLUX OF DIFFERENT NITROGEN FORMS INTO GROWING TOMATO PLANTS

S.E-D.A. FAIZY

Department of Soil Science, Faculty of Agriculture, University of Tanta, Kafr El-Sheikh, Egypt

A model for the carrier mediated active flux of anions, cations, and uncharged molecules into plant roots has been proposed. The model was applied on tomato plants growing for 57 days in a complete nutrient solution, supplied with nitrate, ammonium, or urea nitrogen, the plant system was therefore made to take up more anions than cations, more cations than anions, or uncharged molecules, respectively. The net active flux of anions, cations and uncharged molecules are balanced by an efflux of endogenous OH^-, H^+, or H_2O, respectively.

Whilst the net flux of ammonium showed a hyperbolic dependence, the nitrate and urea showed a sigmoidal dependence on the external concentration. Two expressions were formulated, one to be applied to the hyperbolic and the other to the sigmoidal response.

With the three nitrogen forms, two mechanisms of uptake were found to exist, one operating at low and the other at high external concentration of nitrogen. With all the three nitrogen forms, the shift from the first to the second mechanism occurred at the same external concentration.

7. NEMATODES AND MICROFLORA IN THE ROOT RHIZOSPHERE OF FOUR DESERT SHRUBS

DIANA W. FRECKMAN and R. MANKAU

Department of Nematology, University of California, Riverside, California, USA

The relationship of nematodes, fungi, bacteria and actinomycetes to root systems was studied in soils of the Mojave desert, Rock Valley, Nevada, USA. Parameters examined included spatial and seasonal distribution of organisms and roots and the abiotic factors of soil temperature and moisture. Soil samples were collected weekly from four perennial plant species, *Larrea tridentata*, *Lycium andersonii*, *Krameria parvifolia* and *Ambrosia dumosa*, which represented 75% of the dominant vegetation at Rock Valley. One shrub of each species was examined weekly. Soil samples were taken at 3 depths and 3 distances from the shrubs. An identical set of samples was collected to determine root biomass.

Nematodes were extracted from soil by a modified sugar flotation technique developed for use in desert soils. The nematodes were counted, identified and placed in the following groups: 1) microbivores - mainly Caphalobidae, 2) omnivore-predators - Dorylaimina, excluding *Xiphinema*, *Longidorus* and *Trichodorus*, 3) fungivores - *Aphelenchus avenae*, *Aphelenchoides* sp. and *Ditylenchus* sp., 4) phytophages - mainly Tylenchorhynchinae. Damaged or unidentifiable nematodes comprised a fifth group. Roots were separated into two groups. Small roots (0.5 - 2 mm) were separated by screening and flotation in $MgSO_4$ solution from 0.3 liter of soil and larger roots (2 mm) by manual separation from 1.5 liter soil. Dilution plate counts were made to quantify and differentiate the fungal, bacterial and actinomycete populations of the soil samples.

The spatial distribution of all nematode trophic groups was positively correlated with root biomass. Nematode density, fungal and bacterial populations decreased with depth and distance from the plant. The abundance of microflora in decreasing order was actinomycetes, bacteria and fungi. Actinomycetes were not influenced by depth or distance from the shrubs. The distribution of nematodes was closely related to food sources. Microbivores were the most numerous trophic nematode group at all depths and distances followed by omnivore-predators, fungivores and plant feeders. Plant feeders represented only 14-17% of the total population and were most numerous in association with *Lycium* and *Larrea*. The density and biomass of trophic groups of nematodes other than phytophages was intimately associated with the root rhizosphere in desert soils where food sources were available for even the detritus-based nematode trophic groups.

8. ADAPTIVE RESPONSES OF PLANT ROOT SYSTEMS TO NUTRIENT STRESS

A. L. FLEMING

U.S. Dept. of Agr., SEA, FR PPHI, Beltsville, USA

Plant genotypes differ in their response to nutrient stress (deficiency) and in their tolerance to mineral elements. These responses involve modification of enzyme activity, ion uptake, plant-induced pH changes, accumulation of nutrients, and changes in the reductive capacity of plant tissues. Although the metabolic processes induced by nutrient stress have not been fully characterized, investigations of the physiological indicators of efficiency and their interactions should aid in the development of improved varieties.

This study was conducted to describe the relationship between efficiency in the absorption of Fe and Al tolerance in selected varieties of wheat and soybean.

Soybean varieties, differing in Fe efficiency, and wheat varieties, representing extremes of Al tolerance were grown in nutrient solutions containing suboptimum concentrations of Fe or excess Al. These plants were then transferred to complete nutrient solutions and allowed to recover. The magnitude of the stress and the plant responses were monitored by determining plant-induced pH changes, activity of root surface enzymes, reduction (Fe^{3+} to Fe^{2+}) at the root, and the uptake of Fe, Ca, P, NO_3^- and NH_4 by the plant.

Fe stress in wheat and soybean was characterized by decreases in pH of the nutrient solution, leaf chlorosis, and increased reduction at the root surface. There was also a release of "reductants" into the solution that was significant only for the soybean. Al sensitive wheat varieties were more Fe efficient than Al tolerant varieties. This relationship did not apply to several varieties of soybean.

Al injured roots were severely reduced in growth. In addition, Ca and NO_3^- uptake was reduced, acid phosphatase activity increased, and external P accumulations increased in response to Al treatments.

The adaptation to Fe stress or Al toxicity involves specific systems which change in response to the particular stress.

9. PHOSPHATE CONTENT IN RHIZOCYLINDER AND BULK SOIL RESPECTIVELY AS INDICATED BY SOIL TEST EXTRACTION METHODS

A. JUNGK and L. HENDRIKS

Institut für Pflanzenernährung der Technischen Universität Hannover, Hannover, West Germany

The ability of soil test methods to indicate changes of available phosphate in the interface of soil and plant roots is considered.

In pot experiments maize and other seedlings were grown for three weeks and then the soil strongly adhering to the roots was separated from bulk soil. Several soil test methods used in Germany to determine phosphate fertilizer requirements were applied. Results were as follows.

Only small differences of soluble soil phosphate were found between rhizocylinder and bulk soil by methods based on acidic extracting solutions like Ca-lactate/lactic acid (DL-method, pH 3.6) or sodium formate/formic acid (pH 3.5). Much bigger differences occurred by using water as the extracting agent according to van der Paauw *et al.* With this method P concentration near plant roots decreased by about 50% whereas bulk soil did not change significantly compared to uncropped soil.

Plant species markedly influenced the degree of P depletion near roots. Maize plants, which have higher P uptake rates per unit of root length when grown in nutrient solution, reduced P concentration in rhizocylinder soil more than wheat and green bean plants. This effect increased with plant age and root density.

In mixtures of sand and loess the degree of P depletion strongly increased with increasing percentage of sand.

Increasing soil moisture reduced the degree of P depletion in rhizocylinder soil even though P uptake rates per unit of root length increased. It is assumed that the increase of soil moisture extends the distance of P transport and thus decreases the gradient of P concentration within the depletion zone.

Application of soluble calcium salt ($CaCl_2$) reduced P uptake of plants and also the degree of P depletion in the rhizocylinder. Therefore it appears that calcium ions reduced the availability of soil phosphate more than the solubility in water.

10. WATER RELATIONS OF WHEAT ALTERNATED BETWEEN TWO ROOT TEMPERATURES

M.B. KIRKHAM

Department of Agronomy, Oklahoma State University, Stillwater, Oklahoma, USA

Water potentials, osmotic potentials, and stomatal resistances were measured daily in a growth room during a 16-day period, in growing leaves of winter wheat (*Triticum aestivum* L. em. Thell.), which had roots maintained under one of three root temperature treatments: 1) constant temperature of 24°C (about the optimum); 2) constant temperature of 34°C (10°C above the optimum and near the high temperature limit for growth); and 3) alternating temperature of 24 hours at 24°C followed by 24 hours at 34°C. Air temperature was constant at 24°C.

Water and osmotic potentials, stomatal resistances, and height of plants maintained at the 34°C root temperature were intermediate between those grown at 24°C (highest potentials, lowest resistances, tallest) and those alternated daily between the two temperatures (lowest potentials, highest resistances, shortest). Previous work showed that potentials, stomatal resistances, and growth of wheat subjected to constant root temperatures of 25°, 29° and 33°C were *linearly* related to root temperature with lowest potentials, highest resistances, and least growth at the high temperature. Even though in this experiment the average root temperature of plants exposed to alternating root temperatures was 29°C, they grew the poorest. The wheat plants, therefore, did not integrate a root temperature stress in the same manner over "time" as they did over "space".

The results showed that the stomatal resistance, water and osmotic potentials, and height of wheat seedlings, maintained at a constant root temperature 10°C higher than the optimum, were about halfway between those of plants grown at the optimum and those alternated daily between the optimum and the optimum-plus-10°C root temperature.

11. SIMULATION OF SOIL-ROOT INTERACTIONS

JERRY R. LAMBERT and DONALD N. BAKER

Clemson University, USDA-ARS, Clemson, Mississippi, USA

A dynamic simulation model of the interactions between soil and roots has been implemented in a Fortran programme. A synthesis of known information necessary to quantify root growth and water and nitrogen uptake from the soil is included.

A two-dimensional profile of soil and the contained water and nitrogen root mass, classified by age, is maintained under a row crop of variable spacing and root zone depth. Root growth in any soil cell in the matrix is assumed to be proportional to the roots already present and capable of growth, and to depend on soil water conditions, temperature, soil strength, and on the amount of carbohydrate supplied from the aerial portion of the crop for root growth. Water uptake occurs differentially throughout the profile dependent on root length and age, water diffusivity, and the demand impressed by the environment and aerial crop. The consequent water potential gradients cause capillary redistribution of soil water; a finite-difference form of Darcy's Law is used to calculate the flow rates. Nitrate is assumed to be completely in solution in the soil water and to move by mass flow during uptake and redistribution. Boundary conditions include zero horizontal flux beneath the row and midway between the rows, zero nitrate flux across the soil surface, and field capacity being maintained in the bottom soil layer. Feedback to the aerial crop simulator includes sink strength for carbohydrate and nitrate supply for top growth.

The required initial values are two-dimensional profiles of water, nitrate and root distribution at emergence, and a one-dimensional temperature profile. Information is required on the following soil characteristics: the relationships between water potential and water content; conductivity and water potential; water diffusivity and water content. In addition information on the diffusivity of air in soil would be necessary to take account of variations in aeration. If soil strength is included in the model, information would be required on the relationship between mechanical impedance to root penetration, bulk density and water content of the soil. During the simulation period estimates are needed of the rainfall infiltrated daily, the maximum and minimum air temperature, the total radiation, and the carbohydrate supplied for root growth. The programme is modular so that it can be included in a growth model of any row crop; with modification to one-dimension, it could be used for pastures.

The results include maps of water and nitrate content and root distribution throughout the growing season. The effects of environmental and cultural management variables on behaviour of the root zone can be quantitatively visualized.

12. SOME FACTORS WHICH INFLUENCE THE ACCUMULATION AND ACTION OF MICROBIAL METABOLITES AROUND SEEDS AND ROOTS

J.M. LYNCH

ARC Letcombe Laboratory, Wantage, Oxon, UK

Plant residues and root exudates are the two major sources of carbon substrates for the soil microflora. Residues may contribute the greatest quantity but root exudates are usually in a form which provides more readily available substrates for microbial growth. The total carbon leaving roots can account for up to 20% of the total photosynthate produced and this can influence the formation of toxic substances around the roots, the pattern of pathogen invasion and the formation of microbial polysaccharides which contribute to the stabilization of soil aggregates.

The effect on plants of microbial metabolites depends on their proximity to the roots and the nature of the compounds formed. Studies on the formation of acetic acid, which can be produced in phytotoxic concentrations from straw residues in anaerobic soils, showed that the decline in concentration (c) with distance (d) from straw fitted an exponential curve described by the equation $c = c_0 e^{-0.5d}$, where c_0 is the concentration of acetic acid at the straw surface. The accumulation of the toxin is governed by temperature ($Q_{10} = 2.5$) and redox potential (maximal around zero). The rate of straw decomposition influences the amount of toxin which is produced and preliminary studies involving chemical treatment and application of animal slurries indicate that it may be possible to manipulate this.

Uptake of ^{14}C-labelled acetic acid ($PK_a = 4.75$) by germinating barley seed was found to be about four times greater at pH 3.2 than at pH 7.0 and there was a concomitant increase in toxicity. The application of powdered chalk to the seed could partially alleviate the phytotoxicity, presumably by increasing the pH close to the germinating seedling. Under conditions of poor aeration, root extension is reduced both by the low oxygen concentration and by the toxin. As is common with many toxic materials, low concentrations (1 mM) of acetic acid at pH 6.5 stimulated the extension of young roots from seeds and some evidence suggests that this is due to a stimulation in the repair of the membranes in the seed embryo on rehydration.

13. THE IMPORTANCE OF ROOTING IN THE NUTRITION OF CROPS

CLAUDE MAERTENS

INRA - Station d'Agronomie, BP 12, Auzeville, Castanet Tolosan, France

The paper describes an attempt at the development of a method based on rooting characteristics which could define the potential supply of mineral elements by a soil to a crop.

Knowledge of both uptake by the plants and the root surfaces which develop allows estimation of the flux of potassium and phosphorus at the soil-root interface. These fluxes are then compared with the potential fluxes obtained elsewhere in controlled media.

Determination of root length and the quantities of available nutrients in the different soil horizons, allows calculation of a theoretical cylinder of soil around the root from which all available nutrients have been removed. These results, compared with the average distance between roots in the soil gives for phosphorus and potassium an estimate of the degree of exploitation of the soil.

In a simultaneous field trial with soybeans, an uptake profile was determined using Rb. This improved the study of potassium by indicating root activity in each horizon.

In conclusion, the value of the results which can be obtained using the above approaches, and improvements in the methods, are discussed.

14. ROLE OF THE RHIZOSPHERE IN UTILIZATION OF INORGANIC IRON-III-COMPOUNDS BY CORN PLANTS

H. MARSCHNER and S. AZARABADI

Institute of Plant Nutrition, Universität Hohenheim, Stuttgart, West Germany

In water culture experiments the effect of varied supply of iron-III-hydroxide and phosphate on growth rate, chlorophyll and mineral content of corn plants has been studied.

Using a nutrient solution with nitrate-nitrogen, there was a strong interaction between iron and phosphorus uptake. Severe iron chlorosis could only be prevented by very high supply of ferric hydroxide or very low phosphorus supply, both of which resulted in a normal chlorophyll content but produced phosphorus deficient plants. However, also in water culture, ferric hydroxide was a satisfactory source of iron for corn plants when iron and phosphorus were supplied to separate root zones (split-root-technique).

In contrast to water culture, in sand culture (quartz sand) with the same nutrient solution, corn plants used ferric hydroxide as well as iron chelate, even when high phosphorus concentrations were simultaneously present in the substrate. Using Fe-59 and circulating the nutrient solution through the sand culture it could be demonstrated that the mobilization of iron from ferric hydroxide is restricted to the root-sand (ferric hydroxide) interface (rhizosphere) without increasing the amount of soluble iron in the bulk substrate. The exhaustion zone of phosphorus around the roots in sand seems to be particularly responsible for this "substrate effect" in utilization of ferric hydroxide.

The uptake of phosphorus and iron in sequence along a growing root in a solid substrate could be important in the iron nutrition of iron-inefficient plant species like corn in soils with high pH.

15. IS THE EFFICIENCY OF NUTRIENT UPTAKE GENETICALLY CONTROLLED?

NIELS ERIK NIELSEN

Department of Soil Fertility and Plant Nutrition, The Royal Veterinary and Agricultural University, Copenhagen, Denmark

Under conditions in which the rate determining step in nutrient uptake is located in the root, the efficiency of nutrient uptake was shown to increase if the root parameters: root length per unit of plant weight and maximal net influx per cm root (I_{max}) increase and Michaelis Menten constant (Km) and minimum concentration (c_{min}) decrease. At low nutrient concentrations in the soil solution, I_{max}, Km, and c_{min} were of equal importance in simulated phosphorus uptake by a transport-kinetic model.

In water culture and field experiments, genotypes of maize showed in respect to phosphorus uptake a two- to three-fold variation in the root parameters. These variations seem to be genetically controlled.

At low and high phosphorus levels, correlation coefficients 0.99 and 0.95 were found between phosphorus uptake by maize genotypes in the field and phosphorus uptake simulated from root parameters of maize grown in water culture. This would seem to justify the use of kinetic studies based on water culture experiments in characterizing roots with respect to nutrient uptake and in connection with selection and development of maize genotypes with improved efficiency of phosphorus uptake from soil.

16. COMBINED EFFECT OF LINEAR AND NON-LINEAR TRANSPORT RATES ON SOLUTE UPTAKE INTO BIOLOGICAL TISSUES

J.J. OERTLI

Botanical Institute of the University of Basle, Basle, Switzerland

Strong evidence for an effective rate-limiting boundary layer in nutrient uptake by plant tissues stems from studies of iron nutrition.

Whereas the equilibrium solution can be entirely inadequate to provide iron to roots, precipitated iron adhering to root surfaces either as ferric hydroxide or ferric oxide will support excellent plant growth. This author, at one time, suggested the existence of microcavities between iron particles and root surfaces. In such compartments, pH values, oxygen tensions, or organic acid compositions would be sufficiently different from the equilibrium solution to account for the iron nutrition.

Further support for the existence of an effective boundary layer, which of course must not necessarily be enclosed by solid particles, comes from observations that iron uptake from solid or chelated sources is stimulated by cation uptake in the order $K^+ > Na^+ > MG^{++} > Ca^{++}$ and is retarded by anion uptake in the order $NO_3^- > Cl^- > SO_4^-$. Apparently, rapid cation uptake reduces the boundary layer pH whereas a fast anion uptake has an opposite effect. Solid precipitates of nutrients close to roots also represent capacities of nutrients. Such capacities reduce the effective thickness of the boundary layer. The effectiveness of such a capacity in reducing the boundary layer depends on conditions that govern ion uptake.

A steady state model is presented in which non-linear Michaelis-Menten type solute transport is combined in series with one or more preceding linear diffusion type transports. The former may be typical for solute uptake across biological membranes, the latter for the transport across one or more boundary layers. It follows from this model that the transport across the boundary layer becomes more rate limiting (1) with decreasing concentration of the external equilibrium solution, (2) with decreasing effective diffusion constant in the boundary layers (i.e. the linear phases), (3) with increasing thickness of the boundary layer, (4) with increasing maximum uptake rate (V_{max}) in the Michaelis-Menten term, and (5) with decreasing Michaelis-Menten constants (K values). These conclusions are supported by experimental findings.

17. SOME EFFECTS OF SOIL WATER CONDITIONS ON THE ROOT GROWTH OF GRASS IN THE PEAT MUCK SOIL

WENANTY OLSZTA

Land Reclamation and Grassland Farming, Lublin, Poland

The effect of moisture potential on the dry weight of roots and on the vertical distribution of roots in the soil profile was examined in five plots with the water table held at a different level in each plot (50, 60, 70, 80 and 90 cm depth). Observations were made, every 10 days during two seasons, of the weight of roots to 50 cm depth; soil moisture potential, the depth of the water table and soil temperature were recorded daily. Rainfall was also measured.

The percentage distribution of roots in the soil profile is dependent on the soil moisture potential to the full depth of 50 cm. The depth of the water table also affects the proportional distribution of root weight down the soil profile and on the total root weight. The relationship shown between the distribution of roots and soil moisture potential may be useful in the simulation of root growth during the vegetative phase.

18. MASS FLOW TRANSPORT TOWARDS THE ROOT SYSTEMS OF TWO FOREST ECOSYSTEMS IN RELATION TO THE UPTAKE BY THE ABOVE GROUND TREE PARTS

JÜRGEN PRENZEL

Institute for Soil Science and Forest Nutrition, Göttingen, West Germany

The concept of mass-flow transport of ions to the root system has been used in an ecosystem study of a beech and a spruce forest. The transpirational water uptake of the stands was calculated as a function of time and of soil depth by a soil physical simulation model. In this calculation field measurements of the soil water tension profile were used. The concentration of 11 chemical elements in the soil solution were measured by suction lysimeter plates and funnel lysimeters in the field. In addition the concentrations in the equilibrium soil solutions at different depths were determined.

Results for several years were used to calculate the mass-flow of 11 chemical elements to the root system. In a simple model of uptake this mass flow was combined with a factor, called the Mass Flow Coefficient (MFC), to obtain the actually measured uptake into the above ground stand. MFC values greater than unity indicate that mass flow transports less ions to root systems than they absorb; the reverse is true if they are less than unity. The MFCs vary between less than 0.1 for Al and more than 100 for P. They are lower than 1 for Al, Cl, Na and higher than 1 for Mn, Ca, K, N, P.

19. DYNAMICS OF WATER CONDUCTANCE IN SORGHUM ROOTS

J.T. RITCHIE and W.S. MEYER

USDA, Agricultural Research Service, Temple, Texas, USA and Soil and Irrigation Research Institute, Pretoria, South Africa

We have found evidence that the water potential of covered leaves can be regarded as the potential operative in the root axis at the crown. Assuming a negligible axial resistance, we believe the covered leaf potential can provide a reasonable value for the xylem potential at the point of water uptake within the sorghum root. Thus, knowing the relevant potentials and flow rates in a system, the conductance of roots can be calculated.

Both nutrient solution and soil-grown plant root conductances showed a strong dependence on flow rate. Experimental measurements also demonstrate that the conductance per unit length of a root is not a unique value. The same roots on a sorghum plant were found to have conductance values ranging from about 10^{-3} to 10^{-4} cm^3 bar^{-1} h^{-1} cm^{-1}. The upper limit of conductance to solution-grown roots was about 10^{-3} cm^3 bar^{-1} h^{-1} cm^{-1}. Above that limit, plant top functions were strongly affected.

These results demonstrate that physical models used to calculate water uptake must consider the dynamic nature of root conductance if they are to approximate to reality.

20. THE EFFECT OF INOCULUM DENSITY ON THE DEVELOPMENT OF VESICULAR-ARBUSCULAR MYCORRHIZAS OF CROP PLANTS

F.E. SANDERS and L.E. REED

Department of Plant Sciences, University of Leeds, UK

Vesicular-arbuscular mycorrhizal infection of the roots of crop plants may increase the rate of phosphorus uptake from soils where phosphorus availability is low and consequently increase crop growth rate and yield in favourable circumstances. The growth rate response to mycorrhizal infection is linked closely in time to the development of the mycorrhizal root system.

In the period of vegetative growth of annual crops, the development of mycorrhizal infection appears to follow a set pattern. Infection establishment is followed by a phase of extending infection where the proportion of mycorrhizal roots increases rapidly. Eventually an equilibrium is reached where the proportion of mycorrhizal roots becomes constant although the root weight of the crop may continue to increase. Many factors appear to have effects on this pattern of development. These are reviewed using examples drawn from the literature.

One factor of potential importance, but which has received little attention, is the inoculum density of mycorrhizal fungi present in the soil in which the crop is sown. The effect of varying inoculum density was investigated in a series of pot experiments using maize. Variation of inoculum density was found to produce large effects on infection development. Similar effects occurred on barley in the field.

21. INPUT AND TURNOVER OF ORGANIC MATTER IN ROOTS AND RHIZOSPHERE

D.R. SAUERBECK

Institute of Agricultural Chemistry, University of Bonn, West Germany

From seasonal changes in the carbon content of agricultural soils it has already been concluded earlier that the formation and breakdown of plant root substances during the growth period is considerably larger than the amount of plant residues at harvest time. Recent investigations with ^{14}C-labelled plants have confirmed this conjecture. The plants are grown under a continuous supply of $^{14}CO_2$, so that the photosynthethate consumption and turnover of their roots can be measured both in soil and in nutrient solution.

The results indicate that plants may require about one quarter of their assimilatory power for synthesizing their root system and maintaining its functions. Much of this organic root matter is already mineralized during the growth period, and therefore the root residues at harvest time represent only a relatively small fraction of the plant's total investment. The frequently found difference between the CO_2 output of soils and the conventionally measured plant residue input may thus be explained.

The primary root respiration appears to constitute only a minor portion of this energy turnover, whereas the secondary microbial decomposition represents the bulk of it. This enormous rhizo-deposition of readily decomposable organic matter and as its consequence the most intense turnover processes in the rhizosphere may be of considerable influence on plant-soil relationships and on plant nutrition.

Data are available from wheat, mustard, and corn plants. Further experiments are under way in order to extend these studies to other plants under various growth and stress conditions.

22. USE OF THE FLUORESCENT ANTIBODY TECHNIQUE TO IDENTIFY N_2-FIXING BACTERIA AT THE ROOT INTERFACE IN TROPICAL GRASSES

S.C. SCHANK, REX L. SMITH and DAVID A. ZUBERER

Institute of Food and Agricultural Sciences, University of Florida, Gainesville, Florida, USA

Direct observation of whole root segments of three tropical grasses was accomplished using fluorescent antibody (FA) techniques. Antisera had been prepared which identified isolates of N_2-fixing bacteria of the genus *Azospirillum brasilense* (formerly *Spirillum lipoferum*) strains 13t from Brazil and JM125A2 from Florida. Roots of *Panicum maximum* (guineagrass) had been inoculated for 3 consecutive years with *A.brasilense* strain 13t, whereas *Pennisetum americanum* (pearl millet) and *Digitaria decumbens* (digitgrass) had been inoculated with a mixed culture of *A.brasilense*, strains 13t and JM125A2 in replicated field plots in early summer 1977. Root samples collected during the year were assayed for acetylene reduction and by FA analysis. The FA tests included bacterial counts immediately prior to, and immediately after the acetylene reduction assay.

The guineagrass plots located on Arredondo fine sand (Grosarenic Palendult; loamy, siliceous hypothermic) had a significantly higher number of *A.brasilense* strain 13t bacteria than strain JM125A2. These plots had been inoculated with 13t bacteria for three years, but had not received inoculum of strain JM125A2. Average numbers of 13t bacteria per 59 nm^2 of root surface were 12.2 while counts of strain JM125A2 were 6.48 per 59 nm^2 on guineagrass.

Pearl millet and digitgrass plots inoculated only in 1977 with the mixed culture were located on Myakka fine sand (sandy siliceous, hyperthermic Aeric Haplaquad). In pearl millet, the average number of 13t bacteria observed per 59 nm^2 was 0.55 and JM125A2 was 0.72. The digitgrass had 2.11 of 13t and 3.26 of JM125A2 in equally scored areas of root. In counts using sectioned roots and non-specific staining we found only 1% of the total bacterial population at the root interface was composed of FA reactive *Azospirillum*.

23. INFLUENCE OF SHORT-TERM OXYGEN STRESS UPON THE TRANSLOCATION OF ZINC IN NAVY BEANS (*PHASEOLUS VULGARIS*, L.)

A.J.M. SMUCKER, B.D. KENEZEK and G.R. HOOPER

Michigan State University

Treatments of N^2 were applied to roots in solution culture to determine the adverse effects of short-term oxygen stress upon ^{65}Zn distribution, ultrastructural changes in root cells and ethanol concentration in xylem exudates. Beans (var. Seafarer) were grown in solutions containing 0.1 and 3 ppm Zn for 28 days, then ^{65}Zn was added and aeration gases were replaced by N_2 for 0, 6, 12 and 24 hours. Tissue Zn was measured by atomic absorption and ^{65}Zn by gamma counting. Ultrastructural changes were observed by electron microscopy. Ethanol concentration of xylem exudates was measured by gas-liquid chromatography. Oxygen stresses greater than 6 hours, significantly reduced the absorption and translocation of ^{65}Zn.

Low tissue concentrations of Zn increased the absorption and translocation of ^{65}Zn when plants were aerated. Zinc concentration had no influence upon the movement of ^{65}Zn in plants subjected to oxygen stress. Complete plasmolysis of root tip cells occurred after 24 hours of oxygen stress. Changes in the nucleolus and endoplasmic reticulum were also observed in root cells subjected to short-term oxygen stress.

24. DENITRIFICATION AS RELATED TO POTASSIUM NUTRITION OF RYE GRASS AND MOISTURE CONTENT OF SOIL

GÜNTER TROLLDENIER

Landwirtschaftliche Forschungsanstalt Büntehof, Hannover, West Germany

Rye grass was grown at 3 different potassium levels in pots filled with 2 kg gray brown podsolic soil. One treatment did not receive any additional potassium, while two others were treated with 25 and 50 mg K/100 g soil respectively. Other nutrients were given in adequate amounts. The waterholding capacity (whc) of the soil was adjusted first to 50% and raised consecutively till to the second cut up to 70%. After the second cut moisture content of the soil in half of the pots was maintained at 70%, the other half was adjusted to 90% whc. Two weeks after the second cut each pot received 0.2 g $Ca(NO_3)_2$ with 11% ^{15}N excess. Eleven days later with the third cut the experiment was terminated.

Yield differences due to varying K supply became greater with each cut. The two different moisture contents after the second cut did not influence the yields much. After the final harvest the fate of the added labelled nitrogen was assayed. Nitrogen balances were established by analysing ^{15}N content of shoots, roots and soil. Most fertilizer ^{15}N (74-87%) was found in the shoots indicating a very rapid nitrogen uptake. The amount remaining in the soil was 2-3% only. At 70% whc in all potassium treatments ^{15}N was recovered completely. At 90% whc complete recovery was found at the two higher potassium levels. At the lowest K level, however, only 89% of the added ^{15}N were recovered. The lower nitrogen recovery indicates that oxygen was depleted thus allowing denitrification. The result is in concordance with those of nutrient solution experiments. These experiments revealed higher bacterial numbers and higher microbial activity in the rhizosphere of potassium deficient plants, resulting under suitable conditions for denitrification in higher nitrogen losses. The nitrogen losses are not very serious. It should be taken into consideration, however, that the rapid translocation into the plants under the experimental conditions prevented probably higher nitrogen losses.

25. EFFECT OF DRY SOIL CONDITIONS AROUND THE BASE OF *LOLIUM PERENNE* PLANTS

ARTHUR TROUGHTON

Welsh Plant Breeding Station, Aberystwyth, Dyfed, Wales, UK

During periods of drought the first portion of the soil to become dry is that near to the surface surrounding the shoot bases. When this occurs *Lolium perenne* plants cease to produce new main root axes. This results in a considerable decrease in the growth of the root system as a whole. Provided the lower layers of the soil contain an adequate supply of water and minerals, shoot growth is little affected for some time afterwards. This results in a decrease in root growth relative to shoot growth, due not so much to a change in environment but lack of 'sink' capacity by the root system. Over a period of months the growth of the shoot system gradually decreases and the plant dies. Death is apparently due to lack of calcium. This may be related to a reduction in the transpiration rate, caused by the smaller root system, and results in less calcium being transported by mass flow through the soil to the root surface.

26. THE INFLUENCE OF NATURAL ORGANIC COMPLEXING AGENTS ON THE UPTAKE OF HEAVY METALS BY PLANTS

MADELIJN MARQUENIE - VAN DER WERFF

Biological Laboratory, Free University, Amsterdam, The Netherlands

In studying plant communities the soil profile is of particular interest. The chemical composition of different soil horizons can vary greatly, so that neighbouring species with different root lengths can be subjected to completely different nutritional situations. Metal-complexing components of the soil, e.g. fulvic and humic acids, are of great importance as they may change the behaviour of metal ions in the soil and their availability to plants. Not only the quantity of the metals, but also their chemical quality must be studied. So the influence of such compounds promoting or preventing metal uptake and the possibility of their influencing metal activity or toxicity to plants, makes them interesting for the relationship of plants to micronutrients (Fe, Mn, Cu, Zn) in situations of deficiency as well as surplus, especially in heavy metal pollution (Cu, Pb, Zn, Cd). The chemical constitution of humic acids in comparison with fulvic acids is rather different. In a podzol soil the proportions and composition of and ratio between fulvic and humic acids vary between the horizons.

In this study humic and fulvic acids are extracted from peat soil and from the O- and A-horizons of a podzol soil. As Zn-, Cu-, Pb- or Cd-complexes they are added to nutrient solutions. In these experiments the uptake of heavy metals by plants is compared with the results from experiments with metal complexes of artificial chelating agents (EDTA) as well as with metal ions. The behaviour of these complexes can be considerably different, depending on metal species and kind and origin of the complexing agent. This results in contrasting relationships between metal uptake, translocation and toxicity.

27. SIMULATION OF CHANGES IN NUTRIENT CONCENTRATIONS IN THE ROOT ENVIRONMENT. I. LONG TERM MODELS BASED ON EXPERIMENTS WITH GRAVEL CULTURE

F. VAN DORP, J. SINNAEVE and M.J. FRISSEL

Association EURATOM - ITAL, Wageningen, The Netherlands

To obtain data for simulation models describing uptake processes by roots and changes in nutrient concentrations around roots caused by this uptake, experiments with maize were carried out on nutrient solutions in coarse sand as a supporting medium. Fresh and dry weight of shoot and roots, dimensions of the root system at different depths, transpiration and amounts of K, Ca, Mg, N and P in the plants were determined as a function of time.

In the models, the selectivity coefficient, being the ratio between the actual uptake of ions and the uptake which can be calculated from the amount of water absorbed by the roots multiplied with the concentration of that ion in that water, plays an important role. The following hypotheses were tested: All roots absorb water and ions simultaneously; the calculated selectivity coefficient indicates if the relative uptake of ions is higher or lower than the uptake of water, resulting in depletion or accumulation of ions around the roots. Another model takes into account that part of the roots are suberized and no longer contribute to absorption. Furthermore the influence of the diurnal rhythm in transpiration versus a constant water uptake (per 24 h) has been tested.

During exponential growth, N and K were absorbed actively by the roots, causing depletion around the roots, while Ca and Mg, as expected, were accumulated around the roots. The amounts of the different nutrients, which were accumulated or depleted, changed during the growth of the maize. The hypothesis that roots, due to suberization actively absorb water and ions only during a limited period, could be confirmed when their period of activity is linked to the physiological stage of the crop at the moment of their initiation. The hypothesis that ion uptake is a function of a constant root activity and of ion concentration, could neither be confirmed nor rejected. The diurnal rhythm of transpiration has important consequences only for short term changes of nutrient concentrations around the roots and is discussed in part II.

28. SIMULATION OF CHANGES IN NUTRIENT CONCENTRATIONS IN THE ROOT ENVIRONMENT. II. MODELS OF SHORT TERM CHANGES AROUND THE ROOTS

F. VAN DORP and J. SINNAEVE

Association EURATOM - ITAL, Wageningen, The Netherlands

The long term simulation models, discussed in part I, indicated growth stages during which the greatest depletion or accumulation of a particular nutrient around the roots occurred. The short term models describe the variation of nutrient concentrations in those growth stages from hour to hour and evaluate concentration gradients around single roots. Depending on the stage of development of the crop, on the period of the day and on the considered nutrient, the concentration in the root environment changed considerably. About 40 days after sowing the concentrations of K and N around the roots were depleted with a factor 2 within 6 hours. P and Ca, however, were accumulated around the roots during the second half of the growing period up to 1.5 times the original concentration within 6 hours.

Two general conclusions which may be drawn from the experiments and the simulation models are:

- Transport of water and ions and exchange processes of ions in the rhizosphere can be simulated without problems, provided that enough data on the growth medium are available.
- A complete picture of the uptake of nutrients by plant roots as a

function of the growth processes of the shoot (as they are influenced by climatic conditions), as a function of the geometry and growth of the root system and as a function of the environment of the root is not yet available. At the moment it is possible to calculate in a simulation model changes in ion concentrations around roots as a function of time, provided that sufficient data on the dimensions of the root system, on the amounts of nutrients in the plant and on the transpiration during the whole growing season are available. Reliable data are a prerequisite for such a model.

An experimental set-up with separate conditioning of the aerial and subaerial environment, which is particularly suited for the quantification of the soil-plant-atmosphere interaction will be presented at the poster exhibition.

29. ROOT EXUDATES AND MICROBIAL GROWTH IN THE INTERFACE

A.W.S.M. VAN EGERAAT

Laboratory of Microbiology, Agricultural University, Wageningen, The Netherlands

It is a well-known fact that roots exude organic compounds which can be used by micro-organisms. As a consequence the microbial population close to the root is higher than at some distance from the root. The zone in which stimulation of micro-organisms occurs is called "the rhizosphere". The exudation of organic compounds by roots of several leguminous plants (with emphasis on pea) has been studied. It was shown that the composition of root-tip exudates of seedling plants was quite different as compared with exudates from older parts of the root. In the latter case the release of compounds was due to damage of the main root, mainly caused by the emergence of lateral roots. The composition of these exudates equals those of the soluble fraction of the root itself.

In the root-tip exudates of pea seedlings two very peculiar compounds were identified: γ-L-glutamyl-D-alanine and 2-alanyl-isoxazolin-5-one. These compounds are not found in other plants. During the emergence of lateral roots mainly homoserine is released.

30. MICROBIAL COLONIZATION OF ROOTS OF WHEAT SEEDLINGS

J.W.L. VAN VUURDE and B. SCHIPPERS

Phytopathological Laboratory 'Willie Commelin Scholten', Baarn, The Netherlands

A soil-root model system was developed to study the microbial colonization of roots of wheat seedlings in a controlled environment. The percentage of root surface covered with bacteria was estimated on daily formed root segments ('day-segments', 2-3 cm long) from root tip to root base, both by plate counts and by direct observations using fluorescence microscopy. Two zones of dense colonization of ten-day-old roots were recognized, one on day-segments III to V from the root tip and the other one on day-segments VII to IX. This root colonization pattern is discussed in relation to the most important sources of nutrient release: 1) the zone of cell elongation behind the root tip (root tip exudation), 2) the senescent and dying epidermis and cortex cells, 3) the emergence of lateral roots. The latter two are considered to be the most important in developing the microbial colonization pattern of the roots. The degree of senescence of the root tissue was estimated by stainability of the nuclei with acridine-orange.

31. SOIL ORGANIC MATTER AND ROOT SURFACE ENZYMES IN WINTER WHEAT

D. VAUGHAN

The Macaulay Institute for Soil Research, Craigiebuckler, Aberdeen, Scotland

Although plant roots are in constant contact with soil little is known about the effects of soil organic matter on enzymes at the root surface.

Humic acid (HA) and fulvic acid (FA) were obtained by alkaline extraction. HA was further fractionated by boiling in water or acid to yield water-insoluble (WBS), water-soluble (WBF), acid-insoluble (ABS) and acid-soluble (ABF) fractions. All SOM fractions at 500 mg/l concentrations inhibited root-surface invertase activity of 3 day-old wheat plants, their % order of effectiveness being ABS(67)>WBS(50)>HA(44)>FA(23)>ABF(13)>WBF(8). Root surface phosphatase activity was also inhibited, but for this enzyme the order of effectiveness was ABF(52)>WBF(40)>ABS(31)>WBS(24)>HA(20)>FA(15). Even at concentrations as low as 10 mg/l the effects of HA on both enzymes were always greater than those of FA. The inhibition of both enzyme activities was similar to that produced by water extracts of soil. Indeed FA has a similar C, H and N content to the coloured components in the water extracts of soil.

In all cases the inhibition was due to an effect on the V_{max} of the reaction and the K_m values (4.4×10^{-3}M invertase; 2.7×10^{-4}M phosphatase) were not substantially altered. Magnesium ions stimulated root surface phosphatases and partially abolished the inhibitions produced by SOM.

Experiments using ^{14}C-labelled HA showed that some 70% of the radioactivity was removed from the root surfaces with a brief rinse in 0.02M NaOH indicating that it is adsorbed onto the root surface. All ^{14}C-labelled SOM was adsorbed onto the root surface and there was no connection between the amount of ^{14}C associated with the roots and the effect of the fractions on enzyme activity. Furthermore using gel filtration it appears that the younger, lower MW components of HA are more effective inhibitors of phosphatase activity than the older, higher MW HA components.

These observations demonstrate that SOM can influence enzymes at the root surface, adding weight to the current view that these soil fractions can influence plant metabolism directly.

32. CATION ACCUMULATION BY *GLYCINE MAX*. (L) MERR. AS RELATED TO MATURITY STAGES

JEAN VINCENT, J.E. LEGGETT and D.B. EGLI

Agronomy Department, University of Kentucky, Lexington, Ky., USA

The yield of soybeans (*Glycine max* (L.) Merr.) is a function of the rate and duration of dry matter accumulation in the seed. An increase in the duration of pod fill would lead to an increased yield as a result of an increase in seed size. Recent results suggest that the duration of the pod filling period and yield are limited by the destruction of the leaves as a result of the redistribution of nutrients from the leaves to the seed (Sinclair and de Wit, Agron. J. 68, 319-324). It is not known whether the redistribution of nitrogenous components, K, etc. from leaves occurs because of inability of the root system to absorb nutrients (Barber, Agron. J. 68, 973-975) or for some other reason.

Soybeans (cv. Fiskeby) were germinated and grown in 0.5 x Hoagland solution. Rubidium accumulation was measured after a 24 h absorption period at five developmental stages spanning vegetative and reproductive growth. After a 24 h pretreatment in minus K nutrient solution, the plants were placed in a minus K, mM^{86}Rb nutrient solution. After treatment the tops were removed and separated into parts. Exudate was collected for one hour from the detopped roots. Representative samples of root segments were removed from the older roots and from the root apex region. All samples were oven-dried, ground in a Wiley mill, and analyzed for ^{86}Rb using a gamma counter.

The accumulation of Rb was greatest at the first observation, during vegetative growth. Rubidium accumulation by mature roots decreased from 11 to 1.5 μeq/g fresh weight at late podfill; that of root tip segments dropped from 21 to 0.6 μeq/g fresh weight. Over the same time period, Rb content of the leaves decreased from 50 to 5 μeq Rb/g dry weight. The volume of exudate per plant decreased from 0.7 to 0.1 ml/h, and exudate Rb fell from 1.4 to 0.2 μeq Rb/h at late podfill. However, the total Rb accumulated by the tops during this time was only reduced from 125 to 90 μeq Rb/plant.

Although Rb accumulation by soybean root segments decreased during reproductive growth, the total Rb accumulated by the tops was not reduced as much. The data suggest that the soybean plant can accumulate cations from a nutrient solution during much of reproductive growth. In a nutrient solution the process of mass flow would likely predominate; this would not be the case in a soil situation.

33. DEPLETION OF POTASSIUM CLOSE TO ROOT SURFACES AND ITS IMPLICATIONS FOR SOIL-GROWN PLANTS

A. WILD, P.J. WOODHOUSE and M.J. HOPPER

University of Reading and Grassland Research Institute, Hurley, UK

In order to calculate the limitations imposed by soil properties on nutrient uptake by plants, it is necessary to know the threshold concentration at the root surface below which the flux into roots is too low for optimal growth. This information can be obtained by growing plants in flowing nutrient solutions in which the concentration of the nutrient under investigation is maintained constant.

Using this technique, three crop species were grown for about five weeks, the external concentration of potassium, $[K^+]_e$ was maintained constant in the range 1 to 100 x 10^{-6}M, and other nutrients were non-limiting. During the first two weeks after germination the relative growth rate (0.2 - 0.3 day^{-1}) of two species increased with $[K^+]_e$ up to 20 x 10^{-6}M, but not above, and with the third, growth was not affected by $[K^+]_e$ over the entire range. After two weeks the growth rate of all three species was unaffected by $[K^+]_e$ down to 1 x 10^{-6}M.

From analyses of plants at 2 - 4 day intervals the flux (F) of K^+ into the plants was calculated. Both F, and the effect of $[K^+]_e$ on F, decreased as the plants grew older:

	Mean flux of K^+ for 3 species ($mol\ cm^{-2}\ s^{-1} \times 10^{-12}$)	
	100 x 10^{-6}M	1 x 10^{-6}M
Highest value during first 2 weeks of growth	19.7	7.4
"Steady state" value between 3 to 5 weeks	3.8	3.4

A minimum flux, Fm, for plants to maintain potential growth rates is defined by:

$$Fm = K_{crit} \frac{dW}{dt} \frac{1}{W} \frac{W}{2\pi^{-}rL}$$ where K_{crit} is concentration of K^+ in plant tissues at 95% potential growth rate. The observed values of F were below the calculated values of Fm only during the first two weeks of the experiment.

With an initial $[K^+]_e$ of 100 x 10^{-6}M in depletion experiments, F at 1 x 10^{-6}M was less than half the value obtained in constant $[K^+]_e$. Unless the internal K^+ concentration is the same, depletion experiments give different values of F to those obtained with plants grown at constant $[K^+]_e$.

The findings that, after the first two weeks of growth, plants achieve their maximum potential growth rate at low $[K^+]_e$ is relevant to interpretation of the behaviour of soil-grown plants.

34. ULTRASTRUCTURE OF THE RHIZOPLANE OF SELECTED MARINE AND FRESHWATER VASCULAR MACROPHYTES

D.A. ZUBERER

Dept. of Microbiology and Cell Science, University of Florida, Gainesville, Florida, USA

The structure of the rhizoplane of naturally occurring plants of three Florida mangrove species (*Rhizophora mangle* L., red mangrove; *Avicennia germinans* L. Stern, black mangrove; and *Laguncularia racemosa* Gaertn., white mangrove) and the freshwater floating macrophyte *Lemna* sp. (duckweed) was investigated by transmission electron microscopy. These investigations were undertaken to obtain information as to the nature of the association between the plant roots and epiphytic bacteria in an effort to explain the nitrogen fixation activity which has been observed with washed roots of these plants.

Results of these studies indicated that roots of these emergent and floating macrophytes were well colonized by bacteria and other microorganisms. The associative bacteria were generally observed in the mucigel layer at the root surface. Rarely were bacteria observed within the cortical cells of the macrophyte root. When bacteria were seen in a cortical cell, those cells were invariably senescent. Bacteria were frequently observed in the outer intercellular spaces of the root tissue. Roots of the floating macrophyte *Lemna* sp. were also colonized by blue green bacteria. Samples from duckweed mats in Florida exhibited significant acetylene reduction. The observed activity was about eight fold higher in the light than in the dark and was only partially inhibited by DCMU.

These studies demonstrated the presence of a mucigel layer on the roots of these plant species and the presence of diazotrophic bacteria as a component of the rhizoplane microflora. The nature of these plant-bacterial associations is largely that of root surface colonization with numerous bacteria embedded in the mucigel layer and outer intercellular spaces. Nitrogen fixing bacteria in the rhizosphere are dependent on photosynthate (exudates) and/or sloughed cell debris as an energy source for growth and nitrogen fixation.

INDEX